GIS for Environmental Monitoring

D1698161

Dedicated to our colleague
Prof. H. ANNONI (†),
one of
the main initiators
of TEMPUS project JEP-2013

This book was created within the framework of the TEMPUS project JEP-2013 supported by the European Commission (EC)

ꭆ-Ⅻ-27

GIS
for Environmental
Monitoring

edited by

Hans-Peter Bähr and Thomas Vögtle

with 187 figures and 35 tables

E. Schweizerbart'sche Verlagsbuchhandlung
(Nägele u. Obermiller) • Stuttgart

Editors' address:

Prof. Dr.-Ing. habil Hans-Peter Bähr,
Dr.-Ing. Thomas Vögtle,
Universität Karlsruhe (TH)
Institut für Photogrammetrie und Fernerkundung
Englerstr. 7, 76128 Karlsruhe
Germany

Frontispiece: Reconstruction of the Battlefield of Gyôr, in 1809, by combining satellite images, aerial photographs and old military maps. The dark line shows the old hungarian fort. Small picture (upper right): Positions of the hungarian (red) and french (blue) militaries before the Battle of Gyôr.

Die Deutsche Bibliothek – CIP-Einheitsaufnahme

GIS for environmental monitoring : with 35 tables / ed. by Hans-Peter Bähr and Thomas Vögtle. - Stuttgart : Schweizerbart, 1999
  ISBN 3-510-65191-X

Inv.-Nr. 20/1738389

Geographisches Institut
der Universität Kiel
ausgesonderte Dublette

ISBN 3-510-65191-X

All rights reserved, included those of translation or to reproduce parts of this book in any form.

© 1999 by E. Schweizerbart'sche Verlagsbuchhandlung (Nägele u. Obermiller), D-70176 Stuttgart

⊗ Printed on permanent paper conforming to ISO 9706-1994

Printed in Germany by Strauss Offsetdruck GmbH, D-69509 Mörlenbach

Geograpnisches institut
der Universität Kiel

Preface

1. Background of this book

One of the dominant challenges of the European Community (EC) is integration on all levels. In the academic sector, a lot of programs are available for both students and professors, e.g. for exchange, common research projects, and for technology transfer (*Erasmus, Comett, Lingua, Esprit...*). One of these programs is especially designed for cooperation between Universities from East and West Europe: the TEMPUS-Program. It has been established in the context of the new horizons opened through the political changes in East Europe.

The TEMPUS-Projects require cooperation between one university of East Europe and at least two universities of West Europe. The philosophy of these programs, initiated in 1990, was to support East European universities by hard- and software (hightech) and to accompany this support by establishing new intimate links between universities, students and professors, from East and West. This goal was to be achieved by developing new curricula for advanced subjects.

The authors of this book are happy to have been selected by the European Community Administration for designing a curriculum in the field of "Geoinformation Systems for Environmental Monitoring". The work was completed between 1992 and 1995.

2. Authors and methodology

The idea for this book was initiated by the Technical University of Budapest. This institution belonged, without any doubt, politically to East Europe, though not necessarily geographically. Traditional contacts to the University of Karlsruhe/Germany, already established about 25 years ago during the era of the "Cold War", served as a first link for the TEMPUS-Program. In addition, new links were established to the Technical University of Delft and the University Louis Pasteur together with the ENSAIS in Strasbourg . The representatives of the mentioned institutions were all enthusiastic about the TEMPUS philosophy. Beside the "political" challenge to collaborate in an advanced field of technology like Geoinformation Systems, the composition of the different disciplines involved was highly animating. Geodesy (Budapest/Karlsruhe), Civil Engineering/Hydrology (Delft) and Geography/Topography (Strasbourg), all play an important role in advanced geoprocessing. Consequently, the TEMPUS-Program "Geoinformation Systems for Environmental Monitoring" of the EC gave an additional push to all members.

We have to point out that the individual curricula at the mentioned universities had to be restructured with regard to the new field of Geoinformation. Therefore, the work performed within the TEMPUS-Program had to be done in any case.

The result is presented in English. From the beginning, there was no "language problem". This might be due to the fact that none of the four institutions involved use English as native language. English was adapted without any discussion from the start as the communication idiom between the groups. The written text, however, required a little more effort than it would have been for the native languages. Having in mind "European students" as a target group and addressing an advanced technological field, we think that the English language is a sort of a modern *"lingua franca"*.

To develop a curriculum for "Geoinformation Systems for Environmental Monitoring" was not an easy task, since every discipline has its own specific perspective. Therefore, no "monolithic textbook" can be expected. On the other hand, an international and interdisciplinary output was feasible. The cooperation was achieved through numerous meetings and discussions between the participants, generally two or three co-workers from each place. The methodology applied during these discussions followed the latest findings of group dynamics: A "moderator" had to structure the ideas of the participants who wrote their comments on small cards which the moderator pinned on a board. This process evolved into a "problem tree", a "tree of objectives" and finally into a matrix of "modules", which then resulted in the elements of the curriculum.

3. Structure and contents

The working units (modules) on which the participants agreed, were freely grouped under "Data", "Data Preprocessing", "GIS Database" and "Environmental Monitoring". For the present book, the chapters were restructured taking these parts as basic units. Interestingly, there is no clustering in the respective chapters by just one university. In nearly every one of the four principal parts of this book, we find contributions of *all* members of the four groups involved. This proves the perfect interrelation of the different disciplines.

The listed contents fall into three theoretical parts and one part which is devoted to Environmental Application.
Basic representation discloses the GIS philosophy of the group. In general, GIS depends very much on the individual viewpoint of each discipline, of each individual. In particular, however, this book presents a common scope from Geodesy, Geography and Hydrology. The part on *Information Sources and Data* provides many particular inputs from Geodesy (Surveying, Photogrammetry, Remote Sensing, Cartography). *Data Processing and Information Extraction* mainly treats methodological aspects. This part, like the preceding one, is not restricted to "Environmental Monitoring" since *Data and Methodology* is applicable, in principle for many different tasks.

The last part of this book deals with *Environmental Monitoring Applications*. Here, the respective participants present their experience in special application projects. For land, *erosion* and *waste control* is investigated, whereas for water, *hydrologic measuring*, *runoff irrigation in Sahel* and *process water* is discussed. The international REKLIP-Project in the Upper Rhine Valley deals with *climatic questions*.

4. Target groups and "how to use the book"

In the EC (TEMPUS) contract, target groups are defined as "personnel and students of universities of environmentally oriented specializations (Geodesy, Civil Engineering, Hydrology, Ecology, Geology) and practising engineers (refreshment courses)". The scope of this book is very broad. It refers to a politically and ecologically very challenging question: Especially in countries of Middle and Eastern Europe, the threat of the environment is an enormous problem. Remote Sensing is a powerful mean to monitor the effect of environmental pollution and eventually to detect sources of pollution. On the other hand, Geoinformation Systems offer different possibilities to store Remote Sensing and other georeferenced data so as to combine and to analyse them. This is why experiences indicate that there is a world-wide demand of Remote Sensing, Digital Image Processing and powerful Geoinformation Systems in the field of Environmental Monitoring. A combination of these techniques will offer very useful information to locate and assess the extent of pollution and to take measures to stop and to avoid pollution now and in future. Essential is, however, the availability of knowledge in instrumentation for Digital Image Processing and for Geoinformation Systems to handle and to process adequately the great number of data. The exactly characterized situation of the ecological environment and the available techniques present an important base for this book. It is definitely not a conventional textbook. It addresses both students and professors, both learning and lecturing individuals. However, a certain professional base from Geography, Hydrology or Geodesy is required as a prerequisite.

With regard to this objective, the book offers an ideal base for advanced studies in the field of Environmental Monitoring by modern methodology. Every author and every university present their specific experiences through their most successfully applied examples. As it was mentioned before, this is not possible in the form of a "monolithic textbook". The respective contributions of the different participants will necessarily leave some gaps. On the other hand, redundancy was avoided. Comparison with many textbooks show the particular quality of the presented book :

- Interdisciplinary Perspective: Geodesy, Cartography, Remote Sensing, Geography, Civil Engineering, Statistics, Hydrology, Agriculture involved and presented by professionals
- International Cooperation: Contributions from France, Germany, Hungary, The Netherlands
- Pragmatic Approach:"Engineering Viewpoint": focusing *problem solving instead of academic research*
- Didactic Challenge: Structuring the topics according to logical sequence observing formal criteria for all modules

The book presents a large amount of data and experience. For students, text and examples will be very helpful to understand the interrelation between environment and technology. The lecturer, on the other hand, will find a lot of information and material useful for being included in his individual courses. The book comprises much information displayed from various points of view, which the reader would otherwise only find distributed in many different books. In whole Europe, there is a lack in specific information described by the "key words" *GIS, Database, Geo-referenced Information, Geo-Processing* and *Environmental Monitoring*. Consequently, this book is an appropriate tool for both advanced courses on academic level and for self-study as well.

The subtitle of the book exactly reflects this object: *selected material for teaching and learning.*

Karlsruhe, December 1998   H.-P. Bähr   Th. Vögtle

CONTENTS

INHALTSVERZEICHNIS

I. EINFÜHRUNG

II INFORMATIONSQUELLEN UND DATEN

III DATENVERARBEITUNG UND INFORMATIONSDISTRAKTION

Methodik
Handwerkszeuge

IV. GIS ANWENDUNGEN FÜR UMWELTÜBERWACHUNG

Land
Wasser
Klima

INHOUDSOPGAVE

TARTALOMJEGYZÉK

SOMMAIRE

GIS Introduction

Main Concepts

Hans-Peter Bähr, Karlsruhe

1. Terms

There are three lines which have mainly contributed to the development of Geoinformation Systems :
- availability of high-level computer hardware and software
- progress in digital imaging sensors
- rapidly growing demand for environmental information.

All these aspects will be considered in this book. GIS is a part of the big family of "Information Systems", a specific "georeferenced" information system. This means that the information is geometrically related to the Earth (including the atmosphere and the lithosphere).

"GIS" stands for *Geoinformation Systems* or *Geographic Information Systems*. It is used differently, depending on the respective country and on the different meanings and traditions of the term "Geography". In the Anglo-American countries and in France for instance, Geography opens a broader field than in Hungary or in Germany. To avoid confusion, we will therefore apply here the term *Geoinformation Systems* which in our opinion seems to be more general than *Geographic Information System.*

Within the concept "Georeferenced" Information Systems, a large number of related terms are constantly coined and used, "Land Information System (LIS)" being by far the most important. In 1984, the Fédération Internationale de Géomètres (FIG) defined LIS (see Section 2) as a "large scale, cadaster oriented GIS". Besides LIS, "Environmental Information System" (EIS) has also come into use and some others are appearing. Both "Land" and "Environmental" classifiy the *thematic* content of the Geoinformation System. Among all these, GIS is definitely the dominant concept, the thematic specifications indicating the subordinate terms.

New technological developments are very often accompanied by new terms. "Geoinformatics" is such a new term which tries to bring forward the fact that the conventional analogous processing of information in the "Geo" field has now been substituted by digital techniques. Though it seems logically consistent to "polish up" the good old geographic/cartographic disciplines, the impression prevails that a completely new field of activity has been born. This, of course, is not fully true.

One of the most important objectives of this book is to show what the new advanced technologies offer beyond the classic methodology and to point out which problems in geoprocessing remain unsolved (see Section 5).

2. Definitions

A Geoinformation System is, no doubt, a very complex tool. Therefore, a definition seems to be of particular interest to clarify the concept. Following the FIG definition of Land Information System (LIS), a Geoinformation System is:

"a tool for planning development and environmental control as well as an instrument of decision support. On the one hand, it consists of a Georeferenced Database, on the other hand of techniques for data acquisition, actualisation, processing and visualization of the results. The semantic data are geometrically related to a homogenous georeferenced coordinate system allowing controlled interrelation of the information."

This definition includes not only *data acquisition* and *data processing* but also *decision making*. It takes into account that *data processing* offers alternatives for decisions. Decisions have to be taken by a human being because only he is able to assume responsibilities. Therefore, GIS is, in our view, not a system which will finally substitute the human operator, but a welcome support to his more and more complex work like development planning and environmental control .

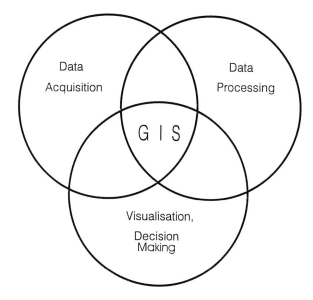

Figure I.2-1: GIS in the context of data acquisition, data processing and
visualization/decision making

In Fig. I.2-1, the three main components of a GIS (data acquisition, data processing and visualization of the result together with alternatives for decision making) are graphically displayed. The three areas are not fully separated because advanced GIS Systems allow a per-manent dialogue-based interaction of the different fields. That is also why **experts from each of these three areas define a GIS according to their very specific and different viewpoint:** a surveyor, photogrammetrist or expert in remote sensing refers primarily to "data acquisition" and "data storage techniques" in database systems. In "data processing", the geoexperts from the various disciplines contribute their particular geomodels. In the "visualization and decision" field, the "real user" is the expert in governmental administration.

3. Political and socio-economic aspects

Worldwide there is an urgent need for Geoinformation Systems. This is due to the circumstance that the *Earth's surface is a constant in man's environment*. It is the limiting factor for all human activities in the Geo-field, since it cannot be enlarged. Enormous population growth on one hand and draught, erosion and inundation as well as general desertification on the other, require large areas of natural environment to be transformed into arable land. Geoinformation Systems are considered to be the best suited tool to attain optimal benefits and achieve sustainable results in this context.

The necessary information for social and technological management has to be **faster, more reliable and more specific** than before. These three aspects still demand extensive research, to meet the requirements of Public Administrations.

Although the technological progress generally advances faster than expected, the development of operational concepts takes considerably more time. The situation is further complicated by the fact that the information that will be required is often not predictable: Who would have thought of a management action plan for all necessary preventive measures *before* the Chernobyl disaster? This example shows the role of operational GIS, and demonstrates the joint responsibility of international politics.

The main task of politics is the a-priori anticipation of future problems. Simulation and prediction in the course of planning procedures is one of the new possibilities offered by GIS. The following examples will reveal the wide range of possible applications:

1. Inundation

   Endangered areas may be detected by simulating floods as a function of precipitation, topography, vegetation, soil type, etc.

2. Draught

   GIS-based simulation may discover endangered areas as a function of climate, soil, land use, precipitation probability, etc.

3. Pseudo-Krupp

   This allergic reaction of children (aged between 3 and 8) has not yet been fully investigated and explained. There is no doubt that climate, air pollution and ventilation as a function of topography and humidity are contributing factors.

4. Unemployment

   Here, of course, economic parameters play a dominant role, like gross national product, available capital, investments, etc. but geofactors must not be neglected. As we all know, there is a clustering of unemployment in the lesser developed regions of a country. Regional planning has to take into account a large amount of information which is generally provided by data bases within a Geoinformation System and these may play a highly important role within unemployment control plans.

These examples show that, in principle, many applications are just waiting for an adequate GIS interface. Though there is obviously a benefit in the application of GIS, we should not ignore the dangers involved, some of which are, for instance:

1. Inadequate application:

   As stated above, the final decision has to be made by the human being, not by the machine.

2. Overly high expectations

   Many people expect "wonders through hightech" forgetting that the output is a function of the input (which is prepared and entered by a human operator).

3. Blackbox syndrome for the target group

   This is exactly the opposite to "high expectations". Many people are very critical towards advanced technology. This is often not justified and a mere emotional reaction.

4. Blackbox syndrome for the operator

   The digital working place of an operator has changed dramatically compared with what it was before. The problems involved should not be underestimated. For some elder people (but not only for them), the challenge may be very high, even excessively high, occasionally.

It is important to realize that a GIS does not only solve problems, but at the same time presents new ones.

4. Example

An example will demonstrate better than a definition or long discussions what the nature and the essential structure of a GIS is. In Fig. I.2-2, the two components (data acquisition and data processing) in a GIS are displayed.

The application presents the evaluation of a 50 km × 50 km territory in the Sahel Zone for runoff irrigation. This very simple technique allows a better use of the available water resources by so-called "micro or macro catchments" (see Chapter IV.6 and [Vögtle, Tauer, 1991]).

Fig. I.2-2 shows which areas in a specific region are suitable for this type of irrigation. The hydrologist, in cooperation with the agronomist, has the know-how to select the conditions which allow irrigation. Their individual expertise is based on "data sources", "data" and "models" (see Fig. I.2-2). For this specific application, satellite images, field work and topographic maps 1:200.000 are required. From such data sources, the particular data for this specific area were taken. Information about soil, vegetation and water was provided by the satellite image. Pedological parameters such as field capacity had to be checked by field work.

Finally, cartographic parameters (mainly geometric parameters) were extracted from topographic maps 1:200.000. We see that again different parameter types (biological, pedological and cartographic parameters) interact in order to find a solution to the irrigation problem.

The central part of the GIS is what we call a "model". Certain data are extracted from the data sources and introduced into the model to obtain a result (see right sector of the diagram). As far as the biological parameters are concerned, only shrubland might be taken into consideration. All other land use classes have to be rejected. Data extraction from the satellite image was done by digital multispectral classification. Here, field work was obligatory, too, not only for checking the field capacity. A capacity factor of more than 15 volume % is acceptable; lower values do not allow irrigation. Finally, the topographic map 1:200.000 was applied to locate villages: it makes no sense to irrigate land that is very distant from people´s dwellings. A distance of 4 km is considered the maximum, fundamentally because women are responsible for field work in Africa. Slopes with inclination of less than 10% and grades steeper than 10% are acceptable for irrigation, but they require the application of different techniques: Micro catchments for inclinations of less than 10% and macro catchments for surfaces steeper than 10%.

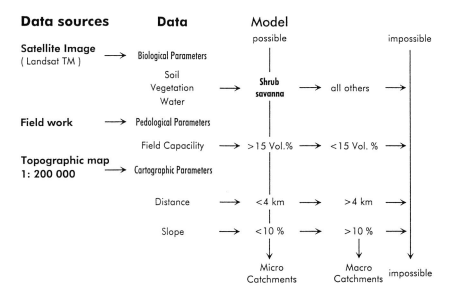

Figure I.2-2: Structure of a GIS: Runoff Irrigation in Sahel

Every parameter in the model (land use, field capacity, distance and terrain inclination) produces a "data layer" for the area under consideration. By combination of the different layers according to the "model", three classes of surfaces are defined: favourable for micro catchments, favourable for macro catchments and areas wholly inadequate for irrigation purposes.

This example shows very clearly the necessary cooperation between different disciplines. The computer specialist, the surveyor, the hydrologist, the agricultural engineer, the remote sensing specialist and the ethnologist have to work together and cooperate.

The result obtained through the above explained GIS application excludes areas which are in-adequate for irrigation. In this respect, the system helps the operator. But the decision about the areas to be chosen for Runoff Irrigation is the responsibility of the decision maker. He is the person to make the decision, in very close cooperation with the local authorities.

5. GIS and Cartography – a comparison

When talking about GIS, one should be aware of the fact that it does not represent a "new" subject: the conventional analogous "Geo Data Base" (GDB) was represented by both topographic and thematic maps. This should be taken into account instead of considering it a completely new terrain. In this sense, GIS might be understood as **"Cartography by advanced technology"**. (This is, of course, the surveyor´s viewpoint). Digital techniques offer many new opportunities and challenges, and we may expect brillant results which would never have been achieved by conventional methods. GIS is only used properly if its full technological potential is consequently taken advantage of.

On the other hand, it is logical and helpful to bear in mind the experience contained in long years of conventional cartography. In this chapter some topics are discussed which are common to a digital GDB and a conventional analogous map.

5.1 Generalization

The excellent term "level of abstraction" is valid for both analogous and digital GDBs. Full computer-aided generalization in cartography, however, is an "ill posed problem" and will never be achieved. This should be considered when talking about generalization in digital GDBs. The rigid scale sequence for topographic maps was conditioned by the prevailing technology (each product reflects the technological environment of its historical period). The scale sequence will become more flexible in the future, though it will remain being a matter of standardization.

5.2 Visualization

Here, conventional cartography offers much expertise. The product itself, the map, is not separable from the (analogous) data base. Visualization is a central issue for cartographers, and in this field computer-based performance must still learn a lot about how to represent spatial data conveniently ("readability").

In our days the colour monitor offers almost unlimited new possibilities which were not available in the times of analogous presentation of data; perspective views and time series, should also be taken into consideration ("Multimedia GIS"). The basic rules of visual perception, however, as formulated by cartographers, are still valid. This does not invalidate the evident need and chance for common research work to be done by cartographers and CAD-experts (Computer-aided design experts).

5.3 Updating

Similarly to what has been stated under "generalization", the problem of how to update conventional maps has not yet been solved in a satisfactory manner. This is particularly problematic because the urgency for up-to-date information is constantly growing. Updating is very much linked with data acquisition. As far as updating a digital data base is concerned, a new level of information is offered through "information sequences", i.e. the record about the *development* of an area. This really requires a fourth dimension (supposing we are working in 3,0 D-Geometry), or at least a 3,5 th. dimension if the time sequence is not continuous. Updating a digital GDB is a fundamental research field.

5.4 Data Quality

Here again cartography has a lot to contribute as long as map making is carried out by engineers, like geodesists and surveyors. For conventional maps, well-defined standards exist for both geometric and semantic data quality. The quality standards related to the applied technology, e. g. when topography is to be highlighted, are strictly tailored to the requirements of the product to be produced. Propagation of errors is carefully taken into account.

   The rigid conventional map making procedure allows a complete error theory, at least for the *geometric* data base. For the digital procedure, it is not yet available. Consequently, no consistent theory exists to describe GIS accuracy, precision and reliability. With regard to this important, still absent feature, the GIS accuracy performance has to be investigated with the highest urgency. This is an extremely complex task which requires a number of effectively collaborating research groups.

GIS Introduction

Additional Remarks

Christiane Weber, Strasbourg

As explained above, a Geoinformation System is definitely a tremendous means to deal with natural or human problems. It has to be noted that all kind of GIS applications can be found from management to scientific use in research and application (Fig. I.2-1).

   "If GIS packages usually include some analytical tools.… they have rather limited analytical power" ([Kilchenmann, 1990], [Bailey, 1994]). This leads to link some analysis and statistical systems with the GIS package. This implies, in fact, going to scientific uses through the usual tools of GIS packages. This book will present some main elements of a GIS through sources, processing and analytical tools and finally examples.

   As presented above, GIS is at the crossroads of different disciplines: Cartography, Computer science, Imagery. Of course as it gathers the abilities of experts in various domains, GIS is also the beginning of an understanding exercise between partners. This is not the least difficulty. With their own surroundings and language habits, pedagological backgrounds, partners have, since the very beginning, to understand each other, to build the necessary approach according to a collective formalized culture.

   Giving marks accepted on a common basis is the best way to avoid misunderstandings, inappropriate use and unreproducable and incomparable results; this is the reason why some concepts and definitions will be given along the chapters.

1. What is in a GIS except reality ?

One of the main issues in GIS development is the transcription of the World into numbers, figures, maps, etc.

How to transcribe a traffic jam into flows or a landslide after a thunderstorm into maps?

Most of the difficulties amount to this operation; most of all, because this has to be understood by the largest community. Modelling the real world which can be observed, is an essential step in the implementation of a GIS. The main point here is the fundamental view of the location. All the knowledge has to be associated with a point, a line or an area, represented by coordinates. Data acquisition is thus the beginning of a dynamic process which implies a maximum of consideration (see chapters on *Field Survey, Topographic Maps, Imagery* or *Questionnaires and Other Sources)* in order to represent the studied reality (space) in an understandable and accessible form.

GIS has to fit into an operational constraining model which step by step narrows the possibilities to acquire information in order to get attribute data. This model implies access and potential restrictive rules (legal restrictions or economic ones); the information (nature, updating, measurement level and reliability, completeness), the dedicated use (pre-processing according to the goal of the study). As a result, the chosen information will be transformed through coding, most of the time into data.

All pieces of information we need must be transformed according to two aspects:
- the semantic sense: the meaning regarding the context;
- the syntactic structure: the coding, or the form to operate.

This transformation has of course to preserve the maximum of sense in order to generate knowledge.

Of course reality cannot be reduced to details; decisions on what has to be to kept, what is essential, what is worthwhile etc. must be made by the researcher. Approaching simplicity and avoiding caricature is the deal, all the more since the final goal is communication. This highlights the importance of the methodological process run by GIS, the result has to be understood and interpreted by a final user. Therefore, the whole process could be compared to a hierarchical tree where each branch identifies a somewhat "irreversible" choice until any hypothesis is verified.

In this context generalization is necessary and applied both to the geometric features and the thematic information, to make it easier to understand and interpret the respective phenomena. The chapter on *Cartographic Principles* presents the main keys of understanding and the rules to be used in order to manage such issues from the cartographic point of view. Chapters on *Image Classification, Geostatistics, Interpolation Schemes* (Regionalization), deal with data processing, i.e. means to synthetise and transform the information by different methods for diverse purposes. Some common spatial analysis methods are missing in this chapter (spatial auto-correlation, variograms, spatio-temporal models or multi-dimensional scaling analysis for multivariate studies). Some applications of these methods can be found easily in the literature or references.

What has to be specified is that "sophisticated statistical techniques in GIS are not likely to materialize in the short term ... but this was probably never a practical proposition" ([Bailey, 1994]). But even if GIS gives the possibility to deal with a large amount of data and extra-topological details, it is necessary to create a linkage to special analysis software.

No chapter deals directly with data structures (raster, vector or hybrid), in this book, but an extended literature does exist on these topics and it is not worth reproducing it here.

The technical aspects of the GIS information are described in the chapters on data conversion ('Digitizing Maps', 'Imagery', 'Digital Terrain Model' etc.) which review all techniques needed to acquire and integrate data.

The *Environmental Monitoring Applications* are a good way to introduce these specific aspects in association with thematic issues (Land, Water, Climate) and to present logical and methodological developments.

The *European Projects* show the interest of the European Community in managing environmental and agricultural problems on a large spatial scale. The power of GIS lies also in the possibility to erase limitations and differences (provided that there is not too much incompatibility between the data used).

These European projects also prove that there is great interest in such developments at the various decision levels, from local entities to European organizations.

2. GIS Future

Indeed GIS is a means to gather experts, but also to tie strong relationships between people or organizations. This does also infer the actual difficulties encountered by the GIS communities. In fact now exchangeability, access to numerous data-bases, integration of a growing number of data coming from very diverse sources etc., are the actual issues of the GIS use. This implies some specific problems: (1) *technical* (standard format, networks, protection against virus propagation or even worse delinquency), (2) *legal* (on property, rigths of use, royalties) and (3) *economical* because it is not usual that information is considered as a value; from the structured good to the value of it, it is not really easy to define market rules. What is a reasonable prize for data with regard to the organization which has constituted the data-base (private or public)? Should differences be made between users, according to the use of such data?

Political issues are involved too, citizen rights to liberty and privacy facing the numerous files which are constituted around the world, citizen and administration relationships with regard to the use of national means to collect information and the sale of it to private organizations (census for instance). These questions are tackled in Chapter II.5 ('Questionnaires and Other Sources'), because they are more important for socio-economic data than for environmental or physical data.

The actual confrontations between different legislations from the national to the international level, each arguing according to opposing political principles "free access market against national protection" are typical for the economic and political issues which are part of the social and technical development and which will become more important in the future.

References

Baley, T.: A review of statistical spatial analysis in geographical information systems. In: Fotheringham, Rogerson (eds.). Taylor & Francis, 1994, p 281

Bill, R., Fritsch, D.: Grundlage der Geo-Informations-Systeme. Wichmann Verlag, Karlsruhe, 1991

Burrough, P.A.: Principles of Geographic Information Systems for Land Resources Assessment. Oxfort Science Publications, Monographs on Soil and Resources Survey No.12, 1985

Buttenfield, B.P., Mackaness, W.A.: Visualization. In: Maguire et al. (eds.): Geographical Information System: principles and application. Vol. 1, Longmann, London, 1991, pp 428-443

Clarke, A.L.: GIS specification, evaluation, and implementation. In: Maguire et al. (eds.): Geographical Information System: principles and application. Vol. 1, Longmann, London, 1991, pp 477-488

Dangermond, J.: The commercial setting of GIS. In: Maguire et al. (eds.): Geographical Information System: principles and application. Vol. 1, Longmann, London, 1991, pp 55-65

Detrekői, Á: Térinformatika és az elsődleges adatnyerés. Geodézia és Kartográfia 44, 1992, évf. pp 340-343

Kilchenmann, A.: Geographical Information Systems and Maps analysis using factor analysis and character analysis. European Conference on Geographical Information Systems, 1990, pp. 585-593

Maguire, D.J.: An overview and definition of GIS. In: Maguire et al. (eds.): Geographical Information System: principles and application. Vol. 1, Longmann, London, 1991, pp 9-20

National Center for Geographic Information and Analysis (NCGIA): Core Curriculum, Santa Barbara, 1988

Schweinfurth, G.: Geo-Informationssysteme. In: Bähr, Vögtle (Hrsg.): Digitale Bildverarbeitung. Wichmann Verlag, Karlsruhe 1991

Vögtle, T., W. Tauer: GIS-Anwendung: Sturzwasserbewässerung in der Sahel-Zone. In: Bähr, Vögtle (Hrsg.): Digitale Bildverarbeitung. Wichmann Verlag, Karlsruhe 1991

II.1 Reference Systems

Ákos Detrekői, Budapest

Goal:

The goal of this chapter is to give an overview of reference systems of geometrical data in a GIS. Various kinds of reference systems and the transformations of data from one coordinate system to another will be discussed . Some aspects of the selection of reference systems will be given, too.

Summary:

The first topic of the chapter is the shape of the Earth. Then the various coordinate systems will be discussed. The elements of map projections are given too. The transformation of data from one coordinate system to another and the discrete georeference system will be summarized. At the end of the chapter examples for reference systems in various countries and some aspects of the selection of reference systems will be given.

Keywords:

Geoid, rotational ellipsoid, geodetic datum, coordinate system, map projection, transformation.

1. Introduction

Before geometrical data can be used in GIS, they must be referenced to a common system. This chapter presents an overview of available reference systems. There are various reference systems which describe the real world in different ways, and with varying degrees of precision. These systems are related to the Earth, so they are often called "georeferences".

In this part the following items will be discussed:

1. shape of the Earth.
2. geocentric coordinates.
3. geographical coordinates on the surface of the Earth.
4. map projections.
5. georeference systems of rectangular coordinates.
6. the transformation of data from one coordinate system to another.
7. discrete georeference systems.
8. examples of reference systems in various countries.
9. some aspects of selection of reference system by environmental monitoring.

2. The shape of the Earth

By the shape of the Earth we mean the physical surface of the Earth and its mathematical description. The physical surface of the Earth is the border between the solid or fluid masses and the atmosphere. The physical surface of the Earth is normally given by the coordinates of discrete points of this surface.

The mathematical surface is given by various models. The most general mathematical model is given by the geoid. The geoid can be illustrated as the level surface of the Earth's gravity field. The equation of the geoid is given by [Torge, 1991]:

$$W = W(r) = W_0 \qquad (II.1\text{-}1)$$

where W is the gravity potential of the Earth.
As a reference system for horizontal positioning the rotational ellipsoid is used in geodetic surveying. The shape of the ellipsoid is described by two geometrical parameters (e.g. the semimajor and the semiminor axis). The placing of a rotational ellipsoid is given by the geodetic datum. The geodetic datum defines the orientation of the ellipsoid within the global Cartesian system. In the various countries different kinds of ellipsoids are used, e.g. *Bessel, Hayford, Krassowski, WGS84*. The geometrical parameters and the geodetic datum of these rotational ellipsoids are different for each model. In plane surveying, the horizontal plane is generally sufficient. In GIS activities we normally use the horizontal plane as Earth model, too.

As various kinds of coordinates are used the knowledge of the reference systems is necessary.

3. Geocentric coordinates

The geocentric coordinate system is an Earth fixed spatial Cartesian system (X,Y,Z) whose origin is the Earth's center of gravity. The Z axis is coaxial with the axis of the Earth's rotation and is positive in the direction of the North Pole. The X axis intersects the zero meridian at Greenwich; the Y axis is orthogonal to the left of the positive X axis (Fig.II.1-1.).

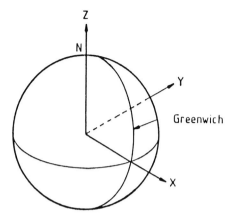

Figure II.1-1: Geocentric coordinate system

Geocentric coordinates are useful mostly for GPS positioning. The distances between two points P and R can be calculated as Euclidean distances in space.

4. Geographical coordinates on the surface of the Earth

The geographical coordinates on the surface of the Earth are given by geographical latitude and longitude. At a point P of the ellipsoid, the geographical latitude Φ is the angle between the point and the equator along the meridian. (A meridian is a line connecting the N and S poles and the point of interest.) The geographical longitude Λ of the point is measured by the angle on the equatorial plane between the meridian of the point and the central meridian (passing through Greenwich, England). The geographical coordinates of a point P are given in Fig. II.1-2.

The latitude and longitude (both angles) are measured in degrees, minutes and seconds. The geographical coordinates are useful for larger areas, e.g., topographic maps of a country. The distance between two points can be calculated using the equations of spherical trigonometry.

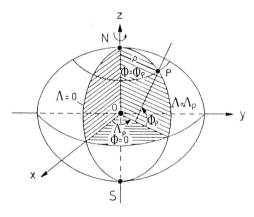

Figure II.1-2: Geographical coordinate system

5. Map projections

A map projection is a system in which locations on the curved surface of the Earth's model (normally on the ellipsoid) are displayed on a flat sheet or surface according to some set of rules.

Mathematically, projection is a process of transforming a coordinate pair (normally Φ,Λ) on a mathematically defined surface to a planar (normally x,y) position. The projection equations can be given generally by:

$$x = f_1\ (\Phi,\Lambda)$$
$$y = f_2\ (\Phi,\Lambda)$$

(II.1-2)

The reverse calculation is the determination of the global location from planar coordinates. The general form of the equations in this case is:

$$\Phi = g_1\ (x,y)$$
$$\Lambda = g_2\ (x,y)$$

(II.1-3)

Angles, areas, directions, shapes and distances are distorted when transformed from a curved surface to a plane. All these properties cannot be kept undistorted in a single projection. Usually the distorsion in one property will be kept to a minimum while other properties are severely distorted.

Maps very often use a conformal (orthomorphic) projection. A projection is called conformal when the angles in the original features are preserved. In this kind of projection over small areas the shapes will be preserved, too. Preservation of shape does not work for large areas.

Various projections are used to represent the curved surface of the Earth on the plane. They fall into two groups:

    - geometric projections,
    - non-geometric (mathematical) projections.

Geometric projections can be conceptually described by imaging the surface to be developed which is a surface that can be made flat by cutting it along certain lines and unfolding or unrolling it. This surface can be a plane, a cone and a cylinder ([Bernhardsen, 1992]). So we can classify:

    - planar or azimuthal projections (e.g. the stereographic projection),
    - conical projections (e.g. the Lambert projection),
    - cylindrical projections (e.g. the Universal Transverse Mercator projection (UTM)).

The three kinds of these projections are illustrated in Fig. II.1-3.

6. The georeference systems of rectangular coordinates

The rectangular coordinates are used on the plane. Planar representations of the Earth are used in the following cases:

    - when we assume the Earth is flat (no projection, small areas),
    - when we use maps with coordinates projected from the ellipsoid,
    - when we work with aerial photos.

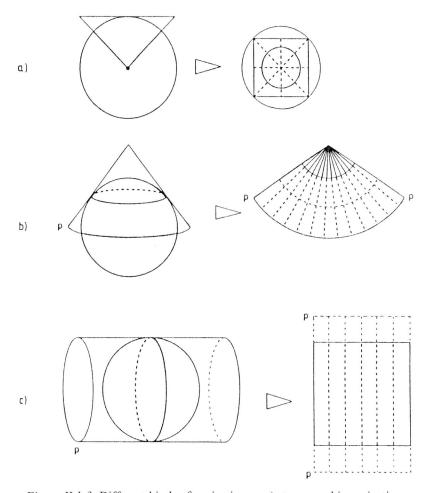

Figure II.1-3: Different kinds of projections: a) stereographic projection
b) conical projection   c) cylindrical projection

The rectangular (Cartesian) coordinates are determined as follows ([NCGIA, 1990]):
    - an origin is defined for the coordinate system
    - two axes passing through the origin are set in fixed directions, at right angles to each
      other,
    - the linear displacement from the origin in the directions defined by the two axes, is
      measured,
    - an ordered pair (x,y) is determined.

In the GIS activity national plan coordinates and local plan coordinates are used. The origin and
the axis of the national coordinates are determined by governmental agencies. In the various
countries the direction and the name of the axis may differ, for example:
    x is horizontal and y is vertical,
    x is vertical and y is horizontal,
    x is east, y is north.

Often more than one coordinate system is used in a single country. For example in USA both the Universal Transverse Mercator (UTM) system and the Lambert Conformal Conic projection system are used ([NCGIA, 1990]).

The origin and the axis of a local coordinate system can be set in various ways. They can be located to specified objects in the terrain, too.

Precision and resolution of coordinates are different elements.The number of significant digits required for a specific project depends on the size of the study area and on the resolution (accuracy) of measurement.

The distances, the areas in the plane coordinate systems can be determined by using trigonometrical equations. However, for large areas, the distorsion of the projection must be considered, too.

7. Transforming data from one coordinate system to another

Larger areas are best represented by several reference systems. In areas where neighbouring systems overlap, one must often work in two systems, so that it may be necessary to transform data from one coordinate system to another.

There are various types of transformation equations. The transformation of geocentric coordinates into geographical or planar coordinates can be done using the space similarity transformation. The number of coefficients for this transfomation is 7. The transformation of geographical coordinates into planar coordinates is possible using Eq. (II.1-2) of the map projection. The most frequent used transformation is the transformation between rectangular systems. There the following types of transformations are used ([Janza, Vozikis, 1984]):

- indirect transformation,
- interpolation methods.

An indirect method should be used, when the relationships between the original system (x,y) and the new system (x',y') are only given by geographical coordinates (Φ,Λ). In this case we first use Eq. (II.1-3) of the original system, and Eq. (II.1-2) of the new system:

$$\Phi = g_1(x,y)$$
$$\Lambda = g_2(x,y)$$

and

$$x' = f_1'(\Phi,\Lambda)$$
$$y' = f_2'(\Phi,\Lambda)$$

When there is no mathematical relationship between the two planar systems and, provided that a sufficient number of points common to both coordinate systems can be identified, the interpolation method can be used. Initially, the coefficients for interpolation are determined by means of common points identified. The most important interpolation methods are the following:

- similarity transformation (4 coefficients),
- affine transformation (2×3 coefficients),
- curvilinear transformation (2×4 to 2×6 coefficients).
- interpolation by least squares.

The transformation equations of the similarity and of the affine transfomation are:

- similarity transformation:

$$x' = a_0 + a_1 x - a_2 y$$
$$y' = b_0 + a_1 y + a_2 x$$

- affine transformation:

$$x' = a_0 + a_1 x + a_2 y$$
$$y' = b_0 + b_1 x + b_2 y$$

The similarity transformation is suitable for small areas with accurate data. The affine transformation is frequently applied when developing spatial databases: data will be provided on map sheets which use unknown or inaccurate projections. The curvilinear transformation is useful in larger areas. The interpolation by least squares is a method for large areas with systematic residuals.

To determine the unknown coefficients of transformation equations (e.g. a_0, a_1, a_2, b_0 by the similarity transformation) a set of common points must be identified that can be located on both systems. Coefficients can be determined, for example, by least squares adjustment.

8. Discrete georeference systems

Positions in the discussed georeference systems are given by coordinates. These coordinates have a continuous character. In discrete georeference systems the position of phenomena is given by discrete, limited units of the surface of the Earth. Such units are, for example, ([Bernhardsen, 1992]):
- postal codes,
- address and street codes
- cadastral number of parcels,
- statistical units and other administrative zones,
- grids (e.g. "A7" on a city map).

The accuracy of positioning is determined by the unit size. Normally the accuracy of the positioning by discrete units is lower than the accuracy of positioning by coordinates. Discrete reference systems are sometimes easy to use but data update is very complicated.

9. Examples of reference systems in various countries

The geocentric coordinates are normally used in all countries. The GPS coordinates refer to the World Geodetic System 1984 - WGS84. The geographical coordinates in different countries refer to different rotational ellipsoids with different geodetic datums.

Rectangular coordinate systems differ according to the countries. Very often, different state coordinate systems are used in the same country and the reference system in the small and middle scale domain differs from the reference system in the large scale domain.

For example, the Federal Republic of Germany uses Gauss-Krüger coordinates in the small and middle scale domain, in the large scale domain also Soldner coordinates are used beside the Gauss-Krüger coordinates. In the Netherlands, UTM and stereographic coordinates are commonly applied. In France the state coordinates are given in Lambert or in Soldner projection. In Hungary two systems exist for the small and middle scale domain:

- Gauss-Krüger,
- Unified National Map System (cylindrical projection), and various systems for large scale domain.

10. Some aspects of the selection of reference systems for environmental monitoring

The selection of the reference systems for environmental monitoring depends on various factors. The most important criteria are:
- the size of the area or region to monitor,
- the methods of the data acquisition,
- the kind of existing data.

There are global, regional and local environmental monitoring systems. For global monitoring geocentric or geographical systems are suitable. For regional and local monitoring planar systems are usually adequate.

Systems/instruments for data acquisition are based on various reference systems. Surveying instruments are working commonly in planar systems. Global Positioning Systems (GPS) determine geocentric coordinates. Remote sensing users get satellite images in UTM system. Therefore, existing data are given very often in different reference systems.

Generally in environmental monitoring data based on various reference systems are applied. So it is necessary to select the reference systems of the data analysis (normally as a function of the size of the studied area). Data of other reference systems must be transformed into the reference system selected for data analysis.

Questions

1. What are the most important models for the mathematical surface of the Earth?
2. What is the difference between geocentric and geographical coordinates?
3. Which kinds of map projection do you know? What are the typical map projections in your country?
4. What is the difference between similarity and affine transformation?
5. Which kind of reference system would you use for environmental monitoring of a nuclear power station?

References

Bernhardsen, T.: Geographic Information System. VIAK IT and Norwegian Mapping Authority, 1992

Bill, R.; Fritsch, D.: Grundlage der Geo-Informationssysteme 1. Wichmann Verlag, Karlsruhe, 1991

Jansa, J. Vozikis, E.: SORA-MP a program for digitally controlled map transformation. ISPRS Congress Commission III, 1984

NCGIA: Core Curriculum - Introduction. Tecnical Issues in GIS Curricula, 1990

Torge, W.: Geodesy. Walter de Gruyter, Berlin, New York, 1991

II.2 Field Surveying Data

Ferenc Sárközy, Budapest

Goal:

Spatial data can be acquired from several sources. One of the most important methods of collecting spatial data is field surveying. The chapter gives the main ideas of the two fundamental methods of field surveying (traditional and space techniques) with emphasis on the characteristics of their products. As a result of studying this chapter, students should be able to assess realistically the overhead and accuracy of the method considered, they should be able to choose correctly the right way of data capture for their LIS or GIS tasks.

Summary:

In Section *Introduction* the basic concepts of field surveying are outlined. The *traditional surveying methods* are divided into *plane surveying, height measurements* and *tacheometry*. All methods are discussed in four parts: geometrical (mathematical) principles, instrumentation, organization of fieldwork, computation and plotting. The basic objectives and some methods of hydrographic surveying are briefly outlined. We shortly discuss the role of the Global Positioning System (GPS) in field data acquisition. Absolute and relative positioning as well as static, kinematic and dynamic point determinations are sketched out.

Keywords:

Surveying, plane surveying, control densification, height determination, levelling, electronic tacheometry, total station, surveying computations, mapping, tide gauges, sounding, global positioning system.

1. Introduction

Spatial data can be characterized by two types of parameters: the first group describes the position of an object, the second some of its features. The first group consists of coordinates, the second one of attribute values.

As a matter of fact, the methods of field surveying produce coordinates of points located on, over or below the Earth's surface. This means that if we want to determine the data of an object for a LIS or GIS, we should find the object on the Earth's surface first and then perform the following sequence of activities: decompose the object into points, survey the points and compute their coordinates, record its topology (sequence) of points and obtain one or more characteristic properties of the object. For example, by surveying an industrial plant we should first of all select such objects as roads, buildings, parks, lines, etc. Secondly we should decompose all the objects into characteristic points (points of deflection of lines, corners of buildings, etc). These should be measured then. We should include in the surveying process the record of the way in which the points should be connected on the map by setting up the object from points in the information system. The third step is the collection of attributive values of the objects, for example the voltage of lines, the diameter of pipes, the destination of buildings, etc. Strictly spoken, this last step is not the subject of field surveying.

Consequently, **field surveying determines the coordinates of points and their sequence to build up an object.**

Now we should answer the question: what do the coordinates of a point mean?

To answer this question we should refer to Chapter II.1 dealing with the reference systems. For clarity and completeness of our discussion we will repeat some basic ideas of that chapter here.

We can speak about two fundamental types of reference systems called **Global Reference Systems** and **Local Reference Systems**. The Global Reference System serves as basis for the **geodetic surveying** that determines control points over large territories and, therefore, needs such a kind of reference that approximates well a generalized Earth surface. Local Reference Systems satisfy the demands of **plane surveying**. In plane surveying relatively small areas are taken into consideration for one selected reference system; the actual size of the area depend on the distortions of the selected projection. Local Reference Systems are not independent from each other but are connected by a choosen Global Reference System.

Fig. II.2-1 shows the principal components of a Global Reference System. We can use a rectangular spatial XYZ coordinate system to fix the position of a point P on the surface of the Earth. As a rule the Z axis approximates the mean rotational axis of the Earth, while the origin of the system (point O) should be close to the center of gravity of the Earth. We can make three remarks about the rectangular systems.

We use the plural in connection with rectangular systems because there are a lot of them. It is to be noted that only coordinates belonging to the same system should be used for each task.

The Global Positioning System (GPS) uses spatial rectangular coordinates for internal computation but delivers geographic coordinates to the user.

The most commonly used Global Reference Systems describe the position of a point on the surface of the Earth by means of the ellipsoidal coordinates ϕ, λ and h. The *latitude* of a point (ϕ) is the angle built by the ellipsoid normal passing through the point and the intersection of the point's meridian plane with the plane of the equator. The *longitude* of a point (λ) is the angle formed at the equator by the intersection of the XZ plane and the meridian plane of the point. The *height above the ellipsoid* of point h is the distance from the point to the ellipsoid surface measured along the ellipsoid normal passing through the point.

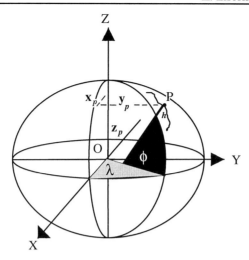

Figure II.2-1: Principle components of a Global Reference System

The ellipsoidal reference system approximates the Earth's surface by a rotational ellipsoid that is created by rotating the meridian ellipse around its minor axis. The shape of the ellipsoid is thereby described by two geometric parameters (distances): the ***semimajor axis a*** and the ***semiminor axis b***. This ellipsoid should be fixed to the Earth so that its center coincides with the Earth's center of gravity and its semiminor axis with the Earth's mean rotational axis. The relation of the ellipsoidal coordinate system to the rectangular spatial geocentric coordinate system is called the orientation of the ellipsoid. There are several ellipsoids devised by different geodesists and given by their semimajor axis ***a*** and semiminor axis ***b***, but the same ellipsoid also may have different orientations. The particular ellipsoid together with the chosen parameters of its orientation is called the ***geodetic datum***.

Due to practical reasons, neither the rectangular nor the ellipsoidal reference systems are used directly for height determinations. Theoretically, the reference system of height determinations is the ***geoid***, a level surface which intersects everywhere the direction of gravity at right angles and part of which coincides with the surface of the world oceans (assumed homogeneous and quiet). To determine one point of this surface physically, the countries with shores bordering the sea measured their mean sea level to remove the influence of water motion. From these starting points (***datum of height measurement***), by geometric levelling in combination with gravity measurements, the geoid was extended to the mainlands. The connection of levelling networks with different datums showed that the mean sea levels are not identical and can differ more than one meter. Therefore, there are several height reference systems based on different starting points.

The reference systems of height determination are directly suitable for local surveying tasks. ***Local Reference Systems*** for plane surveying are planar rectangular coordinate systems (so-called *grids*) connected by projectional equations to one of the Global Reference Systems. The projections transform a part of the ellipsoidal surface into a plane or a surface evolvable into a plane (cylinder or cone). The projections used in surveying are conformal which means that the angular distortions are negligible. The surface of the projection coincides, as a rule, with the ellipsoid (or sphere-approximating ellipsoid) in one point or in one or two lines depending on the projection selected. The distortion of distances and areas grows while moving away from the matching elements. To avoid mismatches between the measured (on the ellipsoid) distance and the distance computed from planar coordinates, we should add corrections to the former. These corrections can be computed using the approximated coordinates of the end points of the line. In most cases these corrections are negligible for the surveying of details, but should be applied for control network densification.

The principal axes of the planar coordinate system are usually directed to the North (X axis) and to the East (Y axis) and, therefore, the English technical jargon calls the x coordinates *northings* and the y coordinates *eastings*.

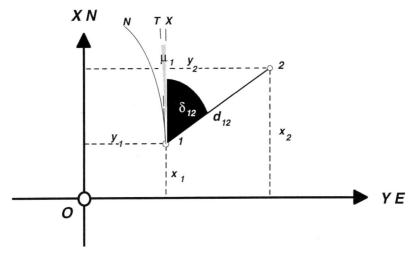

Figure II.2-2: Fundamental components of a surveying coordinate system

Fig. II.2-2 shows the fundamental components of a surveying coordinate system. We have two points, 1 and 2, given by their coordinates x_1, y_1 and x_2, y_2. The two points are connected by the line 12. The clockwise angle of the line 12 with the line passing through the starting point and parallel to the X axis of the system is called **bearing** or **angle of direction** and is denoted by δ_{12}.

While the X axis of the system is the projection of the meridian passing through the origin of the system, the other lines parallel to the X axis cannot be projections of meridians since they do not intersect each other in the pole. That is why the bearings differ from the angle with the meridian called *azimuth* or **true bearing** denoted usually by α. The value of the difference depends on the coordinates of the line's starting point. The name of this angle difference is the **convergence of meridian** and it is denoted by μ.

In connection with two points, we can perform two basic operations in the surveying coordinate system. In the first case the coordinates of the starting point x_1, y_1, the bearing of the line δ_{12} and the distance d_{12} between the two points are given, the unknown values being the coordinates of the second point x_2, y_2. The solution is:

$$x_2 = x_1 + d_{12} \cdot \cos \delta_{12} \qquad\qquad (II.2\text{-}1)$$

$$y_2 = y_1 + d_{12} \cdot \sin \delta_{12} \qquad\qquad (II.2\text{-}2)$$

The inverse task is to determine the bearing and the distance between two points x_1, y_1 and x_2, y_2, the coordinates of which are given. We can solve the task by means of the following equations:

$$\delta_{12} = atan\, \frac{y_2 - y_1}{x_2 - x_1} \tag{II.2-3}$$

$$d_{12} = \sqrt{(x_2 - x_1)^2 + (y_2 - y_1)^2} \tag{II.2-4}$$

We have to emphasize the importance of the order of indices in Eq. (II.2-3). A change of indices results in an error of 180 degrees in the bearing.

In most countries the horizontal and vertical control networks are developed and maintained by the state. Their density may differ according to the size of the country and the development of its infrastructure. The average distance between horizontal control points varies from 7 km to 1.3 km. The points are monumented as a rule by a concrete plinth with a metal point mark (hole or cross) inset in it. The national control points also have supplementary markings under the plinth in the ground. The density of the vertical control points (also called bench marks) is similar to the horizontal control points. They are located along the levelling lines. For their monumentation bolts are often cemented in buildings and other constructions especially in bridge abutments, or, in lack of them, in concrete posts. The top surface of bolts should be spherical.

The points of horizontal and vertical control networks are separated until now. This separation will probably disappear in future. The coordinates and heights as well as the sketches of the points are available for the public in the surveying (cadastral) offices.

2. Methods of control surveying

Detailled surveying performed by any traditional method determines the coordinates of the object points in relation to the control points. These methods, except levelling, require a direct sight between the control point and the related point of the object. The points of national control as a rule do not satisfy this requirement. Therefore, in almost all cases a densification of the control network should precede the actual detail surveying. The most commonly used method for horizontal network densification is called **traversing**.

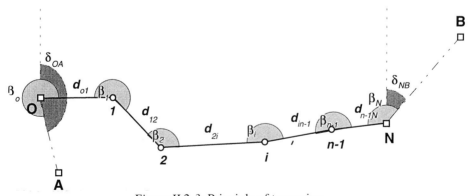

Figure II.2-3: Principle of traversing

Fig. II.2-3 shows the principle of traversing. The known points of the national control network are denoted by *A, O, N, B,* the new points of densification by 1, 2, *i, n*-1. To determine the coordinates of the latter we measure the distances $d_{O1}, d_{12}, d_{2i}, d_{in-1}, d_{n-1N}$ and the angles

$\beta_O, \beta_1, \beta_2, \beta_i, \beta_{n-1}, \beta_N$. From the known coordinates of points O and B and by using Eq. (II.2-3), we can calculate

$$\delta_{OA} = atan\frac{y_A\text{-}y_O}{x_A\text{-}x_O}$$

It is easy to see that the bearing of the line $O1$ is the sum of the bearing calculated from the known points and the angle measured on the point O. That means $\delta_{O1} = \delta_{OA} + \beta_O$.

If we consider that $\delta_{10} = \delta_{O1} + 180$ or, in general, that $\delta_{ij} = \delta_{ji} - 180$, we can calculate all the bearings of the traverse in a similar way. After that we can use Eq. (II.2-1) and (II.2-2) and, consequently, calculate the coordinates. At the end of the computations we will find that we have two values for the coordinates x_N, y_N as well as for the bearing δ_{NB}: one from the traverse computations and one from the archives. The archive data should be considered as unchangeable and with the help of the differences between the true and measured data, we can control our measurements and calculations. If the differences (called misclosures) do not exceed the error limit depending on the circumstances of the traversing (type of instruments, method of centering the instrument and target, number of repetitions, number of new points, average length of lines, etc.), they are used for the adjustment computations.

The first highly responsible task of the traversing procedure is the reconnaissance, the checking of existing control points and the choice of new points. The national control points are often damaged or destroyed and, therefore, many more points than planned must usually be visited. Especially in urban and rural regions, surveyors have to face sometimes more than 50% point loss. The new points should be chosen with the requirements of the survey in mind, aiming at good visibility between points and ensuring a proper shape of the traverse from the point of view of error propagation. This means that the points should be located at nearly the same distance from each other and that the angles of deflection should be close to 180 degree.

If the points of the national control network are too sparsely distributed (when their average distance exceeds 1.5 km) it is preferable to perform the traversing in two steps. As a rule the points of primary traverses are marked by concrete posts, the points of secondary traverses are usually marked by pegs or steel nails.

The procedure of measurement is performed by a surveying team that consists of an engineer, a technician and two or three helpers. The observer centres the theodolite or total station (combined angle and distance measuring instrument) over point O. The technician draws the sketch of the station with measurements from close natural or artificial objects and displays the visible sights to old and new control points marking differently the lines with direction-, distance- or combined measurements. The engineer measures the angle β_o by reading to the back station (A) and to the fore station (1) and subtracting the former from the latter. The recording of readings in the field book is the duty of the technician. The total stations have devices that can record the readings in a solid state storage unit.

After finishing the readings the car takes the surveyor and instruments to the second station (point 1). On this station both angle and distance measurements should be carried out. The helpers with the van remove the tripod from point A, put combined target reflectors in the sockets on points O and 2 and centre the tripod with tribrach onto the next point (i). If the team has separate distance measuring instruments the engineer measures, first of all, the angle β_1 in the same way as he measured the angle β_o on the preceeding point. For computations we need the horizontal distance between two points. In the field the tripod heads are on different levels. To reduce the sloping distance into a horizontal one we should measure the ***vertical angle*** α. Fig. II.2-4 shows the interpretation of the vertical angle. In most of the cases the instrument measures the zenith angle denoted by ζ. These two angles are connected by the equation $\zeta = 90 - \alpha$. Now we can easily reduce the slope distance by means of the equations

$$d_h = d_s \cdot \cos\alpha \qquad \text{or} \qquad d_h = d_s \cdot \sin\zeta .$$

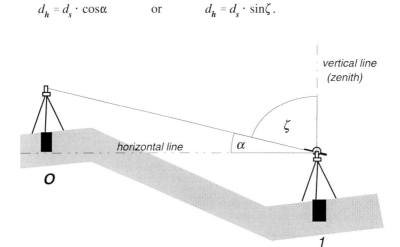

Figure II.2-4: Vertical angle α and zenith angle ζ

From point 1 it is expedient to measure both distances and therefore, both vertical angles have to be measured, too. After finishing the angular measurements the theodolite should be changed for the distance measuring instrument. Meanwhile, the technician reads the thermometer and barometer and calculates (or determines from a diagram) the actual refractive index of the atmosphere that should be set in the distance measuring instrument. With the measurement of the two distances the activities on the station are finished. It is to be noted that distance measurements can only be performed on numbered stations.

If the surveying team uses a total station with data recorder, the procedure is simpler; there is no need to book the readings, the instrument has to be directed to each of the target points only once, the horizontal distance is automatically computed and recorded. Similar operations must be performed on all points including the known control point at the end of the traverse.

It is easy to compute a stand alone traverse by means of a calculator and by applying a method of approximative adjustment. For more complicated network structures manual computation is not efficient enough and, therefore, in such cases various computer programs are in use. With such structures additional directions and distances are measured to ensure the homogenous accuracy of the network. In an up-to-date network of horizontal control densification the relative accuracy of neighbouring points should not exceed 2 cm.

Sometimes one has to determine only a small number of new control points. In such cases the methods of *intersection, resection* or *triangulation* have to be used (Fig. II.2- 5).

The *intersection* method uses two known points as stations. At station *A* the directions *AP* and *AB* are measured and the angle α is calculated as their difference. Similarly, the angle β is determined at station B. The computations aiming at the coordinates of the new control point *P* are easy as well. First we should calculate the bearing δ_{BA} and $\mathbf{d}_{BA}$ applying Eq. (II.2-3) and (II.2-4). On the basis of the sine theorem we obtain the unknown distance

$$d_{BP} = d_{BA} \cdot \frac{\sin(\alpha)}{\sin(\alpha + \beta)}$$

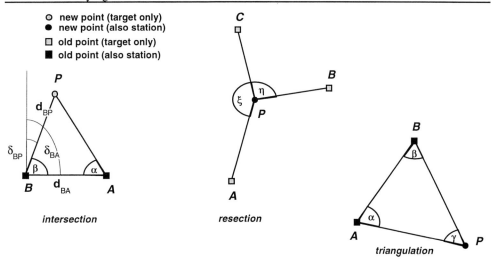

Figure II.2-5: Principles of intersection, resection and triangulation

Considering that $\delta_{BP} = \delta_{BA} - \beta$ we can substitute the actual values into Eq. (II.2-1) and (II.2-2) and obtain the wanted coordinates $\quad x_P = x_B + d_{BP} \cos\delta_{BP}\,, \qquad y_P = y_B + d_{BP} \sin\delta_{BP}\,.$

For ***triangulation*** almost the same procedure is used. The only difference to the intersection method is that the third angle of the triangle is also measured (the point ***P*** is not only target but also station) and consequently the sum of measured angles may differ from 180 degree. The angles should be corrected with the third of the difference (misclosure) in such a way that the corrected angles sum up to 180 degrees.

The densification of the ***vertical control*** is a less complicated task. Aiming at the determination of a few temporary bench marks to provide the vertical survey based on physically realised levels, usually the method of ***geometric levelling*** is applied. The basic principles of geometric levelling are shown on Fig. II.2-6.

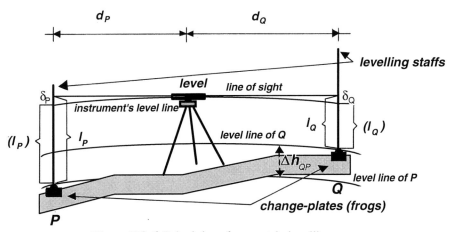

Figure II.2-6: Principles of geometric levelling

Consider two points ***P*** and ***Q*** as shown in Fig. II.2-6. The height difference of these points is the distance along the normal to the level surfaces passing through the points. The level lines are the sections of level surfaces with vertical planes. The telescope of an adjusted level instrument determines the line of sight or more precisely the line of collimation. This line is tangential to the level line passing through the point of its intersection with the instrument's vertical axis. Let us set up scaled levelling staffs on points P and Q. If we could make readings on the staffs at the points of intersection with the instrument's level line using the readings (l_P) and (l_Q), we could calculate the height difference $\Delta h_{QP}=(l_P)-(l_Q)$. But in reality the readings on the staffs l_P and l_Q are made by the line of collimation. It is easy to understand that the height difference calculated from the real readings will deliver the true value only in case the differences

$$l_P - (l_P) = \delta_P \qquad \text{and} \qquad l_Q - (l_Q) = \delta_Q$$

are equal. This is the case when three conditions are fulfilled:
  a) the level lines can be considered parallel;
  b) the instrument is adjusted;
  c) the distances from the level to the staffs are equal, i.e. $d_P = d_Q$.

The condition a) is fulfilled when the points of the national control network are dense enough (the distance between the points does not exceed a few kilometres). The conditions b) and c) are fulfilled by keeping the geodetic process (rules) of levelling.

The distances d in engineering levelling must be less than 100 metres, in precise levelling less than 40 meters, i.e. if the distance between the points to be levelled is more than 200 metres or 80 metres respectively, the levelling should be performed on several stations (Fig. II.2-7).

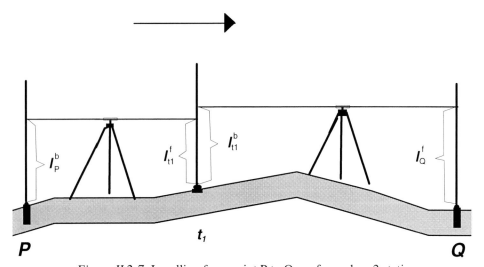

Figure II.2-7: Levelling from point P to Q performed on 2 stations

In Fig. II.2-7 there are two stations between the points P and Q. The arrow shows the succession of the measurement along the line. On each of the stations we make at least two readings: the *backsight reading* and the *foresight reading*. The staff point at the end of the first station is known as *change point*, because it is the staff position during which the position of the level is being changed. The change points are marked either with spherical headed nails driven in a peg (in the case of precise levelling), or with change plates ("frogs"): iron castings with clawed bottom and spherical head (in other cases).

On all stations we can calculate the height difference of the two points making the station. If the foresight is larger than the backsight the difference represents a *fall*, in the opposite case the difference represents a *rise*. The height difference of an optional point of the levelling in relation to the starting point can be calculated by summing up the height differences of the preceding stations, not forgetting that the sign of the falls is negative. The same result can be achieved by the use of the following equation:

$$\Delta h_{PQ} = \sum_{i=1}^{n} l_j^{backsight} - \sum_{i=1}^{n} l_j^{foresight} \tag{II.2-5}$$

where P and Q = the first point and the last point of the levelling (by densification of the vertical control the height of P is given and the measurements aim at the determination the height of Q),

n = number of stations,

l_j^i = readings on the staff.

As a rule the levelling is performed in two directions: from the starting point to the end point (from P to Q) and then from the end point closing back to the starting point. Obviously the height difference of the closed circuit should equal nil. In the reality, because of the errors of measurements, the height difference can differ from nil. This difference is called **misclosure**. For the *engineering* levelling the misclosure should not exceed $\pm 10\sqrt{K}$ mm. The corresponding error limit for *precise* levelling is given by the equation $\pm 2\sqrt{K}$ mm. In both equations K denotes the length of the circuit in [km].

To conclude we have to summarize some basic awareness about the **traditional surveying instrumentation**. We use the word "traditional" meaning "terrestrial" to distinguish from the satellite positioning methods. The latter will be dealt with in Section 4.5.

Surveying determines coordinates by measuring **angles** and **distances.** The angles are measured by **theodolites,** the distances by **tapes, optical distance meters** and **electronic distance meters (EDM)**. Both angles and distances are measured by **tacheometers.** Instruments combining an EDM with an electronic theodolite and a data recorder are called **total stations**. The **levels** obtained by readings on levelling staffs provide data for the determination of height differences. **Theodolites** are mechanical-optical precision instruments suited for measuring angles (directions) in horizontal and vertical planes. Fig. II.2-8 shows these angles.

Fig. II.2-8 shows three points A, B, C on the terrain. The directions **BA** and **BC** form at the point B the spatial angle β. The theodolites measure the horizontal projection of this angle denoted by β. For this purpose the vertical axis of the theodolite is set over the point in such a way, that it coincides with the vertical line passing through the point. A graded glass circle is mounted on the axis at right angles, i.e. the circle of a levelled theodolite is in horizontal position. The upper part of the theodolite also called **alidade** rotates around the vertical axis.

This axis is set in vertical position by means of the spirit levels mounted on the alidade. The **telescope** is fixed on the alidade by a horizontal axis that allows rotation in the vertical plane. When directing the telescope towards a point, the horizontal circle remains unmoved, the telescope rotates first together with the alidade around the vertical axis and then the telescope alone moves around the horizontal axis. By means of a **reading device** located on the alidade, it is possible to determine the reading on the horizontal circle corresponding to the direction to a point. For example on station B we can direct the sight of the line of the telescope towards the target on the point A and make a reading l_A; similarly, after directing towards the point C we can make a new reading l_C. The difference of both readings forms the horizontal angle of the points in question, i.e. $\beta = l_C - l_A$.

The circle which is rigidly fixed on the horizontal axis is called **vertical circle**. The **index level** (or the **inclination compensator**) controls the reading device of the vertical circle. Therefore, a reading on the vertical circle of a direction is an absolute quantity (opposite to the reading of a direction on the horizontal circle) and shows the angle of the direction with the horizontal or (vertical) line.

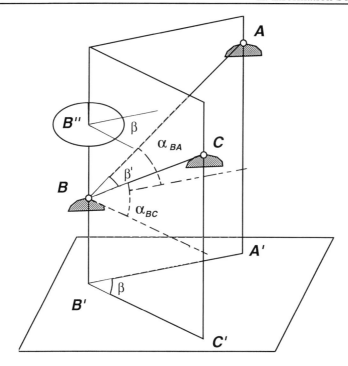

Figure II.2-8: Horizontal and vertical angles measured by theodolites

On the basis of attainalbe accuracy of readings, theodolites can be divided into three groups. First class theodolites have the accuracy of 1 to 2 seconds of arc, the corresponding quantities for second class theodolites are 5 to 10 seconds of arc. Third class theodolites, with reading accuracy of 0.5 to 1 minute of arc are not used in control network densification.

Most modern theodolites are equipped with electronic reading devices and, therefore, are called *electronic theodolites*. Electronic readings are suited for recording which can be realised internally by *memory cards* or externally by *data collectors*. The built-in microprocessor controls the reading device and by running programs increases the functionality of the instrument. External recording (use of data collectors) is less convenient than internal recording and is only applied when a lot of other data (not provided by the instrument, e.g. land use classes etc.) have to be acquired additionally or when sophisticated computations are to be carried out in the field. In fact, data collectors are field computers and supplied with proper software they can offer much more functionality than the electronic theodolites themselves. On the other hand, a significant decrease of the speed of the field work results, if the operator of the instrument also operates the data collector. The proper solution is to entrust the data collector to a separate, technically well-trained helper.

For a very long time, distance measurements were more difficult and less accurate than angle measurements. Nowadays the situation has changed. The *electromagnetic distance measuring (EDM)* instruments operate with amplitude modulated infrared light. The basic idea of this method is very simple: if we measure the time of propagation of a marked light spot from the instrument to the target and back to the instrument and, knowing the velocity of light propagation in the atmosphere, it is easy to calculate the instrument-target distance. For 'marking' the light spot the amplitude modulation is used. The light beam is normally reflected at the target station by a prism. The EDM instruments operate in different ranges: short range instruments, 1-2 kilometres; medium range instruments, 5-7 kilometres and long range instruments, 15-20 kilometres. The accuracy of the measured distances depends on the precision of the instrument as well as on the method of determination of the atmospheric influence. For short distances, the latter can be neglected. In this case the error of a measured short distance should not exceed 2-3 millimetres.

EDM instruments measure   sloping distances. The coordinate determinations require horizontal distances; therefore, the measured distances should be reduced. The reduction is possible if we know either the height difference of the measured points or the vertical (zenith) angle of the measured line. The first is available if the end points of the distance have levelled heights. In the second case we should use a theodolite and consecutively or simultaneously measure the distance with the vertical angle. As a rule the end points of lines have no levelled heights and, therefore, the use of stand alone EDM instruments is considerably limited, except of the ones with built-in sensor measuring the vertical angle. This sensor inputs the reading to the instrument's microprocessor so that the sloping distance, the horizontal distance and the height difference between instrument and reflector can be displayed at will.

Modern EDM instruments are small and of very light weight (1-2 kg). This makes it possible to mount them either on the telescope or on the telescope's supports of the theodolite. In the first case the target point should be directed only once (the telescope and the EDM instrument are moving together); in the second case the target should be sighted separately in the vertical plane (the EDM instrument moves with the alidade but not with the telescope), hence the EDM instruments mounted to the supports or operating in stand alone mode, should have their own telescope, too.

The availability of *'add-on'* EDM as a hybrid system has facilitated the transition to the homogeneous electronic distance and angle measuring system called *'total station'*. The new system consists of the following four main parts: *electronic theodolite, EDM instrument, microprocessor with wired in programs, internal* or *external storage unit.*

Total stations have at least two main advantages compared to the hybrid system: both, distance and angle readings are carried out electronically. That allows their automatic recording. The micro-processor, depending on the type of built-in programs, can solve several surveying tasks in the field using as input the measured angles and distances. As disadvantages we can mention in the first place the significant weight of most systems and secondly, frequent problems with their power supply. While in the hybrid system the theodolite does not need any power source and the EDM instruments are turned on only for the time of distance measurements, the total stations cannot work without electricity and they often exhaust more than one battery during a long working day in the field.

The accuracy of distance and angle measurements of a total station should be in mutual correspondence which actually is not always the case. The technical solution of the high precision distance measurement is easier than that of angular measurement and, therefore, we find total stations on the market with high accuracy of distance measurements (3 to 5 mm) and low accuracy of angular measurements (10" to 30"). Such instruments are adequate only for detail surveying. The control densification demands at least 5" angular accuracy, but the use of instruments with higher accuracy (2") is advisable.

The instrument for the most precise determination of height differences is called a *level*. The principle of a level is shown in Fig. II.2-9. The level should direct the line of sight tangentially to the level surface. This objective is achieved in the *tilting levels* by a spirit level fixed to the telescope. This level is adjusted to the line of sight if its axis (a tangential line to the origin of its scale) is parallel to the former. In this case the line of sight is horizontal if the bubble of the spirit level is brought to the centre of its scale. The bubble's movement is a result of the slight rotation of the telescope around its horizontal axis. This rotation is steered by a fine setting screw at the eyepiece end. The bubble should be brought to the centre for each reading of the levelling staff.

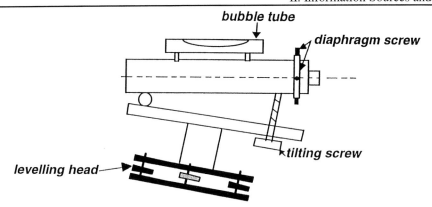

Figure II.2-9: Principle of a level

Automatic levels bring the line of sight into horizontal position by means of a *compensator*. The operating range of the compensator is at about 10 min of arc. Therefore, the instrument should be preliminarily levelled, only coarsely, using the three-screw levelling head and a small target level mounted on the tribrach. Because of the operation of the compensator the surveyor has not to level the instrument manually during the observations.

The first and until now sole *electronic level* the *Wild Na 2000* was constructed in 1990. By virtue of electronic scanning this instrument evaluates *bar coded* staff images. The surveyor should set up the instrument using the three screw-levelling heads, focus the staff and press a button on the control panel; after 4 seconds the staff reading is digitally displayed. An external data collector allows the data to be stored, pre-processed and loaded into computers in the office.

Both, automatic levels and tilting levels can be divided into two groups. The *engineering levels* have technical parameters (magnification, bubble sensitivity, compensator's stability) that ensure an accuracy of about ±2 mm for a double run of levels over 1 km. The accuracy of *precise levels* should be one order of magnitude higher than that of engineering levels. To reach this accuracy, besides the improvement of the technical parameters mentioned above, a new device for readings and corresponding levelling staffs are used. The new device is called parallel plate micrometer and is a glass plate mounted in front of the object lens. By rotating a scaled drum the plate gets a tilting motion that results in a parallel shift of the line of sight. The corresponding levelling staffs have usually 5 mm scaled lines painted on an enforced invar-steel strip embedded in the staff. The direct reading of the equipment is 0.1 mm. It is worth mentioning that readings with engineering levels *estimate* only 1 mm.

3. Detail surveying

There are several points of view when discussing detail surveying. If the basis of classification is the objective and the end product is the survey, we speak about *cadastral surveys, engineering surveys* and *topographic surveys,* respectively.

The *cadastral surveys* produce plans of property boundaries for legal purposes. The scales of these plans depend on the mean dimensions of the sites (parcels) and on the tradition of the particular country; thus they may vary from 1:500-1:1000 in urban areas to 1:2000 in villages and to 1:4000 in uninhabited areas. One of the main features of the cadastral plans is the lack of height data. If thinking in digital plans, the result of the cadastral surveys may be organised in separate coverages. The lots themselves can form different coverages according to the type of crop; other coverages can consist of the roads, buildings, sidewalks and surveying control points.

Engineering surveys are required before, during and after any engineering work. From the point of view of LIS and GIS the *realisation* or *'as built' survey* is especially important. The scale of engineering surveys usually varies from 1:500 to 1:2000 and they always contain some kind of height information. Such survey products as *cross sections* and *longitudinal sections* display height variation with distance. The digital plan of an 'as built' survey can cover area type objects (e.g. buildings), line type objects (e.g. pipe-lines) and point type objects (e.g. lamp-posts).

Topographic surveys produce plans and maps of natural and man-made features. The scale of such **plans** is so large that all the features can be drawn adequately at their "true" scale. On **maps** many features are expressed by symbols that are larger than the features themselves at true scale. The scales of topographic plans vary from 1:500 to 1:1000. They are created for selected areas, as required by the engineering design. Typical scales of topographic maps are 1:10.000, 1:25.000, 1:50.000 and 1:100.000. Nearly all European countries are fully covered by topographic maps at scales of 1:25.000 and less. These maps are of great importance for almost all branches of economy, defence, culture, environmental issues, regional planning and administration. Digital products consist of five main coverage groups: relief, hydrography, railways and roadways, settlements, (pipe)lines, type of crops. In most cases the relief is stored in the form of a **Digital Elevation Model**. Topographic maps are usually surveyed by aerial photogrammetry while topographic plans are produced as a rule by field surveying.

Another approach to classify detail surveys is based on the characteristics of the methods applied such as apparatus used, types of measurements and form of recording. Starting with the latter, we can distinguish electronic recording and manual recording (booking) into a field book and/or by means of a drawing (sketch). It should be mentioned that *in most of the cases, even with electronic recording, the topology has to be supplemented with a manual sketch*. The types of methods differ in the quantities they yield. These measurements can only be distances, distances and horizontal angles, distances and horizontal and vertical angles, only horizontal angles, horizontal and vertical angles, distances and height differences, horizontal angle distances and height differences. The instruments used can range from single tapes to electronic tacheometers. When selecting the surveying method the number of points to be surveyed is of crucial importance. Here, only methods suitable for surveying a large number of points shall be discussed.

3.1 Offset surveying

This is the most commonly used, traditional method to determine a large number of horizontal coordinates. The apparatus used for this method is very simple and consists of a steel band, two tapes, 6 ranging rods, a set (10 pieces) of arrows and an optical square. While the inexpensive equipment can be considered as advantageous, the large number of the surveying team members (e.g. two technicians and six helpers) is, no doubt, a disadvantage. Another disadvantage of this method is its unsuitability for automation.

The basic idea of the method is explained in Fig. II.2-10. The traverse points (e.g. T1 and T2) define the **surveying line**. A steel band lies in the line and the ends of the band are fixed with steel arrows. The surveyor finds the starting points of the rectangular to the surveying line *offset*s passing to the object points. This is possible with the help of an **optical square**. When using the square, the ends of the line as well as the object point should be marked by **ranging rod**s. The distances between the beginning of the line and the starting point of each offset are measured on the steel band. The offsets are measured by tapes. The results of the measurements are recorded on a **sketch** as shown in Fig. II.2-10. The numbers in brackets are the end measurements, the hyphens connect the measurements on the surveying line to the corresponding offsets, the arrows show the succession of line measurements. For checking purposes the sides of the objects are also measured. If there is an obstacle crossing the offset to the object point, it can be surveyed by *ties* i.e. two distances measured between selected points of the surveying line and the object point.

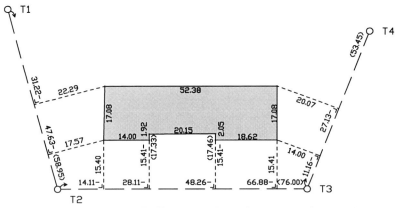

Figure II.2-10: Principle of offset surveying using rectangular coordinates

The first step of the surveying procedure is site reconnaissance: existing points of control of the network must be located and if necessary new control points should be marked and measured. In this phase a sketch should be drawn showing the control points, the surveying lines and the object points. The second step of the procedure is the surveying itself. Two helpers are on the ends of the band with rods ready to move the band forth. The technician with the square moves along the band and finds end points of offsets to object points on both sides of the line. Two helpers mark these object points with rods and two other helpers pull tapes from the object points to the square on the line. The head of the surveying party makes the readings on the band and on the tapes and books them on the sketch. Often it will be necessary to complete the sketch. The third step of the procedure is the office work. Measured distances are then transformed into the coordinate system of the control points and then the coordinates are plotted. The points can then be connected on the basis of the field sketch. For computerised plotting the coordinates are usually used as input data. The connections of object points can be made interactively by using an adequate graphic software (e.g. AUTOCAD).

3.2 Method of radiation

The section above clearly shows that the method of offsets is nothing but the measurement of rectangular coordinates in the coordinate system of the corresponding surveying line. The method of radiation measures the polar coordinates in the same system (Fig. II.2-11).

The polar angles β must be measured by traditional or electronic theodolites. In principle the distances d may be measured in an arbitrary way, but in practice the taping of long distances involves a large amount of difficulties. Therefore, the method is efficient only where distances are measured by EDM instruments mounted on the theodolites or by total stations. The distances should be reduced for the horizontal plane and, therefore, the vertical or zenith angles must be measured, too. If we also measure the height of the instrument h and the height of the target l, then we can determine the height differences between the object points and the station point, too. In this case the method of radiation grows into the method of *tacheometry.*

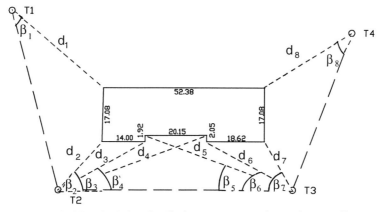

Figure II.2-11: Principle of radiation measures using polar coordinates

3.3 Method of (electronic) tacheometry

We have seen that in instrumental operations there are no essential differences between the modern version of the method of radiation and that of tacheometry. In fact, the instrumental heights should be measured only once on a station, and the target heights are often constant because of the fixed length prism poles. The two methods differ mainly in the goal of the survey. The tacheometry is the universal method of field surveying while the other methods are applicable only for a limited part of all possible surveying tasks.

As a consequence of its universal applicability, we should pay more attention to the preparations of the surveying operations.

Using the method of tacheometry both the situation and the relief can be measured. While the points of the relief usually occupy only one coverage, the objects of situation can be shared by several coverages. The first step in preparations is the assignment of object types to coverages. One coverage can contain different objects and object points; therefore, the second step is assigning feature codes to object points. The code of an object point should refer to the coverage of the object as well as to the individual features of the object and the point itself. For example if - among other things - wells are to be surveyed we can construct the following code: *wwell1s*. The first letter in the code denotes a pointer to the coverage (w=water), well is the name of the object type, the number after the name designates the class of the well and the last letter shows that the point itself is a stand alone object. The feature codes are loaded into the stack of the total station or into the data collector and in the surveying process they are selected and recorded together with the other data related to the point. In the third step the surveyor prepares a detailed drawing about the objects of the area considered. The scale of the sketch should be chosen in such a way that the ID numbers of the surveyed points could be written easily. By drawing the sketch for the relief surveying the lines of valleys and ridges and those of slopes and slope changes, in general all "form" lines and point-like singularities should be fixed. The surveyor imagines the relief as a polyhedral surface composed of plane triangles. The edges of triangles coincide with the form lines, their vertices with the point like singularities. A successful sketch ensures a good quality relief survey in spite of sparse point locations. A good sketch increases both the quality and the efficiency of the survey.

The surveying is performed by a team of two technicians and two or three helpers. Depending on the scale of the survey the helpers holding the prism poles either walk (in the case of large scales) or use some kind of vehicle (bicycle, motorcycle, van) to cope with the larger distances in the case of smaller scales. All members of the team are equipped with "walkie-talkies". One of the technicians operates the total station. He sets up and orients the instrument, measures the height of the instrument, and turns on the power. The circles of the instrument as a rule need indexing which can be performed either automatically or manually. After indexing, the horizontal circle should be oriented to grid bearing. To complete the orientation, the coordinates of the station point and those of another one, visible from the station control point (the so called Backsight BS), should be fed into the total station. After setting up a target over the BS and sighting at it by running a key-press function the horizontal circle is oriented. The next activity of the operator is to type data (among others the date, point number, coordinates, instrument height, feature code, atmospheric correction) in the station. From that moment on the instrument is ready for detail surveying. The operator sights at the prism, selects one of the recording masks (it is advisable to record the originally measured quantities: horizontal angle, vertical angle, slope distance). The entire surveying process is governed by the other technician. He indicates to the helpers where to go with the prisms, queries them about the point features, target height, etc., and draws the point numbers (which have been checked up before with the operator) into the sketch. Often he has to complete a sketch prepared in advance. The operator, in turn, types in the complementary data to be recorded dictated by the team leader.

The following office work depends essentially on the goal of the survey. If the end product is a plan without description of the relief we can use, for example, the well-known CAD-program AUTOCAD. But how to transfer the data recorded on the memory card into the processing program? If the processing program has a direct interface to the total station, we can call the particular module of the program that completes the transfer after connecting the turned-on instrument to the computer. When lacking an interface program module (for example in the case of AUTOCAD) first of all we should transfer the contents of the card in form of an ASCII file to the computer's storage unit. Vendors of instruments usually provide transfer cables and floppy discs containing such kinds of transfer programs with the total stations. Secondly: we should compile a program in AUTO LISP that reads the transferred ASCII file and transforms it into AUTOCAD objects. This program sorts the points into layers on the basis of feature codes, marks them by a corresponding block with point numbers as attribute values, and inserts them into the places determined by the coordinates calculated from the measurements. The objects must be constructed from the inserted points using the sketch on the one hand and the functions of the software on the other. The points determined by ties should be inserted manually, by proper use of the possibilities of coordinate transformation. The insertion does not require any additional calculation. The result can be plotted or transformed in DXF format so as to transfer it into another GIS/LIS software. If relief is also measured, the office work requires a software with a DEM or TIN module. While the problem of data transfer still depends on the particular features of the software (existence or lack of interface, ability to access and manipulate ASCII files), the operations regarding the relief are mostly automated. The end products are coverages of situations and a Digital Elevation Model. In most cases a hard copy, i.e. a paper plan is produced, too. The relief on the paper plan is usually displayed by means of contour lines.

If the scale of the survey requires the production of hard copies in form of maps, the construction of the output is much more complex, since it demands an operator with cartographic skills. To facilitate this task most of the up-to-date GIS software-programs have modules supporting cartographic output.

3.4 Methods using levelling

The only complete method in this section is called *gridding*. As shown in Fig. II.2-12, two classes of squares are set out on the area in question. The setting out can be made using a theo-dolite and a tape or EDM instrument. The corner points of the large squares are marked by pegs, the small squares are marked only along two parallel sides of the large square, usually by ranging rods. Along both a "vertical" and a "horizontal" side of the large squares a second ranging rod is set out on the "horizontal" or "vertical" line passing through each rod on the original side. Each pair of rods determines a "horizontal" or "vertical" line and helps the staffmen to place themselves in the corners of the small squares.

The surveying team consists of two technicians and ten helpers (staffmen). The first step of the procedure is setting out the squares. At the same time another part of the team runs double levelling between the nearest bench-mark and one of the pegs. The proper gridding begins after these preparations. The level is set up approximately in the centre of the first large square. The staffmen occupy the corner points on the line *IH*. The points *I*, *H* (and *G*, *F*) play the role of change points. The operator makes readings on the staffs, the technician books them onto a sketch or into an electronic data collector with adequate software. For electronic recording the staff locations should be coded by the row and column indices and the number of the large square. If the survey aims at the completion of an information system, the national grid coordinates of two of the peg points must be determined, too.

This method is widely used for the survey of flat areas to design drainage, irrigation channels, airports, parking areas, sports-grounds, etc. It has two disadvantages: it demands a lot of helpers and it can not handle the survey of manmade objects effectively.

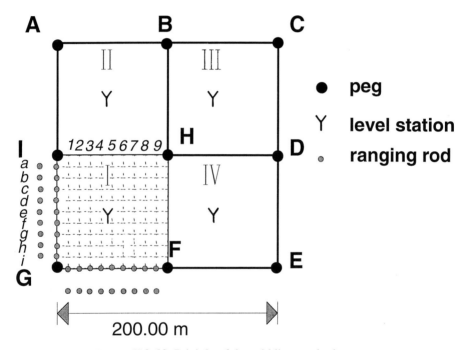

Figure II.2-12: Priciple of the gridding method

A special method of levelling for engineering purposes is to take ***longitudinal sections and cross-sections.*** The basic idea of this method is to exchange the national grid against a reference system of an arbitrarily shaped line that is physically fixed on the terrain. In other words, we can choose a road, a railway or a simple line fixed on the ground by pegs consisting of straight lines with points of deflection as the main axis of this particular reference system. By taking longitudinal sections we measure distances from the beginning of the line and level points (usually marked by pegs or nails) located on round distances (20 m, 50 m or 100 m), on places where the slope changes and where constructions intersect the line. The results of the survey are displayed in a diagram, the scale for the distances (horizontal axis) being different from the scale for the heights (vertical axis). The cross-sections are taken at selected points of the longitudinal section, at right angles to the straight line determined by the two neighbouring points. The points of slope changes are surveyed. The distances are measured from the starting point on both sides of the longitudinal section (to the left, the sign is minus; to the right, it is plus). The distances and staff readings are booked on a sketch. The end product is a diagram with equal scale for the distances and staff readings. The method of longitudinal profiles and cross-sections is frequently used in conventional design processes, but it is of no importance for data acquisition for GIS and LIS.

4. Hydrographic surveying

The methods of hydrographic surveying are methods that provide the information system with attribute data, with positional data, or with both.

A good example for the first category are *tide measurements.* The level of the water surface depends not only on the location of measurement but also on the time of measurement. The location of the measuring instruments - the so-called *tide gauges*, must be determined by some method of detail surveying. The readings on the gauges must be taken frequently at fixed time intervals and, therefore, some method of automatic recording is commonly used.

The most simple gauge is the *staff gauge* a graded staff with a scale that is longer than the interval between the highest and lowest tide. The staffs are fixed to moles, banks or other coastal constructions. The calibration of the staff is achieved with the help of a levelling line to the nearest bench-mark. The readings on the staff are made visually which is often difficult to do because of the waves.

For long-term regular observations *float gauges* with *self-recording devices* must be used. Up- to-date versions of these instruments have remote connections to computers. The float gauge diminishes the influences of waves and the self-recording equipment eliminates the necessity of the continual presence of an observer. The float gauges are located similarly to the staff gauges and their horizontal coordinates and zero levels should be also determined by detail surveying and levelling.

The results of tide measurements have countless applications, one of them is the *reduction of sounding*. Data sounding itself is a kind of depth measurement to determine sea (river) bed reliefs. Strictly speaking the method yields only positional data, but it is easy to find systems where the depth is considered as an attribute. We can distinguish direct sounding methods as the use of *sounding rods*, or *sounding leads on graded lines* and indirect sounding method such as *echo sounding*.

Direct methods are limited to depth measurement (of about 5 m the rods and 30 m the leads) and require a large amount of manpower with high professional skill. The sounding is performed from boats. The horizontal location of the boat must be determined by some method analogous to the methods of land surveying. For river sounding the method of hydrometric check-sections can be used. The end points of the sections are marked by means of concrete plinths and their horizontal coordinates and heights are determined by high accuracy. A graded steel wire is stretched over the river and aligned in the section between its end point and its middle point. The latter is located on a vessel that is moored. The boat with the rod or with a suspended weight incorporated into a sounding machine, moves along the wire and makes depth measurements at regular intervals marked on the wire. The readings on the wire and on the rod (sounding machine's dial) as well as the time of the measurement is booked, usually in a traditional way. For the determination of sounding corrections the record of the nearest tide gauge can be used.

The most modern and universal way of depth determination is the *echo* or *acoustic sounding*. Transducers for emitting and receiving sound pulses are mounted on the craft. The instrument measures the time between the pulse emission and the reception of the *first* reflected signal and, taking into consideration the velocity of sound propagation in (sea)water, calculates the depth. The location of the vessel can be determined by different methods, the most up-to-date being the Global Positioning System (see Section 4.5). In off-shore measurements the use of radio navigation systems is also common.

For river sounding the EDM instruments of both optical and microwave ranges are frequently applied. There are two basic schemes of river sounding. By the first scheme the vessel is steered in a straight line. The endpoints of the line have fixed coordinates. An EDM instrument is set up on one of the endpoints and measures the distances to the boat in fixed time intervals. An electro-optical instrument with data collector records these data at its station (the time should be typed in by the data collector). If a pair of microwave instruments is used, it can be equipped with an adapter to the remote station on the vessel that controls the speed of the paper of the graphical recording unit in such a way that the diagram will show the depth with the distance. Nowadays, electronic storage is frequently used instead of graphical recording. The second scheme allows more freedom in the steering of the vessel. In this case both coordinates of the vessel should be measured on the ground station and recorded together with the corresponding points of time. At the same time the sounding equipment on the vessel records the depths and the corresponding points of time. The two data sets are then matched in the office computer taking time as the common parameter.

A very important characteristic of the rivers and sea currents is the ***distribution of velocity*** in selected sections. The velocity of the water surface can be measured by means iffloats, but for measuring the velocity under the water surface special electric or electronic instruments called ***current meters*** have been developed. These are attached to a rope and let down in the water to the selected depth. The velocity, depth and time are then recorded. The determination of the horizontal position depends on the scheme of measurement and is identical with the corresponding method explained in connection with the sounding.

5. The Global Positioning System (GPS)

5.1 Basic concepts

The initiation of the development of the GPS came from the US, Department of Defense (DoD) in 1972. The system was conceived as a ranging system from the known position of satellites in space to unknown positions on land, sea, in air and space. The primary goals of the system were military ones, but its civil use was also considered.

The ***space segment*** of the system consists of 21 operational and 3 reserve satellites deployed in six planes with an inclination of 55°, 4 satellites per plane. The satellites move in nearly circular orbits at an altitude of about 20.000 km above the Earth, their weight varying about 1500 kg.The fully deployed system will provide 4 to 8 satellites visible over 15° elevation from all points on Earth.

The satellites broadcast two signals with carrier waves L1=1575.42 MHz and L2=1227.60 MHz. These signals are generated by multiplying the fundamental frequency of 10.23 MHz controlled by high precision atomic clocks. Both carriers are modulated by the P-code (precision code) with a wavelength of approximately 30 m. The carrier L1 is also modulated with the C/A code (coarse/acquisition-code) with a wavelength of approximately 300 m. The receiver determines the ***pseudoranges*** using these codes. The P-code provides a higher precision, the C/A-code a lower precision in pseudorange determination.

Because of the primary military goals of the system, the DoD uses a policy of ***selective availability*** (SA), so the full use of the system is casually denied to civilian users. One way of achieving this objective is to truncate the message about the coordinates of the satellites. Another method will be applied after the full deployment of the system. The so-called "anti-spoofing" technique denies the access to the P-code to unauthorised users by means of a hardware key.

GPS uses the WGS-84 reference system. The instantaneous time tagged coordinates of the satellites are incorporated into the navigational message and transmitted by both the C/A and P codes. Naturally, the original processing is also done in this reference system. If we work in another reference system the results of GPS measurements should be transformed in our system.

The transformation is only possible if we have at least three points with known coordinates in both systems. In most of the countries the parameters for this transformation are provided by the national geodetic surveys.

5.2 Types of observables and receivers

GPS receivers measure distances to the satellites and read the coded information about their position. From the geometrical point of view the three unknown coordinates of the station point can be computed if three measurements are available.The measured distances are called *pseudoranges* because they are affected by the receivers' clock delay. This new unknown principally demands an additional (fourth) measurement.

For the determination of pseudoranges the receivers can use the codes. In most of the cases the receivers use the C/A code applying the *correlation technique*. This technique provides all com-ponents of the satellite signal: the satellite clock reading, the navigation message and the unmod-ulated carrier. While the quality of the last two components does not depend on code selection, the satellite clock reading's accuracy depends on the length of the code's period. This length corresponds to a distance of 300 m by the C/A code and 30 m by the P code. The satellite clock reading can be made with a precision of 1% of the length; therefore, the C/A code provides 3 m, the P code 0.3 m accuracy by pseudorange measurements.

The other, more precise method determines pseudoranges by carrier phase measurements. To achieve highest accuracy both carriers' phases should be measured. The receivers that do not know the P code can reconstruct the L2 carrier for phase measurements by the codeless technique based on signal squaring. The phase measurements lead in the expressions of pseudoranges to new unknowns (the counts of integer carrier waves in the receiver-satellite distances) and, therefore, the number of measurements should be increased.

GPS receivers available on the market fall into three groups. *C/A-code pseudorange receivers* are usually battery powered hand held devices, used by hikers, boaters, drivers. Good quality models have four or more independent channels for simultaneous reception of four or more satellites. They display the three dimensional position of the station possibly in both geographical and national grid coordinates. *C/A-code carrier receivers* are used by the geodesists and surveyors for all types of precise surveys. They have from four to twelve independent channels and store the measured code ranges and carrier phases in time-tagged form for post processing. These receivers also perform all functions described in connection with the previous group. Recent models using signal squaring, measure the phase of the L2 carrier, too. This is useful for the elimination of ionospheric influences. The *P-code receivers* are mostly used by the military. The two advantages of this group stated on the basis of a civilian test are as follows:

- the model is capable to measure very long lines (100 km) with an accuracy of a few centimetres;
- the precise measurement of medium length lines (20 km) is very fast (10 minutes).

5.3 Modes of measurement

It is very difficult to use a clear terminology when explaining GPS measuring modes and schemes. The technology is being developed and often a slightly modified old method gets an entirely new name.

The first basic mode is the *point positioning using code pseudoranges*. This mode usually provides coordinates in *real time*. If the positioning is *static* (the receiver remains on the same location for an interval of time) and no real time data are required, the accuracy of the method is at the meter level. This accuracy can be improved smoothing techniques. *Kinematic point positioning* (the receiver moves together with the car) can be used to determine the vehicle's trajectory with an accuracy of 10-100 meters. *Differential positioning* (Fig. II.2-13) requires two receivers and a communication (radio) link between them. The base receiver is stationary, and its coordinates are known, the remote receiver is mounted on the car. The base receiver computes the range corrections and transmits them to the remote receiver. This, later, computes the location of the vehicle, displays it in real time and records the data with an accuracy of 1-3 m. This method is widely used to complete the coverage of transportation networks.

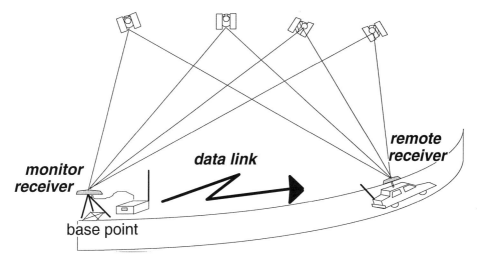

Figure II.2-13: Method of differential positioning by GPS

The second basic mode uses ***carrier phase pseudoranges***. The ambiguity of the range (the number of whole carrier wave lengths in the distance to the satellite) should be determined either before beginning with the positioning by some kind of initialization procedure or during the positioning process. Both cases demand at least two receivers as well as a point with known coordinates. Hence, the methods using carrier phases are called *relative positioning*. Relative positioning provides coordinate differences between the known point and the new ones.

By *static relative positioning,* one of the receivers measures on the station with known coordinates, one or more other receivers measure on the new points. The measurements are performed simultaneously to the same satellites (Fig. II.2-14). The length of an observation period depends on the distance between the old and the new points. For long lines of 15 to 30 km it takes 1.5 - 2 hours. Static relative positioning ensures the highest accuracy (1 - 2 cm for a line of 15 to 30 km in length). The method is solely used for control point determination.

Nowadays new software developments allow *rapid* static relative positioning. This technique uses code and carrier phase combinations for a fast initialization in static mode. The method requires both code and carrier phase measurements, the latter on both frequencies. One observation lasts 5 to 10 minutes, the accuracy is one order lower than that of static relative positioning. The method is used for control point densification.

Kinematic relative positioning by carrier phase measurement is the most productive method (Fig. II.2-15). The first step of the method is the initialization of the two receivers. There are several techniques for this purpose. The initialization means the solution of the ambiguities (introduced by the phase measurement) prior to the survey. This can be done with the help of a distance called ***baseline.*** The methods of initialization differ in the way of solving the baseline. Let us consider the three main methods for initialization.

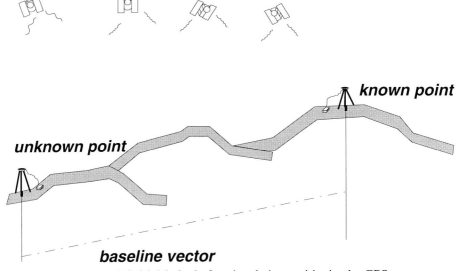

Figure II.2-14: Method of static relative positioning by GPS

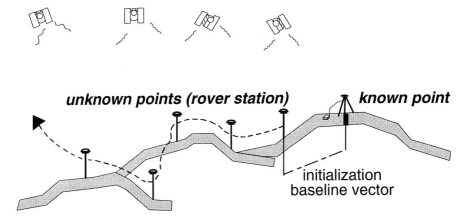

Figure II.2-15: Method of kinematic relative positioning by GPS

The first method uses a known distance between the fixed (base) point and the first point of the survey. The baseline should be short (less than 2 km) and known with precision, at the 3-5 mm level. After a few minutes of observation at both ends of the line to four or more satellites, the initialization is finished.

The use of static relative positioning in relation to the fixed and starting points of the survey can be mentioned as *the second method*. By this method the baseline is determined by the processing of one-hour long simultaneous observations on both points. The *third method* is the so-called **'swapping'**. At the beginning of the procedure receiver *A* works on the fixed point and receiver *B* on the starting point of the survey. After a few minutes of observation, the receivers are to be changed while continuing the observation of at least four satellites. The procedure is completed by another change that results in the restoration of the original situation. After the initialization, receiver A continues the observation on the fixed point while receiver B (the so-called rover) moves to the points to be surveyed. Through the entire survey (that is also on the route) both receivers should lock on four or more satellites. To ensure this condition careful reconnaissance work must previously be performed since trees, buildings, electricity poles, etc. can obstruct wave propagation. Therefore, the method is convenient only for detail surveying of open areas. The accuracy of the method is in the centimetre range. Methods using carrier phase measurements usually compute the coordinates by post-processing. Post-processing is applied because the computation of coordinates needs the common processing of observations on at least two sites. Nowadays we can read about experimental schemes aiming at real time coordinate determination using field data radio transfer and additional computing units.

Summing up the different GPS techniques we can draw the following conclusions. *Relative static positioning* is the most efficient way of building up a national control network. *The rapid relative static technique* is applicable for control densification, but the high price of the hardware and software suitable for this technique hinders its widespread use. *Differential positioning* is useful for surveying at scales of 1:10.000-1:25.000 especially if there is a road network where cars can move rapidly. *Relative kinematic positioning* can be used for large scale three-dimensional surveying of **open areas**, but nowadays electronic tacheometry is still less expensive and more generally applicable for such tasks.

We can generally state that GPS alone is not a universal solution for all problems of field surveying. The most usable method is differential positioning but it also has definite limitations. As mentioned above, its precision is about 2 m, hence it can not be used in large scale surveying. Although the method does not require the rover to be continuously connected with the satellites, measurements can only be performed in places where the connection (lock) is not lost. After rebuilding the interrupted connection to the satellites the measurements are less precise for a while; in such cases the speed of survey should be reduced. Survey of highways demands continuos data even for the sections where the sight to the satellites is blocked (in deep cuts, in forests, in urban areas, etc.). To avoid gaps in the surveys, surveying cars are usually equipped with supplementary navigation systems (revolution counter, speed sensor, gimballed flux gate heading sensor, etc.) that provide the software with dead reckoning inputs. These less precise data allow interpolation between the points determined by differential positioning.

Differential positioning is frequently used for surveying of points spread over a large area (e.g. manholes of an irrigation network). The method is very efficient when the contact to the satellites is not blocked by trees or buildings. In the case of blocking, however, traditional methods such as radiation or tights should be used. In such cases we use GPS to determine two auxiliary points which form a surveying line and measure the object as described in Section 5.3. The conclusion is, that in general surveying tasks we should have in addition to the GPS at least an EDM instrument and a theodolite to be able to cope with obstacles caused by blocked visual contact with the satellites.

In some countries the role of master stations' transmitters fulfilled by public radio stations which cover an area of about 100 km radius. In this case the surveyor should not care about the siting: operating the master station, he can perform the survey with the only remote station tuned to the particular radio transmitter.

For GIS data collection two practical novelties of instrumental development should be finally mentioned. Some new GPS receivers are equipped with a full keyboard which simplifies typing in the attributive data. The second development is a convenience for the walking surveyor by putting the antenna and other parts of the receiver (except the control panel) into a metal-framed knapsack. The antenna is mounted to the frame so that it rises over the head of the surveyor. To work with this arrangement is much easier than using one hand for lifting the ranging rod with antenna and the other for holding and typing on the control panel built into the instrument or only typing when the instrument is put in a shoulder-bag.

Questions

1. Which datum(s) is (are) officially used in your country?

2. What is the difference between a bearing and a true bearing?

3. What kind of preparations are necessary for a large scale (1:1000) farm survey?

4. What is the most advantageous method for topographic surveying of an open land at a scale of 1:5000?

5. Which is the most reliable way of transferring the surveying data into the computer?

6. How can we record the topology of objects by tape surveying and by topographic surveying?

7. Desribe the complete equipment of a surveying team performing long range traversing!

8. How much time do you think is necessary for a 1:1000 topographic survey of 100 hectare in an open hilly area?

9. Which method would you choose for updating the GIS of the road administration?

10. What do you think is the weight of the surveying equipment for height surveying by gridding?

References

Bannister, A., Raymond, S., Baker, R.: Surveying. 6th edition, Longman Group UK Limited, 1992

Hofmann-Wellenhof, B., Lichtenegger, H., Collins, J.: Global Positioning System. Springer-Verlag, Wien, New York, 1992

Torge, W.: Geodesy. Walter de Gruyter, Berlin, New York, 1980

II.3 Topographic Maps

Christiane Weber, Strasbourg

Goal:

This chapter deals with topographic map products and their associated peculiarities.

Specific characteristics of different countries (France, Germany, Hungary, the Netherlands) are presented.

Summary:

This chapter specifies the characteristics of topographic maps considered as the reference maps in numerous countries. This official map corresponds to the representation of the territory in metric and planimetric dimension. The creation of such maps follows specific steps involving different kinds of disciplines. New technologies like photogrammetry, global positioning systems or remote sensing have provided a major plus in designing those maps. Conventional symbols are added in order to facilitate the representation for everybody. Representation rules have not yet been standardized world wide, different countries use specific scales and/or symbols.

Keywords:

Construction steps, generalization, updating, semeiology, topographic maps

1. Introduction

Cartographic science and maps:
Definition: **Cartography**: A set of concepts, methods and techniques allowing the representation of a part of the Earth (or another planet) on a plan, this plan is called MAP" (C. Cauvin).
Definition: **Topographic Map**: "Partial representation of the Earth at a defined scale, with two explicit components: the location references (X, Y), and the altimetric attributes (Z)".

2. The topographic map - a reference for all derived maps

Man has paced up and down the Earth since the dawn of humanity, and locations have been - and still are - the only means to get the measure of human presence on it. Using stars and the horizon or using satellites, human movement is linked to the capacity to control the direction, the distance between points and the duration of the displacement.

Among the cartographic domains three main parts (Basic, Essential and Optional domains - see Chapter III.1: 'Cartographic Principles') shall be defined in order to understand clearly how the world can be represented on a map. The first known cartography can be associated with topographic mapping, due to its specific situation within cartography.

Even if "the distinction seems to be artificial (topography being a theme), this specificity could be historically explained (land representation is deeply rooted in military or legal needs)" ([Foin, 1987]). This explains why the topographic map is considered the basis of all other maps. Thus, reference maps have a great importance in the human development as the basis of locations and land representation. Three main domains can be pointed out: the basic domain and topographic mapping, the essential domain to control map creation and the optional domains gathering thematic and analytic cartography. All these domains are described in detail in Chapter III.1 ('Cartographic Principles').

3. The topographic map: construction steps and "designers"

The topographic map is used to link a point on the surface of the Earth with its locations (reference and altitude) on a map. These locations correspond to concrete, settled and lasting phenomena existing on the Earth ([Cuenin, 1972]).

Two different uses of topographic maps:
- the first one is related to property boundaries, planning, infrastructures, etc, and corresponds to a large scale (from 1:500 to 1:2500). All landscape details are represented, the magnitude being here the metre.
- the second use refers to the compilation of maps covering large areas. The scales used are smaller: 1:5000 to 1:100.000).

To be considered a reference map, a topographic map has to be precise and to imply the knowledge of the inherent transformations from 3D (Earth) to 2D (map). This procedure is split into two parts:
- first, the topometric phase during which measurements are made (distances, angles, direction etc.) ;
- second, the topographic phase during which the topographer draws planimetric curves, hydrographic information and elevation curves.

In fact, the construction of a topographic map involves several specialists from various fields. The first three experts deal with measurements and transformations on a plan:

- the astronomer defines the geographical coordinates using latitude, longitude and altitude.

- the geodesist defines the correspondence between Earth surface and map surface, assessing the reference ellipsoid, point location and altitude definition.

- the "engineer-geographer" chooses the projection system to be used for the map. In order to transform from ellipsoid to map, it is necessary to determine and measure the occurring distortions. According to the studied phenomena, various projection systems can be chosen.

In the end, three experts are involved in the completion of the product in order to put it on the market:

- the topographer collects the details of the studied areas, splitting them into two categories: altimetric and planimetric elements. The data can be obtained by field collection, land survey, and imagery (photogrammetric and satellite images).

- the cartographer must define the generalization level, in order to sort the collected information and decide which is to be kept. He prepares the legend and drafts the document which will be given to the printer.

- the printer reproduces the document by means of processes such as typography, heliogravure or lithography.

Nowadays these different steps are modified due to the technological progress (global positioning system , photogrammetry, remote sensing). The first steps are substituted by quasi-automatic processes and ground measurements are made to densify data coverage.

4. Topographic map construction

4.1 Goals

"A topographic map is created in order to give metric and altimetric knowledge on a region or a country added with complementary information in a general use purpose" ([Foin, 1987]).

Acquisition of topographic data[1] is the preparation of the georeference of the map. It must be a medium which can be transformed. Topographic (geometric) data are linked to the geographic phenomena to be mapped. Two kinds of links can be distinguished:
- a link between the mapping process and the object locations.
- a link between the support and the thematic data, according to its nature (punctual, linear or areal data) .

Topographic data are linked to the three elements which define them: scale, projection system and distances.

Projection system: The geographical grid relies on the comprehensiveness of the projection systems using latitude and longitude. As presented in Chapter II.1 ('Reference Systems'), several projection systems exist and can be used as best suited to the map purposes. The

---

[1]   See Chapter II.2 ('Field Survey') for data collection process.

characteristics of the geographical phenomenon studied, such as its location on the Earth, its size and the anticipated manipulations are crucial for the choice of the projection system.

Scale: The scale is the ratio of the distance between two points on the map and the equivalent distance on the Earth. The dimensions of objects in a map differ through scale variations. Choosing a scale implies the determination of the number of phenomena to be shown on a given sheet. Aggregation of spatial objects must be taken into consideration according to scale and variance of the phenomena at different aggregration levels.

Distances: Distances between points can be geometrically preserved as in reality or be transformed into time-distances, for instance. One of the first tasks entrusted to the designer will thus be to compile existing maps of the studied zone. Projection properties have to be considered (equivalence, conformity, equidistance). Familiarity with the projection system is important for anybody working on this problem

"Designers of thematic maps have a choice of hundreds of projections, but the range is reduced by differences in the suitability of projections for mapping the desired part of the globe" ([Dent, 1985]).

4.2 Restricting and restraining aspects

As a topographic map is considered a reference map, the importance of metric and altimetric precision must be reinforced. Details have to be selected and represented so as to avoid confusion and maximize legibility.

Scale and generalization process: The scale of a map is important; it decreases from large scale to small scale, the degree of detail which can be noted and the portion of Earth that can be described increases. Therefore graphic precision in large scale maps leads to an increase of detail. For small scale maps, only a relative precision is required. Of course, if the result has to be an aggregation of embedded areas it will be more useful to start from a larger scale, keeping details which can be important at a smaller scale, rather than using the smaller scale map to start with. Control on generalization sometimes appears to be more efficient.

Generalization means "*to modify a specific data in order to increase the effectiveness of the communication by counteracting the undesirable consequences of reduction*" ([Robinson, 1978]).Generalization depends on the purpose of the map, the final representation scale, the graphic limits for mapping design, map understanding and the data quality.

In the case of topographic maps, several steps are identified in this generalization process (other information can be found in Chapter III.1: 'Cartographic Principles', in particular algorithms):
  – **Simplification-selection** leads to keeping only the essential characteristics of geometric data to "the retention and possible exaggeration of these important characteristics and elimination of unwanted details" ([Robinson, 1978]).
  – **Structure schematization** which is a simplification of the linear forms, and the outline of zonal elements. Angles and sinuosities have to be deleted to be readable and to allow for conventional signs (double lines for roads, and so on).
  – **Classification** makes it possible "to order, to scale or group data" in order to enhance geographical phenomena. It "reduces the complexity of map image, helps to organize the map information". Conceptual schematization is the result of the different levels of analysis: reduction of the number of categories of objects represented, creation of classes (according to qualitative or quantitative levels) and, if necessary, modification of the spatial locations on the map.

For instance, in France 12 road and path classes exist for the 1:25.000 scale map, only 4 for the 1:50.000 scale map where roads and main roads are represented by the same symbol (whatever its width is > 7 m) while they are separated at larger scales ([Cuenin, 1972]).

Through **symbolization** the real world appears as sets of marks (symbols). These marks can look like the real things (trees, railroads, etc.) or be abstract symbols, generally geometric shapes (circles, squares, etc.). As conventional signs - to remain readable - increase their relative dimensions in the same way as the scale diminishes, choice rules are absolutely necessary. Conventional symbols and signs increase the representation dimensions in different ways. On the „World Map" (1:1 Mio.) the sign for the class "highway", is drawn with 0.9 mm which is equivalent to 900 m on Earth, but represents only a roadwidth of 30 m in reality. In opposite, for a city of more than 100.000 habitants, the contour line is drawn without increase.

A cartographic product necessarily provides a distorted vision of reality. For instance, in reality several objects lie often in short distance to each other, e.g. a road, a railway and a river. To obtain a visual separation between these objects in maps of smaller scales there has to be introduced necessary displacements from the correct location. Only relative positions between them are preserved, and for one object among them the correct location is kept. Geometric relationships are thus to be preserved, when they exist in reality: parallelism, crossing linear features, equality or inequality of distances and surfaces. All these abstractions have repercussions on the whole map.

Making a road relatively wider than it is on the Earth makes it visible in a map; distorting distances on a projection enables the map user to see the whole Earth at once; separating features by greater than Earth distances allows representation of relative positions. Distortion is necessary in order that the map reader be permitted to comprehend the meaning of the map (M. Monmonier, 1977, Map, Distortion, and Meaning. Resource Paper n °74-4, Washington DC, Association of American Cartographers.

Updating the information: Updating processes are less time consuming than before, technological progress allows to update information without travelling up and down the landscape. Photogrammetry provides a large amount of precise and measurable pieces of information in a short time. Aerial images with vertical views are produced in black and white, panchromatic, colour or colour infrared. Imagery can be integrated in the map creation process, according to its specific geometric quality.

4.3 Semeiology and conventional rules

Relief representation:
Two aspects are rather delicate to reconcile, namely the mathematical precision and the suggestion of the third dimension on a plan. In France in the XIXth century, the "Etat-Major" map (1:80.000) provided for the first time a more precise representation: 'minced lines' getting closer and closer as the slope becomes steeper. Later on in the XXth century, relief cartography became exact and precise. Slope levels replace 'minced lines' with equidistance varying according to the slope: for instance in a 1:50.000 map: 10 m in a plain, 20 m in mountains. But the pictorial aspect remains present to avoid too much abstraction: on the one hand some 'blurring process' is added to accentuate the slope and on the other hand high mountain areas for which slope levels are too close to be drawn: a figurative depiction is done, this depiction of rocks specifies the major features of the relief. Remote sensing has supported the development of some interesting products such as spatial maps, the production of which is easier but implies higher precision control, e. g. through Global Positioning Systems.

In a topographic map, the planimetric representation deals with all the details appearing on the Earth surface, with the exception of the relief. Thus different components are represented:
- natural phenomena: hydrologic networks, vegetation etc.;

- man-made constructions: channels, buildings etc.;
- abstract elements: administrative limits, regulation features etc.;

Different categories of objects are usually defined: railways, electric utilities, hydrographic details, concrete limits (enclosure), vegetation and cultivation and administrative limits; for instance, on a French topographic map, 240 kinds of objects are registered.

An ideal map symbol is one which is most understandable to anyone without looking at the legend. It is necessary to evoke the real form of the object since the exact dimensions are impossible to get. A link between cultural knowledge and geometry has to be found. Most of the punctual details have the same importance, therefore, it is necessary to avoid hierarchies. For linear position the shape is imposed by the layout. Order is introduced by the width of the map symbol, value variation (grey or colour or discontinuity).

Some conventional map symbols are rather similar in different countries, even if they change in detail. But an absolute standardization is far from existing.

5. Peculiarities of topographic maps

Different projection systems have been presented in chapter II.1; this section will specify some characteristics of the French, German, Hungarian and Dutch systems.

5.1 Projection systems in use in different countries

- France: The Lambert conical projection is conformal (Clarke ellipsoid) and divided into 3 zones (Corsica has a special one). Each sheet has 3 graduations: according to the Greenwich meridian, in grades regarding the Paris meridian and square tessellation of 1 km (Fig. II.3-1 and II.3-2).

- Germany: The geodetic Gauss-Krüger system is used (Transversal Mercator Projection, Bessel ellipsoid), but will be converted to UTM in the near future.

- Hungary: The largest system of maps is the Union National Map System (UNMS). The reference ellipsoid of this system is the International Union of Geodesy and Geophysics (IUGG) 67 ellipsoid. The map projection is a special cylindrical projection. Another map system was for military purposes, the reference ellipsoid is Krassowskij ellipsoid. The map projection is the Gauss-Krüger projection.

- Netherlands (The): Reference system used by the Topografische Dienst is based on the Bessel ellipsoid, the geoid reference is Von Willigen 1985. The map-series 1:250.000 and 1:500.000 is produced according to UTM Projection and the International ellipsoid.

5.2 Commonly used map scales

- France: The National Geographic Institute (IGN) promotes mostly large scale maps, the 1:25.000 and 1:50.000. Four series are available now, (but they will be replaced soon by the numerical product):
  - the 1:25.000 (blue series) with 2000 maps, 4 colours: blue for hydrology, green for vegetation, orange for relief and black for objects left. It is the reference map, the legend has been coded in 1972.
  - the 1:50.000 (orange series) with 1100 maps, it is a generalization of the 1:25.000, with the same legend.

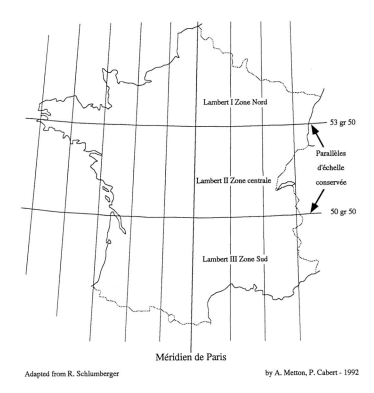

Figure II.3-1: Lambert projection zones

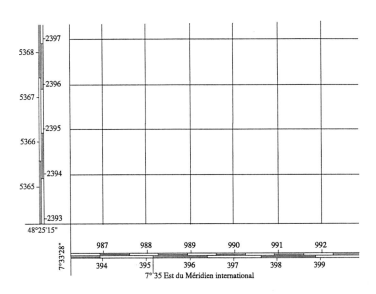

Figure II.3-2: Square tesselation of 1 km (Lambert grid), map of Strasbourg (1:50.000)

Geographisches Institut
der Universität Kiel

- the 1:100.000 (green series) with 74 maps, 5 colours (yellow added for secondary roads).
- the 1:200.000 (red series) with 16 maps and 5 colours, but are focused more on tourism and driving.

- Germany: Maps are produced by different surveying administrations. Scales from 1:5000 to 1:100.000 are processed in each of the 16 states of the Federal Republic of Germany by the "Landesvermessungsämter":
  - 1:   5.000 (DGK5), contents: isolines, buildings, parcel boundaries, traffic systems, water
  - 1: 25.000 (TK25), contents: isolines, water, forest, vegetation, settlements, traffic systems
  - 1: 50.000 (TK50), contents: as above, plus shading
  - 1:100.000 (TK100), contents: as above
  Smaller scales (1:200.000, 1:500.000 and 1:1 Mio.) are produced nation-wide by the central administration *Bundesamt für Kartographie und Geodäsie*, Frankfurt/M.

- Hungary: The maps of the UNMS in the medium scale (topographic) domain have the following scales :
  - the 1: 10.000 (84% of the maps are ready)
  - the 1: 25.000 (12% of the maps are ready)
  - the 1:100.000 (ready)
  - the 1:200.000 (ready).

- Netherlands (The): The Topografische Dienst is responsible for providing medium and small scale geographic data for military and civil use. Several products are available.
  - the TOP500 vector - database 1:500.000 (coverage 50%, completion 1994, updated every 4 years, scale of digitized documents 1:250.000)
  - the TOP250 - database 1:250.000 (coverage 100%, completion 1992, updated every 4 years, scale of digitized documents 1:125.000)
  - the TOP250 gazetteer - database 1:250.000 (coverage 100%, completion 1990, updating to be decided, contents: geographical names, features code, province code and map sheet references UTM and national grid coordinates.
  - the TOP50 - database 1:50.000 (coverage 50%, date of production by 1997, updated 4, 6, 8 years, contents: coded points, linear and area information, line elements defined as open polygons running between nodes, centre-lines of road, scale of digitized documents 1:25.000)
  - the TOP50 Wegen - road database 1:50.000 (coverage 100%, date of production 1986-1994, updated 4, 6, 8 years, contents: centre-lines of the raod and railways, scale of digitized documents 1:25.000)
  - the TOP50 gazetter - gazetteer 1:50.000 (coverage 100%, completion 1987, updating to be decided, contents: geographical names, features code, province code and map sheet references UTM and national grid coordinates.

5.3 Cartographic Numerical Data-Base

- France: The National Geographic Institute (IGN) produces many numerical data:
  - the geodetic database with all geodetic measurements;
  - the topographic database (1:5000 to 1:50.000) with hydrologic network, energy network, circulation network, buildings and agglomerations, administrative limits, vegetation etc.,
  - the TRAPU database is a 3D database which provides a realistic view of the urban areas;
  - the cartographic database built up with the 1:50.000 "Topo. base";

- the Georoute database for circulation, the goal of this one is to promote on-board driving assistance;
- the altimetric database (1:25.000) has a vector data structure;

• Germany: A nation-wide approach for creation of digital databases is ATKIS (Amtliches Topographisch-Kartographisches Informationssystem). The processing is similar to the map production: the large scale part DLM25 (related to a scale of 1:25.000) is produced by the surveying administration of each state, while the small scale parts DLM200 and DLM1000 (related to a scale of 1:200.000 and 1:1 Mio., resp.) is produced by the *Bundesamt für Kartographie und Geodäsie*. The contents (mainly vector data) is adequate to the corresponding maps (7 main object classes), additionally they include digital elevation models (DEM) for relief information. All states also create a (large-scale) digital cadastral database ALK (*Automatisierte Liegenschaftskarte*).

• Hungary: Most of the available maps are in analogous form. However, the digitized information of the 1:100.000 UNMS and the digital elevation model of the 1:50.000 Gauss-Krüger maps are now available on a commercial basis.

• Netherlands (The): The production and providing of topographic databases is now considered the main task of the organization. Three main databases, containing the data of the map series 1:10.000, 1:50.000 and 1:250.000 are to be completed by 1997.
A digital elevation model has been produced by digitizing the Altitude Map (1:10.000) in close cooperation with the Meetkundige Dienst of Rijkswaterstraat.
The topographic map series 1:25.000 is available as raster data on CD-ROM.

6. Conclusion

The importance of topographic maps is crucial in GIS applications. The planimetric and altimetric references used at large scale, according to the specifications of the projection systems, provide all the information necessary to further manipulation with respect to geometric precision. Of course such maps do not exist everywhere and a large number of countries have a very bad (even non-existent) reference map layer. The difficulty is to evaluate the quality of the given information according to the purpose of the project. Updating corresponds to a former aspect of these maps, new technologies introduce a very interesting progress. With technological progress regarding cartographic products, a grouping of the official cartographic representatives of Europe was created in 1982. The goal of the Committee of Representatives of Official Cartography (CERCO) is to promote the sharing of information and the improvement of consultation and co-operation in all cartographic fields ([Mousset, 1992]). Some twenty five countries participate in this committee. Ten working groups are dealing with copyright, map of Europe at a 1:1 Mio. scale, database and standardization, GPS, or the definition of a European Exchange Format. CERCO may play a key role in the future, due to the differences between countries, for instance in the frontier Euregio (Belgique), the German maps are drawn at 1:5000, the Dutch maps at 1:10.000, the Belgian maps at 1:25.000 at the present time ([Donnay, 1991]). So all are around, from plans to maps, showing different objects, different generalization processes and so on. Harmonization can be promoted by a European organization, even if we cannot expect an ideal solution to this problem to be found for the moment.

Questions

1 - Is the topographic map considered a "reference map"? And why ?

2 - What are the main steps of the topographic map production ?

3- What are the main elements defining such a map ?

4 - What is, in this case, called "generalization" ?

5 - What are the benefits of new technologies in the topographic domain ?

References

Cuenin, R.: Cartographie Générale. Tome I et II, Ed. Eyrolles, 1972, 524 p.

Dent, B.: Principles of thematic map design. Addison-Wesley Publishing Comp., 1985, 389p.

Donnay, Jp.: Usefulness and Pitfalls of GIS applications in a frontier regions context. ESF, Davos, 1991

Foin, P.: Cartographie, topographie et thématique. Paradigme, Télédétection Satellitaire, 1972, 127 p.

Metton, A., Gabert, P.: Commentaire de documents géographiques de la France. Sedes 1992, 410 p.

Mousset, J.: The European Committee of Representatives of Official Cartography (CERCO). Mapping Awareness & GIS in Europe, Vol. 6, N°4, 1992, pp 61-62

Richardus, R., Adler, R.K.: Map Projections - For Geodesists, Cartographers and Geographers. North Hollard Company, Amsterdam, 1972, 172 p.

Robinson, A. et al.: Elements of Cartography. John Wiley & Sons, New York, 4th ed., 1978, 448 p.

II.4 Imagery

Hans-Peter Bähr, Karlsruhe

Goal:

1. Image vs. map: differences and common features

2. Systematic classification of different types of image according to spectral range and acquisition techniques

3. Images operationally available to be applied in a GIS

4. Possibilities to enter images into a GIS

Summary:

A broad concept for "imagery" is presented. In more detail, characteristic features are discussed, referring to a matrix, giving the sensors as a function of platforms (Tab. II.4-2).

Finally, imaging systems, both experimental and operational are presented and some pictorial examples are displayed.

Keywords:

Electromagnetic spectrum, photographic imagery, scanner, thermal IR, microwave, LANDSAT, SPOT, ERS-1.

1. On the concept and content of "image"

The word "image" has a very wide meaning in everyday language: it may refer to concrete or abstract objects, to something pertaining to the real world or merely present in man's imagination. When we consider Geoinformation Systems (GISs), however, only images that really exist are actually significant. A further constraint to the use of images within these systems is that they must represent the "real world".

Fig. II.4-1 tries to show graphically the different meanings assigned to the word "image". They are arranged in groups, according to their nature, into "mathematical", "optical" and "manual" images, the latter referring specifically to man's craftmanship, to the fact that these images are produced by man himself. All these images share the characteristic of representing the real three-dimensional world in two dimensions. Images never depict the world as a whole; they merely show a part, a sector "clipped off" the real 3-D world. The representation of this 3-D environment on a 2-D surface is what we call an "image" in the GIS context.

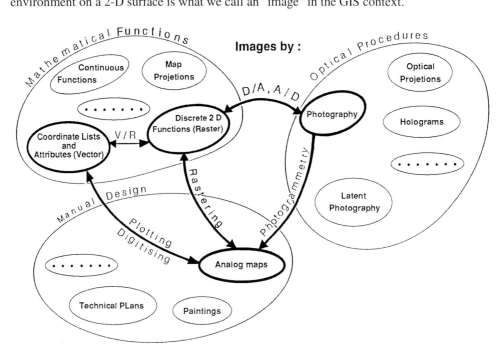

Figure II.4-1: Interrelation of the concept "images"

For GIS only one subset of each group of images shown in Fig. II.4-1 is relevant. These subsets are :

- well known conventional photographs (optical images)

- discrete mathematical functions (mathematical images)

- thematic or topographic maps (manual images) as well as their representation in vector (lists of graphic primitives like points) or raster form ("pixels").

These products in their widest meaning fall under the concept "image". In fact, they are all closely related, especially when they are digitally treated, as normally occurs in GIS's.

To identify the characteristic features of these "images" we shall start with a conventional photograph which can be digitized and thus be transformed into discrete mathematical functions. This process is becoming more and more important and the frequency of its application is growing correspondingly. (For more detail, refer to Chapter III.8: 'Digitizing Imagery'). The reverse process, i.e. the transformation of discrete mathematical functions representing an image into the picture itself so as to make the scene recognizable to the human eye, is equally important. This step, also known as D/A transformation, has been represented in Fig. II.4-1 by lines with 2 arrows.

While analogous photographs on one hand and discrete mathematical functions in digitized form on the other, are still identical, their relationship with the type of image called "map" is totally different. The map also deals with a part of the real world represented on a surface to be looked at, but its contents have been filtered through the basic knowledge of the system "man". A map is no longer a "true copy" of physical reality.

What characterizes manual representations is that objects in them are coded. This means that significance is assigned to certain details within the image "map". There are also links between maps and mathematical representations. These links are the lists of coordinates with attributes. Maps and charts may be derived from photographs; this is one of the fundamental tasks of photogrammetry. In order to gain access to the third dimension, however, stereophotos are required.

After this rather formal description of the different types of images, we are going to analyse the concrete differences between photographic and cartographic representations of the ground in a practical manner. In Fig. II.4-2 the pictorial and the cartographic representations of a sector of the Earth's surface are confronted. We see that while the photograph shows many details, the map does not. There are, however, certain objects which appear continuously on the map even when they are covered by vegetation or shadows on the photographs. Analysing the amount of information contained in both images, it is evident that neither contour lines nor names can be found on the photograph. Both types of representation require interpretative knowledge of the human observer, the photograph obviously posing higher demands on the interpreter.

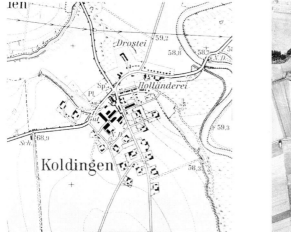

Figure II.4-2: Comparison of a line map (a) and the corresponding aerial photography (b) (approx. area: 1.5 km × 1.5 km)

The reduced information contained on maps compared to photographs presupposes a generalization process. This "generalization during acquisition" has nothing to do with the actual "cartographic generalization" resorted to due to the necessity of representing very small details on maps. The smaller the scale of charts and maps, the more cartographic generalization is required in order to keep the product "readable".

The question: "Which is more useful, the photograph or the map?" is a purely academic one, since in practice both supplement each other. A compromise was arrived at by the use of "photographic maps or charts", i.e. rectified photographs with overprinted names, contour lines, border information, legend, etc. (compare Section 4 of this chapter).

Fig. II.4-1 shows the meaning of "image" in its widest sense, covering also mathematical and manual representations. From now on, when we speak of "image" we shall only refer to that type of representation which can enter the data base of a GIS.

2. Classification of images according to their importance for GIS

There are different ways to classify images, e.g. they may be grouped according to their type into analogous vs. digital[1] images; according to their scale into large scale vs. small scale images; according to the platform from which they are taken into space-borne vs. aerial photographs, etc.

Table II.4-1: Platforms, Sensors and Applications (after [Bähr,Vögtle, 1991])

Platform	Sensor	Applications (examples)
Ground	Metric camera (analogous)	Architectural photogrammetry
Ground	CCD camera (digital)	Robotics - industrial quality control
Aircraft	Photogrammetric camera (analogous)	Topographic maps, environmental impact analysis (large scale)
Satellite	Multispectral scanner	Environmental monitoring (small scale)

Considering the platforms from which images are taken, we shall analyse Tab. II.4-1 where the typical platforms, sensing devices and their corresponding applications are listed. From the point of view of their applications, analogous photogrammetric cameras on terrestrial platforms are dominant for architectural applications while airborne platforms are preferred for conventional topographic and ecological uses. CCD cameras with A/D transformation are nowadays playing a fundamental role in the terrestrial field. They are used in industrial quality control processes and robotics. Finally, the satellite is the typical platform for multispectral scanners which are of special service to ecological monitoring. These are mostly operational uses (see Section 3.5). Good results have also been obtained with tests from shuttle platforms. The latest developments are microwave imaging systems on satellites (see Section 3.6).

Obviously images may also be classified according to spectral range because the applicability of images depends, in first instance, on the spectral range in which they were taken. Fig. II.4-3 gives a general view of the whole electromagnetic range, from gamma rays to short waves, showing that only some few bands are adequate to generate pictures due to the fact, that not all wavelengths are able to pass through the atmosphere. The upper area of Fig. II.4-3 represents a magnification of the range from 0.3 µm to 15 µm. The hatched zones show those parts of the electromagnetic range which do not allow rays to pass through, whereas the remaining zones, the so called "atmospheric windows", are basically appropriate for the generation of images by means of different systems.

---

[1] The term "digital image" sounds strange when speaking of an image which should be in analogous form "by nature"; the international usage, however, tends to accept this designation in spite of its apparent incongruity.

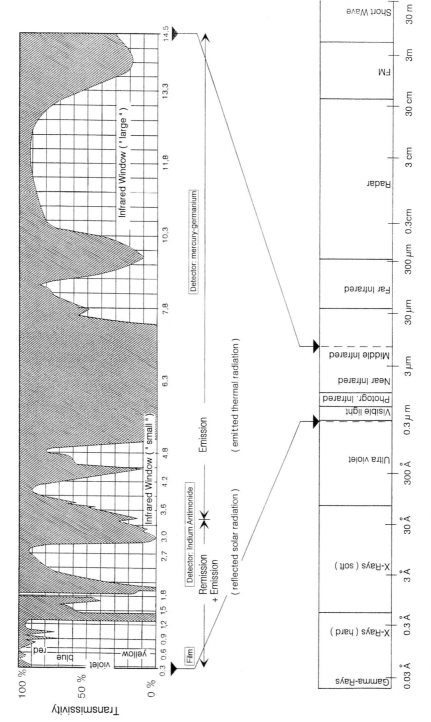

Figure II.4-3: Relationship of imagery and the electromagnetic spectrum (source: [Hirth, 1972])

Four spectral ranges are most important for GIS applications :

a)   **The visible light/VIS (0.4 - 0.65 μm)**
        The information supplied by this spectral range is "qua nature" well known to man. Therefore, it is not necessary to go into more detail. What should be stressed, however, is that the characteristic acquisition is done by photographic cameras and pan-chromatically sensitive photographic films. Panchromatic sensitivity means, in this case, to be receptive from blue to and including red. Panchromatic films are black and white films. For colour films, three emulsions are required, each of which is sensitive to a different spectral range (usually red, green and blue).
        Through the corresponding colouring of these three layers and by substractive mixing of colours the natural colours of the photographed objects are reproduced (For literature on this subject refer to the Manual of Photogrammetry).

b)   **Photographic infrared or near infrared/NIR (0.65 - 1.0 μm)**
        Some photographic films (emulsions) are still sensitive in this range: the human eye, however, cannot perceive it any more. In the NIR healthy vegetation shows very high reflection. Readily recognizable is the steplike increase of reflected energy between 0.7 and 1.0 μm (see Fig. II.4-4). Sometimes colour IR or black-and-white films capable of recording this effect are successfully applied, especially for ecological purposes such as vegetation monitoring.

c)   **Thermal infrared/THIR (7.8 - 14.5 μm)**
        What is registered here is the thermal emissivity of objects. "Infrared", however, should not generally be related to "heat". Photographic emulsions are no longer sensitive to this range; we require semiconductor diodes. The corresponding cameras, also called "scanners", are significantly different from conventional photogrammetric cameras (see Section 3). It should be stressed that semiconductor diodes can be made sensitive both to the photographic infrared range and to the visible light, so as to allow scanners to cover the whole range from 0.35 to 15 μm. When cameras are built to record more than just one spectral range they are called "multispectral scanners".

d)   **Microwaves (in the range of [cm] to [m])**
        In the above mentioned spectral ranges reflected solar energy is used to generate pictures. This does not occur with microwaves. These are radiated by an antenna which receives the returning waves after they have been reflected by the Earth's surface.
        Microwave systems, vulgo "radar (Radio Detection and Ranging)", are active systems compared to the passive systems of the optical scanners previously described here. An outstanding feature which is characteristic of microwaves, is that they are not blocked by clouds. Microwave systems can be used both day and night since microwaves, when emmitted, "illuminate" the terrain.
        Another characteristic feature is the fact that microwaves can, to a certain degree, penetrate vegetation and even enter the soil, depending on moisture and the wavelength of the sensor.
        Till now microwave images have found only few application in GISs. This situation has changed after the launching of ERS-1 in 1991, the European Remote Sensing satellite (see Section 3.6).

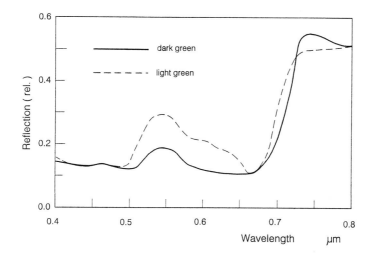

Figure II.4-4: Reflection of healthy vegetation as function of wavelength

3. Characteristic features of images

In the preceding section different kinds of images used in a GIS were introduced as a function of sensor platforms and spectral range. We are now going into more detail and make reference to their importance for actual GIS applications. To this end, sensors and platforms shall be described according to Tab. II.4-2.

Table II.4-2: Sensors and platforms as presented in the following 6 sections (see numbers). Sections marked with (*) indicate the most frequently used configurations.

Sensors	Platforms	
	Aircraft	Satellite
Cameras	**3.1 (*)**	34
Scanners	32	**3.5 (*)**
Microwave Systems	33	3.6

3.1 Cameras on airborne platforms

This is the specific "photogrammetric case". Therefore, we shall exclude in the following all other uses of cameras on aircrafts (e.g. military uses, amateur cameras). The most important devices for GIS are undoubtedly conventional photogrammetric cameras on airborne platforms. One fundamental characteristic of these systems is that they have always been standardized, standardization being a prerequisite for operationally and technically successful systems:

- picture format : 23 cm × 23 cm
- calibrated focal length (in [cm]):
  - 8,5 / 8,8   (super-wide angle)
  - 15   (wide angle)

21   (intermediate angle)
30   (normal angle)
60   (narrow angle)
- known interior orientation (calibrated focal length, distortion, principal point)
- highest geometrical quality (difference of image coordinates from rigorous central pro-
    jection in the range of a few µm).

The lens system most commonly used in photogrammetry has a focal length of 15 cm (wide angle), mainly for geometrical reasons (e.g. precision of height determinations).

Modern photogrammetric cameras are to be considered a mere component within the global photogrammetric data acquisition system. Such an up-to-date system is shown by Fig. II.4-5. The camera itself is here reduced, so to say, to a "peripherical unit", although to a very important one. The data acquisition system includes a navigation system, too, which allows an effective, correct and highly precise flight. These systems are controlled by computers as far as possible. For more detail refer to the literature, e.g. [Lorch, 1992].

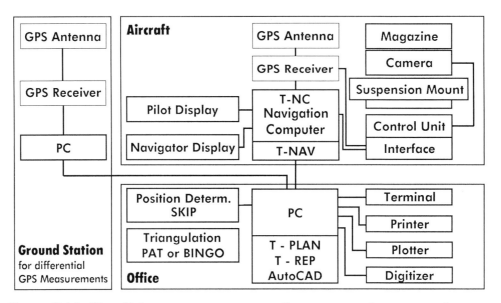

Figure II.4-5: Photoflight management system: Components and system environment (Carl Zeiss)

3.2 Scanner systems on airborne platforms   and
3.3 Microwave systems on airborne platforms

Scanners and microwave systems are called "dynamic" systems. They produce a continuous image strip, different from the individual pictures taken by photographic cameras, each of which could practically be considered a "snap-shot".

Dynamic image generating systems on airborne platforms have the great disadvantage of being geometrically unstable, a feature which disappears with satellite platforms. Nevertheless, scanners and microwave systems on airborne platforms are still used but have so far found little *operational* application in GISs. Among the pure scanner systems, thermal scanners on airborne platforms are the most commonly used, one of their most usual applications being the monitoring of water bodies (thermal emmissions). Military uses of radar on airborne platforms are common.

The detection of microclimates, especially the detection of heat deviations by means of scanner systems on airborne platforms has already been commercially exploited to a limited extent. In general terms, however, it can be stated that the fields described under 3.2 and 3.3 have not yet found widespread use within GISs because of high costs and considerable difficulties for their geometrical rectification (see Chapter III.2: 'Geometrical Models'). It should also be mentioned that GISs include the multisensor concept. The interrelation of diverse data requires their relationship to a geometrical system and this must be attained with utmost precision.

3.4 Photogrammetric cameras on satellite platforms

Many tests have been made with this sensor/platform combination, but none has been too effective, mainly limited by the fact that the used film must somehow be returned to Earth. This results in relatively expensive projects, most of them are shuttle based experimental systems.

Table II.4-3: Photogrammetric cameras on satellite platforms

Type of camera	Spectral range [μm]	Pixel size on ground [m]	Image size [cm] Scale	Calibrated focal length [cm]	Orbit data	Repeti-tion	Remarks
SKYLAB S 190 B - USA -	True colour	approx. 38 x 38	11.5 x 11.5 1:950.000	457	h = 435 km i = 57°	------	1973
SOJUS 2 MKF-6 - SU -	about 0.48 0.54, 0.60 0.66, 0.74 0.84	approx. 20 x 20	5.5 x 8.1, 1:2.7 Mio.	125	h = 200 km .... 400 km	several times	since 1976 Image motion compensation, multispectral
Space Shuttle Metric Camera -ESA-	Pan film: 0.53 ....0.70 col. infrared 0.53...0.90	approx. 20 x 20	23 x 23, 1:820.000	305	h = 250 km i = 57° Spacelab STS 9	------	Nov. 1983 Zeiss-RMK, stereoscopy
Large Format Camera (LFC) - USA -	0.40 .... 0.90	approx. 10 x 10 ... 20 x 20	23 x 46, 1:210.000 ... 780.000	305	h = 240 km .... 370 km i = 57° Spacelab STS 14	------	Okt. 1984 stereoscopy, image motion compensation, reseau
KFA-1000 - SU -	0.57 .... 0.69 0.68 .... 0.81 two layers spectrozon.	approx. 5 x 5	30 x 30, 1:280.000	100	h = 280 km KOSMOS-programme	several times	1983 stereoscopy, image motion compensation
KATE-200 - SU -	col. infrared 0.50 .... 0.90	approx. 30 x 30	18 x 18, 1:1.4 Mio.	20	h = 280 km KOSMOS-programme	several times	stereoscopy
MK-4 - SU -	True colour 0.46 .... 0.69 col. infrared 0.40 .... 0.90	approx. 10 x 10	18 x 18, 1:750.000	30	h = 280 km KOSMOS-programme	several times	1989 stereoscopy, image motion compensation

Tab. II.4-3 lists some systems on satellite platforms for cameras, among which the Soviet systems play an outstanding role. The "western" systems are the Metric Camera (ESA, 1983) and the Large Format Camera (USA, 1984). Both are photogrammetric systems but mainly experimental ones. The strongest argument for cameras in space is that much better resolution on the ground could be attained by films than by electronic scanner systems. Especially analogous storage and processing of imagery on film is operationally done for many decades.

The Soviet Cameras in space were originally developed for military purposes but now many be acquired on the civil market. Most of the films are sent to Earth by parachutes dropped from the platforms. The optical quality of these films - especially their geometric resolution- is excellent, e.g. the KFA-1000 system or newer ones like the KWER-1000 ([Riess, 1993]).

Colour-Page 1a is an example of a photographical picture from space and is a part of a picture showing Karlsruhe taken by a KFA-1000 camera. The recognizability of details is considerably higher on these images than that of its "electronic counterparts" from SPOT or LANDSAT. However, its spectral resolution is extremely small (only two channels). This deficit will be strongly felt in environmental studies.

3.5 Scanners on satellite platforms

These scanners provide the basic image material for GISs. Their evolution started in 1972 with the launch of the first terrestrial satellite LANDSAT (see Tab. II.4-4). The main features of these USA-developed systems were:
- pictorial coverage of the *whole* Earth's surface
- low price of image data (at the beginning)
- 4 colour bands
- data in digital form (magnetic tape)

Images of LANDSAT-MSS (Multispectral Scanner) became "guiding features" ("guide fossil"), to the vigorously evolving Remote Sensing technology (Colour-Page 1b). They also stimulated the further development of digital systems and processing.

Criticism over low geometric and spectral resolution was quietened by the next satellite LANDSAT-TM (Thematic Mapper, see Tab. II.4-4). This system was launched in 1982, 10 years later and might also be defined as being totally operational. Its seven spectral bands (channels), cover not only the whole scale of colours of visible light but also the infrared range as well as two neighbouring bands which are devised mainly for geological applications. It also includes a thermal band within the 10.4 - 12.5 µm range ([Kramer, 1994]).

Colour-Page 2a is an example of an image taken by LANDSAT-TM using colour components of different band combinations. The results of the diverse band combinations can be further quantified with the help of computer-aided processes (see Chapters III.3 and III.9)

Another operational system taking images by means of a scanner mounted on a satellite platform is the French SPOT (Système probatoire de l'observation de la Terre, see Tab. II.4-4) launched in 1986. With only three bands, the spectral resolution attained by this system is far lower than that of LANDSAT but its geometric resolution is much higher: on panchromatic films the size of 1 pixel covers an area of 10 m × 10 m on the ground (Colour-Page 2b).

The coloured SPOT images where 1 pixel covers 20 m × 20 m on the ground also attain higher visual quality than those taken with LANDSAT-TM. Even on a colour hardcopy which is considerably worse than a screen image, it is possible to recognize clearly the superiority of the geometrical resolution of SPOT compared to LANDSAT-TM. Another important feature of SPOT is that it allows the generation of stereo-images by recording one image strip from one orbital path and the second from another path. SPOT images are always preferable when high geometric resolution is required (for example, for topographic map production). For environmental applications, however, where very fine saturation differences must be shown, LANDSAT-TM images are better suited.

Tab. II.4-4 shows beside LANDSAT-TM and SPOT, the ESA-System MOMS (Modular Optoelectronic Multispectral Scanner). This is an experimental system designed in preparation for a possible future operational system of data acquisition. The first shuttle flight in 1992 has shown excellent results with regard to image quality. Among numerous combinations of geometric and radiometric resolution, smallest pixel size on the ground is 5 m.

Table II.4-4: Scanners on satellite platforms

Type of scanner	Spectral range [μm]		Pixel size on ground [m]	Image size	Orbit data	Repetition	Remarks
	pan	multispectral					
LANDSAT 1,2,3 MSS - USA -	0.80 ... 1.10	0.50 ... 0.60 0.60 ... 0.70 0.70 ... 0.80	79 x 56	3200 x 2300 elements 185 x 185 km²	h = 920 km i = 99°	not flexibel 18 days	start 23.7.72
LANDSAT 4,5 TM - USA -		0.45 ... 0.52 0.52 ... 0.60 0.63 ... 0.69 0.76 ... 0.90 1.55 ... 1.75 2.08 ... 2.35 10.4 ... 12.5	30 x 30 120 x 120 (thermal)	7020 x 5760 elements 185 x 185 km²	h = 705 km i = 98°	not flexibel 16 days	start 16.07.82
LANDSAT 6 ETM - USA-	0.50 ... 0.90	0.45 ... 0.52 0.52 ... 0.60 0.63 ... 0.69 0.76 ... 0.90 1.55 ... 1.75 2.08 ... 2.35 10.4 ... 12.5	15 x 15 (pan) 30 x 30 (multisp.) 120 x 120 (thermal)	7020 x 5760 elements 185 x 185 km²	h = 705 km i = 98°	not flexibel 16 days	Stereooption, high resolution 5 m x 5 m
MOMS 01 - ESA -	0.57 ... 0.76 0.82 ... 0.98		20 x 20	6912 elements per line 140 km	h ≈ 300 km i = 28,5° Spacelab (SPAS - 01)	------	first flight: June 83, second flight: February 84, line detector
MOMS 02 - ESA -	0.52 ... 0.76	0.44 ... 0.51 0.53 ... 0.58 0.65 ... 0.69 0.77 ... 0.81	4.5 x 4.5 (pan) 20 x 20 (multisp.)	Scanning width on ground 37.6 .. 78.1 km	h = 308 km i = 28.5°	------	Mission D2, 7 operation modes, stereoscopy
SPOT 1.2 - F -	0.51 ... 0.73	0.50 ... 0.59 0.61 ... 0.68 0.79 ... 0.89	10 x 10 (pan) 20 x 20 (multisp.)	6000 x 6000 elements (pan) 3000 x 3000 (multisp.) 60 x 60 km²	h = 832 km i = 99°	flexibel 26 days	start 22.2.86, line detection, stereoscopy

General observations:
- There is a tendency towards higher geometric resolution. Due to restrictions on data transmission rates this higher resolution is presently limited to the panchromatic, i.e. black-and-white realm.
- The seven bands of LANDSAT have proved to be useful and will continue to be at the user's disposal in the long run.
- There is a tendency towards the production of stereoscopic images.
- The international community of space imagery users must be given the certainty that they may count on long-term image production. This is a very important condition for the successful incorporation of space imagery into a GIS.

3.6 Microwave systems on satellite platforms

After the first, rather experimental microwave system (SEASAT) was launched in 1978, different countries made considerable efforts during the early nineties to develop microwave systems mounted on satellite platforms but up to now only the European ERS-1 has been launched and is in orbit (1991). After ERS-1 much research has been undertaken and is under way. The effects of the image generating microwave sensor of ERS-1 will probably be similar to those that came in the wake of the first LANDSAT-MSS (1972). With good results, eventually applicable to practical uses, ERS-1 might evolve into an operational system ([Winter et al. 1992]).

Table II.4-5: Microwave systems on satellite platforms

Parameters	SEASAT (US)	ERS-1 (ESA)	JERS-1 (Japan)	Radarsat (Kanada)	EOS-SAR (US)
Start	1978	1991	1992	1994	1994
Pixel size on ground [m]	25	30	20	28	10-20
Scanning width on ground [km]	100	80	75	130	50-300
Frequency [GHz]	(L) 1.28	(C) 5.30	(L) 1.20	(C) 5.30	(L) 1.20 (C) 5.30 (X) 9.60
Polarisation	HH	VV	HH	VV	Quad.
Incidence angle	20°	23°	44°	20° - 45°	10° - 55°
Altitude [km]	800	777	568	1000	800

Compared to optical systems, the human observer faces considerably more difficulties with microwaves as they follow other rules than those of the optic rays and depend, among other factors, on the electric conductivity of the surfaces involved. This is why ERS-1 for instance shows ships navigating on the Rhine river in spite of the low geometric resolution of this system (raster size approximately 25 m). The detection of such specific objects (with metallic surfaces) can be better done with microwave images than with the theoretically five times better resolution of the KFA-1000 images. Due to the same reason, i.e. the strong reflectivity of well conducting materials, can railway tracks and even powerlines be distinctly made out.

The fact that weather does not interfere with microwaves cannot be stressed enough since it guarantees the acquisition of images at predetermined fixed times. This advantage becomes particularly significant when considering the enormous difficulties involved in the continuous and temporally equal sensing of extense areas by means of conventional optical satellite systems. Its relevance becomes still more striking when frequently changes of phenological phenomena has to be taken into account. Finally, the combination of optical and microwave systems show promising results ([Bähr, 1995], [Foeller, 1995]).

4. Possibilities to enter images into a GIS

A GIS, as stated in Chapter I.2, rests on three "columns" :
  - original data
  - parameters
  - models

Images are undoubtedly "original data" and, therefore, can enter GIS as a data layer. To do this, some conditions must be fulfilled: image data must be available in digital form (they are, if they are in raster form) and these data must be geocoded. This geometrical reference is indispensable so that images and other data layers can be interrelated by means of geometrical and logical links. This geometrical reference can be done absolutely to a hierarchical coordinate system or relatively to the predetermined geometry given by other data layers.

The practical inclusion of image data into GIS's can be attained in different ways:

a)   **As background information**
        It is very useful, especially for planning purposes, to overlay a large scale vector plan onto an orthophoto. The vector plan may be e.g. the actually "authorized information" according to the Federal Building Regulations. The digital orthophoto underneath offers additional information about the areas in question. This combination is an outstanding

basis for a planner, who might want to derive plans and decisions from such hybrid documents.

b)   Direct use: Colouring according to spectral signature

When image data are *primary* information, contrary to the above example, it should be separately displayed, avoiding overlay of other data sources. The satellite image then directly yields the necessary information, e.g. the type of surface rocks in an arid area for geological interpretation purposes. The legend may be taken directly from the image itself. The additional information is limited to legends.

c)   Base for image segmentation (indirect)

The third case contemplates the inclusion of "raw data" into a GIS. This means that images, although geocoded, are not yet classified. The extraction of attributes will be carried out by digital image processing within the GIS. Examples of this are extraction of roads or other linear elements (e.g. rivers) or a digital multispectral classification to find the attributes for use of the land. The actual use of information contained in images within a GIS remains open and the decision about how to use these information depends upon the intended purpose ([Kaufmann, Schweinfurth, 1986]).

Questions

1. In Fig. II.4-1 "maps" are assigned to "images". What are the reasons for this concept?

2. Try to show systematically the different nature of raster and vector data (advantages/ disadvantages).

3. Problem: advantages and disadvantages of "analogous photography from space"!

4. Which expectations are put from microwave imagery for environmental monitoring?

5. Which sensors are available for thermography? Is there thermal imagery in stereo?

6. Discuss the concepts "operational", "experimental" and "technical standard". Give examples!

References

Bähr, H-P., Th. Vögtle (Hrsg.): Digitale Bildverarbeitung - Anwendung in Photogrammetrie, Kartographie und Fernerkundung. Herbert Wichmann Verlag, Karlsruhe, 1991

Bähr, H-P.: Image Segmentation Methodology in Urban Environment - Selected Topics. GISDATA Specialist Meeting on Remote Sensing and Urban Change, Strasbourg, 1995 (to be published)

Foeller. J.: Kombination der Abbildung verschiedener operationeller Satellitensensoren zur Optimierung der Landnutzungsklassifizierung. Diploma Thesis, University Karlsruhe, 1994.

Fritsch, D.: Synergy of Photogrammetry, Remote Sensing and GIS - the MOMS example. Int. Archives of Photogrammetry and Remote Sensing (IAPRS), Ottawa, Vol. 30, 2, pp. 2-9

Jacobsen, K.: Cartographic Potential of Space Images. ISPRS, Comm. II, Dresden, 1990

Kaufmann, H., G. Schweinfurth: Definition of Processing Parameters for TM based "Spectral Maps" with Respect to Geologic Applications in Arid Areas. Presented Paper, ERIM-Kongreß Nairobi, 1986

Kramer, H.J.: Observation of the Earth and its environment: Survey of missions and sensors. Springer Heidelberg, 1994

Lorch, W.: Zeiss RMK TOP - erweiterte Möglichkeiten eines Luftbildaufnahmesystems. Zeitschrift für Photogrammetrie und Fernerkundung, 1/1992, pp. 19

N.N.: Nouvelles de SPOT. SPOTIMAGE, Toulouse-Cedex

Riess, A., J. Albertz, R. Söllner, R. Tauch: Neue hochauflösende Satellitenbilddaten aus Rußland. Zeitschrift für Photogrammetrie und Fernerkundung, 1/1993, pp. 42

Slama, Chester C.: Manual of Photogrammetry. American Society of Photogrammetry, 1980, pp. 316-318

Winter, R., D. Kosmann, B. Schulz, M. Sties, M. Wiggenhagen: Radarmap Germany - First Mosaic and Classification. Proc. 1. ERS-Symp. 'Space at the Service of our Environment', Cannes, ESA SP-359, 1992, pp. 551-554

II.5 Questionnaires and Other Sources

Christiane Weber, Strasbourg

Goal:

Understanding what "Geographic Information" is.
Nature of information and access to questionnaires and other sources of information.
Limits and constraints in the use of questionnaires and other sources of information.

Summary:

After a precise definition, a general presentation of questionnaires and other sources of information is made. The problem of information access is discussed as well as limits and constraints due to the nature and reliability of information. Specific aspects are developed such as the differences of legal rules among countries and the role of the EC in the exchange of insights. The economic aspect of the geographic information is presented as well as the changing behaviour of different information producers.

Keywords:

Information source, information, data, nature of data, data reliability, questionnaire, legal aspects, economic aspects.

1. Introduction

In order to promote environmental and other studies based on the knowledge of the geographical space, it is necessary to develop methods which make spatial as well as temporal data comparisons of data possible. These can only be developed within the framework of a scientific approach which favours a deductive procedure as a basis for further studies.

However, in order to avoid overemphasizing a certain quantitative empirism which might lead mainly to accumulating data and procedures, it is necessary to take into account a few elements which convey scientific rigour:

- theoretical formalization
- data mobilization[1]

Any theoretical formalization is part of a context which has an impact on the conceptualization of the phenomenon under study. The passage from the "observable" to the mental, abstract representation of reality from which a researcher extracts his knowledge, is influenced by him/her. Nobody can deny the cognitive aspect in any research work and its impact on the perception and even the attitude of the individual, the researcher[2]. His theoretical scientific construct, his hypothesis of interrelations, his selection of information, his geographical background, all these elements have an influence on the results that he will get. Therefore, the analysis is strongly connected to the interpretation and geographical judgement of the researcher, who alone is in a position to "assess to which degree his various measurements really represent the phenomenon under study" ([Béguin, 1979]).

To have a clear idea of what one aims at and to acquire the means to reach this aim - or at least to get close to it - is the key point of the approach. To acquire the means implies identifying and describing those elements available in the observable reality which have to be taken into account. This procedure implies investigating sources of information and potentially relevant data which might provide answers to the following three questions about the phenomenon under study: How is it distributed? What is the quality of the collected data? Are they adequate for the questioning?

Before proceeding further, it is necessary to address some ambiguities which are inherent in the terminology used, e.g. "Information sources, information, data, etc".

Actually the term "information source" can refer both to the original document which contains the information and to the organization which provides and distributes the information. "Information" refers to the geographical pieces of information available to individual researchers, either by direct observation, or indirectly through organizations. "Data" result from the need to structure the information - be it of a qualitative or of a quantitative nature - by identifying groups, categories or classes through existing relationships within collected geographical objects. This structuring is carried out on a conventional, simplified basis (the most common form being coding) with the aim of making information available in an acceptable (understandable by the largest number) form, easy to handle and to memorize (possibly in computerized form). Structuring the information available will allow its distribution.

Therefore, to generate knowledge which adequately represents the structure of the observed geographical space implies minimizing information loss and maximizing the links between data and space.

---

[1] The word "mobilization" includes setting up a data collection plan, collecting the data and structuring the information with the aim of distributing it.

[2] Bailly, Béguin, 1982, p 25.

It is possible to classify sites and their characteristics according to a double-entry table, *the geographical matrix* ([Berry, 1964]), in which the sites can be linked to the values of their attributes. This matrix can moreover be enriched in successive stages, adding a temporal scale to the already existing spatial scales.

Different lessons can already be learned before starting our analysis :

- the obvious non-neutrality of data, because they convey the choices made by the researcher and they imply the beginning of a structure

- the impossibility to apply all sorts of analysis to all data

- the need to measure is justified by comparisons; but measuring[3] implies two types of problems: the problem of the conditions under which the measurements are being done, and the problem of the unit of measurement. According to the types of data (measurable, nominal or even ordinal) the statistical methods are different: descriptive and inferential in the first case, statistical and non-parametrical in the other case.

The conditions of measurement are connected with the quantification of the phenomenon, the repeatable and precise quantification, providing for the identification of a geographical object by its state and/or the intensity of the measured characteristic. The measure cannot be separated from three notions: precision, linked with the quality of the feature measured, accuracy defining the conformity with the observed reality and reliability, corresponding to the notion of confidence, integrity and security.

The general procedure followed in all research works must include as a prerequisite, that "some thought be devoted to the methods to be used, as soon as the data collection starts" ([Ciceri et al., 1977]). The present chapter distinguishes more specifically between the information obtained from administrative or other organizations (e.g. statistical offices of various kinds) and information obtained from individuals or institutions[4] (questionnaires or enquiries), followed by a quick survey of all sources of information, as opposed to other information available and useable in the GIS.

2. Sources of information

Information which supports a better understanding of geographical space, can be obtained - as explained earlier - by direct measurement (field data), or provided by organizations or may be part of a process of data collection directly carried out by social groups.

Field data, some of which correspond to topographic land surveys, others to phenomena which are continuously spread over space and which necessitate specific measurements of the composition of the ground or the air (e.g. pedological, geological, or climatological measurements) and the organization of layers or fluxes. These measurements are carried out within spatial sampling projects characterized by a two-dimensional sampling unit (point, line or area). Different sampling methods are possible: either by randomly distributed sampling units, by systematic or by stratified distribution according to the zones previously selected.

Other data are also based on an implicit geospatial reference, but they are "representations" of the observable reality. They are obtained by analogous or numerical means, without

---

[3] D. Badariotti, 1994.

[4] In this case, it can be any institution, firm or responsible person carrying out activities in the study or collection of information

abstraction or generalization (e.g. aerial photographs[5]), or by interpretation and abstraction in the case of maps and plans. The cartographic reasoning which shows the concept, creation and use of documents of this type is explained in Chapter III.1 ('Cartographic Principles').

A third category of information is provided on different types of media (lists, magnetic tapes, texts etc.), corresponding to sites which are explicitly defined. Statistics, inventories and texts provide information and describe sites by connecting the locations and their attributes, to characterize, identify and classify spaces, distributions of phenomena and links between various elements.

This information may come from different public or private corporations which - according to their scope of responsibility - collect, structure and distribute information, e.g. institutes which provide statistics on a given territory. Some of these data are up-dated more or less regular (e.g. census data). Others are up-dated in a more sophisticated way (e.g. consumption index), and some are up-dated depending on the economic and political circumstances.

But these data may also come from individuals or institutions which choose the medium of survey or questionnaire to go beyond the simple statistical elements available and to obtain more complete information which better describes the behaviour of people, their habits, their wishes and their actions. Of course none of them is sufficient, although some allow comparisons provided the procedure used is the same. Such approaches are interesting because they make it possible to present the spatial distribution of „everyday elements" such as reactions to environmental pollution or decision schemes concerning mobility.

3. Questionnaires and other sources of information

3.1 Questionnaires and surveys

Before dealing with the characteristics of the described sources of information, it is proper to give a general outline of a questionnaire and a survey.

As explained above, this type of information acquisition makes it possible to obtain information which cannot be realized without a designing effort. It is necessary to precisely define the aims of the study and to identify clearly three main points:

- the topics of the survey or questionnaire and the sampling techniques

- the experimental site, localized and defined clearly without ambiguity

- the choice of the collection technique in view of the hypotheses.

Two types of surveys are possible depending on the subject: either the subject is active and carries out an action (direct survey), or the subject is passive and provides information based on memories and perceptions (indirect survey). In the latter case - according to the positioning of the investigator - observations can be made and facts registered without the subject being aware of it, or participation is ensured and the subject communicates with the investigator.

A summary presentation (Tab. II.5-1) of the possible variations of survey techniques shows how important the choice is. This affects the following research work, since it implies the type of questioning, the counting and coding of the survey, as well as the use which is to be made of it. Depending on the principles on which the survey protocol will be based, i.e. the rules to be applied during the survey (information, logical aids for the questioning in connection with the survey, localisation principles), the encoding and its implementation will be easier or more sensitive.

---

[5] Chapter II.4 ('Imagery') and associated treatments

Table II.5-1: The techniques for collecting information (according to [Cauvin, 1984])

Subject \ Investigator	Passive Observation	Semi - passive listening	Active Interview- questionnaire
Active Direct survey	with or without participation of the observer	Requested partial participation by action	Variable according to the scale of the study (laboratory or field)
Passive indirect survey	No contact with the subject	Free Interview	Contacts with or without material participation

If the questioning is appropriate, the questions may still be poorly posed, the encoding may ignore the internal underlying variations and the exploitation may not allow the proper treatment of all possible modalities.

These forms of information acquisition are presently of high interest as they portray the types of behaviour and decisions in a spatial dimension when faced with everyday or extreme situations (such as major hazards), or choices or aspirations, in particular in environmental matters. Using GIS, it is then possible to provide a spatial dimension of an essentially "private" information connected to behaviours as well as culture, education, citizenship, etc.

3.2 Purposes

To deal with "thematic" information, i.e. information which aims at informing on a specific population, implies making a number of distinctions between the types of collected information, considering in particular the objective and type of the information collection.

The aim of this information collection is defined by the organizations (responsibilities in terms of information and follow up: costs, demographic evolution, work structures or state taxes etc.) and the thematic concerns of these organizations which distribute the information collected. The aim will have an impact on the type of information collection: either exhaustive collection (as in the case of taxes), or inference (poll and questionnaire). Questionnaires are to be considered here as a mean of acquiring information on a specific phenomenon, applying mathematical inferences to determine the unknown characteristics of the set under study. To use a survey as a mean of mobilizing information, implies that the observable reality is not finite, and, therefore, exhaustivity is difficult to achieve.

3.3 Use

Another major difference lies in the use of the information collected: either the information is widely distributed and different types of users have access to it (geographers, demographers, sociologists, economists etc.), or it is used only for very specific purposes by specialists. However, in some cases it is possible to have access to dedicated information which is not too "sensitive", i.e. data which does not adversely affect the rights or the social and cultural habits of individuals.

Due to the use of computers, information is increasingly available in computer compatible form and is maintained in data bases by the collecting organizations. To distribute this information by interorganization exchange or by public sale it is necessary to define exchange formats by agreements or associations between the organizations. In many countries - considering the proliferation of data collected by individuals - one of the main concerns at present is to set up laws that protect the freedom of the citizen faced with such practices ([Maisl, 1994]).

4. Sources of information and information providing organizations

Among the organizations which provide information, there are some which offer information at international or European level: UNESCO-GEMS, World Data Center, UNEP-GRID, Master Data Base (NASA), EUROSTAT (the statistical department of the European Community), the European Environment Institute (Corine Land-Cover), IERS (ESA), etc.

Some of the data bases offer dedicated or non-dedicated information: imagery, environmental statistics, demographic data; directories make it possible to access to various data bases (by inter-connections), and to find useful information.

However, in many cases, it is difficult to obtain a clear idea not only of the information content (nature, scale, up-dating etc.) but also of the various means of access and acquisition. Moreover, connection time and costs can be high. Some national institutions, such as the Conseil National de l'Information Geographique (CNIG) in France, publish a catalogue of computerized sources at national level, indicating the information producer, the different types of products and their characteristics, a detailed description, the maintenance principle, the conditions of use and the distributing organizations. It is thus possible to have a general overview of the information potentially available on the French territory.

One of the lacunae which is often mentioned is the difficulty to trace back the source of information and to find out precisely how and by whom it has been collected and structured. As explained above, information is related to a context and can hardly be separated from it, for fear of misinterpretation or even mishandling.

5. Constraints and limits in the use of data

Among the constraints and limits in the use of data, two distinct directions are evident. On the one hand the "practical" aspects and on the other, the more formal ones connected with geographical information.

5.1 Practical aspects

Before using the information provided by an organization, it is necessary to examine it closely in order to assess its value and reliability. This assessment bears on the conditions under which these pieces of information were obtained. These must be read taking into account the precise and operational definitions previously agreed upon. The elements observed (e.g. establishment of natural or migratory balances of a given population) as well as their attributes must be precisely defined in order to avoid wrong conclusions being drawn from the given data. Errors may invalidate the results of the survey.

There may be random errors, in the data resulting from the definition of values on the basis of a representative sample. This errors can be overcome but it is necessary to know how to use correctly the provided data. Other errors concerning observations are possible, especially in census data. They may occur during the collection, coding, storage etc. They can be compensated by syntactical (out of limit answers for questions) or logical corrections (coherence among the answers), if they are really accidental.

This attitude is of course valid for any provided or acquired information and it affects the previous treatments according to the levels of measurement of the data: frequential or mathematical transformations, standardization, class interval selection and statistical treatments to be applied later.

It is necessary to take into account the validity of the information in terms of the subject of the study. Often the user is faced with a dilemma that the data are out of date. Census data are carried out in principle every five to ten years; how long is it permissible to use these data?

Up-dating problems and hence reliability problems are important for information provided by questionnaires or surveys. This is all the more plausible as the possible inference methods are not really convincing. Moreover, according to the national contexts and the purposes, the up-dating principles (information mode, precision, out-dating) can differ. The knowledge requirements may vary from one country to another. This aspect provides for a different orientation to this operation. In developed countries, precision is considered to be essential, even if it implies high additional costs in the acquisition of the information; therefore, surveys or polls are carried out in a detailed way and on a long term basis. In developing countries where information is absolutely necessary on a short term basis, precision is not as important as the urgency to have the knowledge. In cities with a yearly growth rate of about 10%, it is useless to know the exact population every 10 years; but it is important to assess within a realistic margin where these persons are going to settle.

The out-of-date aspect of information has only recently been considered. The use of Geoinformation Systems has implied the integration of various types of information. The quality of data, its relevance and reliability has implied a concern over the limits within which to use this information. It is wise to know the "expiry date" of a piece of information (as is the case for consumables), especially when this information is part of a combination of data, in order to avoid drawing wrong conclusions or making biased reasoning. "Reliability" percentages associated with the information and their results provide the final user with an opportunity to fully play his role as a decision maker.

5.2. Formal aspects

It is difficult to give an overview at European level of what information is available at the present time. It can nevertheless be said that "there is hardly any compatibility between data produced in one country from that produced in another". The field surveys or other kinds of surveys made in any country rely on the legal-administrative concepts of that country, e.g. the concept of "agglomeration" is different in France from that in Germany. The characteristics associated with these territorial partitions affect the modalities and the content of the surveys. Some partitions are strictly defined legally by linking distance and density in a given country but in another country they may be defined taking into account a specific aim which is not necessarily legal.

Such information covers different aspects: whereas in Germany or Sweden it is possible to obtain information on the standards of living through income classes and rents within the data resulting from the census, in France, this is totally impossible. Harmonizing these socio-economic data remains a real challenge among the European countries.

Another point which differs among the various European countries is the access to information; whereas it is possible to obtain information easily on the basis of polls, obviously this method can only be used for a reasonable geographical coverage and a limited objective, unless one wants to leave this in the hands of the national organizations which have this responsibility. Establishing information files presents certain specificities in some European countries; in France, for instance, it is legally compulsory to apply for permission to establish a file with the "Commission Informatique et Libertés", when individuals are involved.

Because of this limitation the census data of the French population of 1990 for urban districts are available only for zones with more than 5000 inhabitants. This type of limitation to the access and distribution of data creates discrepancies between European countries.

It is interesting to note the growing economic importance of geographic information, both in terms of hardware and software and in terms of information itselfe. In France in particular, establishing digital data will force those who own the information (different ministries), to collectively consider the private market, i.e. a competitive market and a free flow of information. The content of the Council directive concerning the protection of the EEC citizens (1992) against the misuse of computerized information during exchanges between member states suggests that the legislation should become more liberal: to reduce the preliminary formalities, to promote exchange between various local services, to provide private enterprises with access to public data, including sensitive ones.

This clearly raises a number of problems[6], such as:

• the value of information which is beyond the costs of acquisition is difficult to estimate

• the justification of the transfer costs or utilisation rights for public information by third parties, when it may be considered a national value

• the relevance of the evolution of responsibilities of organizations which collect or distribute information.

Every country has a different solution to these problems: distribution at the price of the medium, standardization of formats, creation of specific products, different access according to the user (private or public) etc. However, this is just a new line of thought and only joint action will ensure in the future access and utilisation possibilities for all Europeans, while protecting the privacy of the individual. There is a great discrepancy between the various EC countries as far as the protection of non-anonymous data is concerned. Some observers consider that "the work of the Council of Europe will lead to an arbitration and probably a reconciliation of both constraints (protection of privacy/economic liberalism)" ([Aillaud, 1992]).

Actually this is closely linked to the legal problems with geographical information; property rights are not the same in the various countries, copy rights, access rights and utilization rights are considered differently. Interpretations are even more complicated when one does not deal with a file but with a data base or a central computer gathering information coming from various producers. To whom does the information derived from the joint use of 3 or 4 files from different organizations belong? The problem is presently under consideration but in this case also the economic aspects of information have to be taken into account.

6. Conclusion

The new forms of geographical information obviously leads to new progress of the conceptual, methodological and practical aspects. Digital information has completely changed the habits and practices of the users of geographical information. Developments in the fields of hardware and software have had an impact on the ways of working and thinking. The flexibility of use, the speed of realization and the reliability of the media have opened up a new era which coincides with a social and cultural evolution. However, as is the case of all progress, it has brought along a number of problems in various fields at the practical as well as at the conceptual and formal level. The continuously growing use of information in different countries demands comprehensive discussions on the limits of its use. In the near future the proliferation of files should not promote an exaggerated liberalism which might restrict the privacy and freedom of individuals.

---

[6] C. Weber, 1994, GISDATA, European Science Foundation - Malgrate

Questions

1 - What are the main differences between statistical and questionnaire data ?

2 - What is needed to design an enquiry ?

3 - Give some characteristics of the data-bases used around the world ?

4 - What does the "use value" of information mean ?

References

Aillaud, V.: La protection des données personnelles dans les Systèmes d'Information Géographique mis en oeuvre par les collectivités locales. Institut d'Urbanime, Paris VIII. Mémoire de DES, 1992

Badariotti, D.: Les sources d'information de l'aménageur et du géographe, cours de Maîtrise. Institut de Géographie, Université Louis Pasteur, 1994

Bailly, Beguin: Introduction à la géographie humaine. Masson, 1982, p 188

Berry, B.: Approaches to regional analysis: a synthesis. Annals of the Association of American Geographers, 54, 1964, pp. 2-12

Beguin, H.: Méthodes d'analyse géographique quantitative. LITEC, 1979, p 252

Cauvin, C.: La perception des distances en milieu intra-urbain. CNRS-CDSH, 1984, p. 101 et annexes

Chadule (groupe): Initiation aux pratiques statistiques en géographie. Masson, 1987, p. 189

Ciceri, M.F., Marchand, B., Rimbert, S.: Introduction à l'analyse de l'espace. Masson, 1977, p. 173

Maisl, H.: La diffusion des données publiques. L'actualité juridique - Droit Administratif. mai, 1994

Weber, C.: European strategic Review. GISDATA, European Science Foundation, Malgrate, 1994

II.6 Data Quality in GIS Environment

Ákos Detrekői, Budapest

Goal:

An overview of the measurement and measures of data quality.
Discussion of sources of error.

Summary:

The most important measurements for data quality are: the accuracy, precision, reliablity, and the Cohen kappa index.

The data quality has the following measures: georeferencing accuracy, attribute data accuracy, consistency between geometries and attributes, topological consistency, data completeness and actuality, lineage.

The errror sources and the data quality report will be discussed, too.

Keywords:

Accuracy, reliability, misclassification matrix, error propagation, error sources.

1. Introduction

Digital systems are capable of processing data more precisely than analogous systems, but their overall quality still depends on the quality of their source data which in most cases remain analogous. In GIS the situation is similar, the final quality depends on the quality of the sources ([Bernhardsen, 1992]). GIS data quality is a compromise between needs and costs. In practice the choice is often a question of what is currently available or can be acquired in a reasonable amount of time.

Four aspects of data acquisition comprise the criteria for selecting data quality ([Bernhardsen,1992]):

- need
- costs
- accessibility
- time frame

The most important factors of data quality are ([NCGIA, 1990], [Bernhardsen, 1992]):

- georeferencing,
- attribute data,
- consistency of the links between geometry data and attribute data,
- geometry link consistency,
- data completeness,
- data currentness,
- lineage.

In this chapter the quality of the georeferencing and the quality of the attribute data will be discussed. In the quality management it is necessary to estimate the values of the quality components before the realization of the data base (a-priori estimation), and it is required to have redundance in the data for an a-posteriori estimation of the quality components. The a-priori estimation is a tool of the design of GIS. The a-posteriori estimation has an important rule for the control of data bases, it is very often useful to make a data quality report.

2. Measurement of data quality

There are various parameters to measure data quality. The most important are the following:

- accuracy,
- precision,
- reliability,
- degree of the misclassification.

In literature the definitions of these measures are often different. First in this section a possible definition of each measure will be given, afterwards the estimation of these measures in the design and in the control processes will be discussed separately.

Accuracy is defined as the degree of closeness of measurements to the true values. Normally accuracy is characterized by the standard deviation or root-mean-square. If the result is characterized by the random variable x, the standard deviation s is the following:

$$s = \sqrt{E[(x - E(x))^2]} \tag{II.6-1}$$

where E is the symbol of the well-known statistical *expectation value*.

The **precision** is defined as the number of decimal places or significant digits in the measurements. The precision is not the same as accuracy. A large number of significant digits does not necessarily indicate that a measurement is accurate ([NCGIA, 1990]). The precision expresses the repeatability of the measurement.

The **reliability** gives information about the facility to control and about the blunders (gross errors) of the data. The value of the smallest detectable blunder is characterized by the reliability. In the case of normally distributed measurements the value of reliability can be determined ([Baarda, 1968]) by evaluating the statistics of the measurements.

The **degree of misclassification** is a measure of the quality of specific attribute data. It can be determined as a function of the misclassification matrix. Often used functions are the percentage or the Cohen Kappa index ([NCGIA, 1990]). Other measures are given by [Goodchild, Gopal, 1989].

3. The data quality estimation in the design period of GIS

Technically correct and economical data acquisition requires good planning, consisting essentially in selecting the data sources, the method of data acquisition and deciding on the quality of the data to be acquired.

The data quality estimation in the design period is based upon:
   - the experience of the specialists,
   - the literature,
   - the standards.

The data quality can be characterized by the accuracy, precision, reliability, and the degree of misclassification. The a-priori estimation of the various measures requires various methods. Some methods and values from literature will be summarized in this section.

The **georeferencing accuracy** (Fig. II.6-1) depends on the method of data acquisition. Some informative values should be presented here ([Bähr, Vögtle, 1991]; [Bill, Fritsch, 1991]):

Method of data acquisition	Accuracy (approx.)
Surveying	cm-dm
Photogrammetry	
- stereoplotting	0.00001 × photo scale
- DTM (elevation)	0.0001 × object distances
Remote sensing (pixelsize on ground)	
- LANDSAT Thematic Mapper (TM)	30 m × 30 m
- Systeme Probatoire d'Observation de la Terre (SPOT)	20 m × 20 m
- Russian space photograph (KATE)	5 m × 5 m
Digitizing of maps	0.00025 × map scale

Digital map data generated directly from aerial photos or entered during surveying entail fewer processing steps and consequentely are less subject to error than data from digitized maps. Considering the aspect of accuracy, original data are always preferable to maps ([Bernhardsen, 1992]).

In large scale domain a special problem is the uncertainty of the definition of natural points or lines. [Kraus, 1993] gives some values of the uncertainty of different locations in the field (Tab. II.6-1). In the small scale domain the cartographic adaptation may introduce errors, too ([Bernhardsen, 1992]).

The **accuracy of attribute data** also depends on the data acquisition method. In the literature we can find sometimes information about data quality (for direct measurements (e.g. seismic data) or other kinds of sampling (e.g. data of public opinion tests)).

Table II.6-1: Uncertainty of different locations (from: [Kraus, 1993])

Type of point	Planimetry	Height
House and fence corners	7-12 cm	8-15 cm
Manhole cover	4-6 cm	1-3 cm
Field corners	20-100 cm	10-20 cm
Bushes, trees	20-100 cm	20-100 cm

The a-priori estimated value of data accuracy is very useful, if the required accuracy of data is given (e.g. in a standard). We can use a data acquisition method only if the following equation is valid:

$$a \leq A \tag{II.6-2}$$

where a is the a-priori estimated value of accuracy,
           A is the required value of accuracy.

Eq. (II.6-2) could be useful in the case of georeferencing and in the case of attribute data. Normally the values a and A are standard deviations or tolerances.

Precision of data is defined as the number of decimal places or significant digits. For measurements normally of interest in GIS, precision is limited by the instruments and methods used. The precision of data is usually lower than the accuracy.

Reliability gives information about the facility to control and about blunders (gross errors) of data. For determination of reliability we can use the methods of adjustment of survey and photogrammetric measurements ([Baarda, 1968]; [Mikhail, 1981]; [Förstner, 1980]). The preliminary condition for determination of the numerical value of the reliability is the existence of redundancy. In the design period of data acquisition we can estimate the redundancy of data. The redundancy of a data set is given by the following equation:

$$h = N - n \tag{II.6-3}$$

where N is the number of all data,
           n is the number of necessary data for the unique solution.

The ratio $t=h/n$ is a possible measure of the facility to control a data set. If

$0 \leq t \leq 0.01$	the data set isn't controllable,
$0.01 \leq t \leq 0.1$	the data set is poorly controllable,
$0.1 \leq t \leq 0.3$	the data set is sufficiently controllable,
$0.3 \leq t$	the data set is well controllable.

In standards often the number or the precentage of correct data is used as a quality measure. Let this number be defined as H, the following equation must be satisfied:

$$h \leq H \tag{II.6-4}$$

The **degree of misclassification** is very difficult to estimate in the design period. Only the literature and the experience of operating GIS gives some information for this purpose.

4. Data quality estimation after data acquisition

An estimation of the accuracy, precision, reliability and the degree of misclassification is possible by using various methods.

Accuracy of data is characterized normally by the *standard deviation*. The real value of standard deviation of data can be determined in different ways:
- using the least squares or other estimations,
- using more accurate data of other sources for comparison,
- analysis of attribute data.

If we use a *least squares estimation* (e.g. for determination of georeferenced data) the standard deviation values can be calculated using well known equations, e.g. the *covarince matrix* of coordinates using the observations equations is defined as ([Mikhail, Ackerman, 1976]):

$$M = 1/h \; v'Wv \; (A'WA)^{-1} \qquad\qquad (II.6\text{-}5)$$

where h is the redundancy
                   v is the vector of the residuals
                   A is the design matrix of the observation equations
                   W is the weight matrix of the observations.

The standard deviations of the coordinates are the squares of the diagonal elements of the covariance matrix. Sometimes the standard deviation will be determined using *more accurate data* (e.g. in the case of the control of digitized elevation from topographical maps using levelling). In this case we can use the following equation:

$$s = \sqrt{1/n \; (d'd)} \qquad\qquad (II.6\text{-}6)$$

where n is the number of the compared values
                d is the vector of the differences of values.

Attribute accuracy must be analyzed in different ways depending on the nature of data. Quantitative data may be defined in non-numerical terms (ordinal data), as discrete variables divided into classes (interval data) or continuous variables without numerical limits (ratio data) ([Bernhardsen, 1992]).
    The **precision** is determined by the used instruments. The **reliability** r_i of each data can be determined using the methods of adjustment. In the case of normal distributed uncorrelated data the following equation may be useful ([Baarda,1968])

$$r_i = gs_i \sqrt{1/m_i} \qquad\qquad (II.6\text{-}7)$$

where $g = g\,(\alpha,\beta)$ (α and β are significant levels)
                   s_i is the a-priori standard deviation of the measurement,
                   m_i is the square of diagonal element of the matrix

$$[Q - A\,(A'WA)^{-1}A'\,]W$$

where Q is the cofactor matrix of the measurement
                   A is the design matrix of the observation equations
                   W is the diagonal weight matrix of the measurement

This method of calculating the reliability can be used in the case of a least squares adjustment of the measurements.

The **degree of misclassification** is characterized using the misclassification matrix. The determination of the misclassification matrix and the calculation of the function of this matrix (e.g. Cohen Kappa index) was published in the literature (e.g. [NCGIA, 1990]). Various possibilities of the determination of other measures were given by [Goodchild, Gopal, 1989].

5. Conclusion

In the data quality management it is necessary to estimate the values of the quality components before creating data bases (a-priori estimation), and it is required to have redundancy in the data for a-posteriori estimation of the quality components. The a-priori estimation is a tool of the quality design of GIS. The a-posteriori estimation is important for quality control of existing data bases. The results of a-posteriori estimation will be very often used for a quality report. The methods of adjustment of survey and photogrammetric measurements are applicable for the a-priori and a-posteriori estimation of the data quality. Fig. II.6-1 to II.6-6 show some of the error influences on GIS data resp. parameters for quality control.

Questions

1. What are the aspects for selecting data accuracy?
2. What is the difference between precision and accuracy?
3. Give some approximate values for the accuracy of methods of geometrical data acquisition!
4. What are the most important questions in data lineage control?
5. Which kind of error sources do you know?

References

Baarda, W.: A testing procedure for use in geodetic networks. Netherlands Geodetic Commission, Vol. 2, No. 5, Delft, 1968

Bähr, H.-P., Vögtle,T. (eds.): Digitale Bildverarbeitung. Wichmann Verlag, Karlsruhe, 1991

Bernhardsen, T.: Geographical Information System. VIAK IT and Norvegian Mapping Authority, 1992

Bill, F. , Fritsch, D.: Grundlage der Geo-Informationssysteme 1. Wichmann Verlag, Karlsruhe, 1991

Burrough, P.A.: Principles of Geographical Information Systems for Land Resources Assessment. Claredon Press, Oxford, 1986

Förstner, W.: The theoretical reliability of photogrammetric coordinates. XIV Congress of International Society for Photogrammetry, Commission III, 1980, pp 223-245

Goodchild, M., Gopal, S. (ed.): Accuracy of Spatial Databases. Taylor-Francis, London, Bristol, 1989

Kraus, K.: Photogrammetry, Volume 1. Dümmler Verlag, Bonn, 1993

Mikhail, E.M., Ackermann, F.: Observations and Least Squares. IEP Dun-Donelley Publisher, New York, 1976

Mikhail, E.M., Gracie, G.: Analysis and Adjustment of Survey Measurements. Van Nostrand Reinhold, New York, 1981

NCGIA: National Center for Geographic Information and Analysis: Core Curriculum - Introduction, 1990

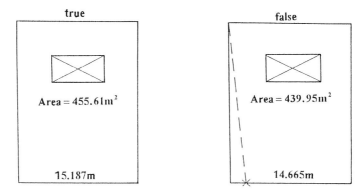

Figure II.6-1: Georeferencing

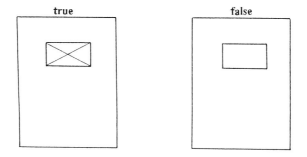

Figure II.6-2: Attribute data

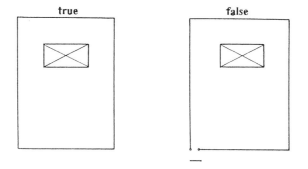

Figure II.6-3: Geometry link consistence

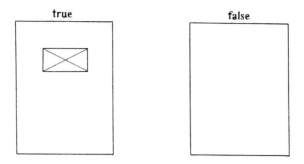

Figure II.6-4: Data completeness

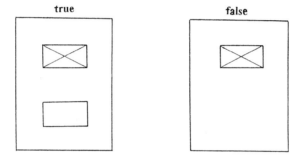

Figure II.6-5: Data actuality

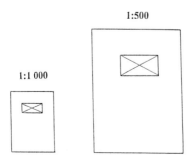

Figure II.6-6: Lineage

II.7 Database Systems I

Joachim Wiesel, Karlsruhe

Goal:

This chapter will introduce the concepts of relational database systems. It will be shown, how to use the Structured Query Language (SQL) for data definition and data manipulation tasks.

Summary:

The principal definitions of relational database management systems are introduced. A coarse overview of interactive SQL (Structured Query Laguage) is given together with some example tables demonstrating the functionality. Base tables are introduced and simple DML and DDL statements are explained.

Keywords:

Database, relational model, SQL, DDL, DML, tables

1. Introduction

What is a Database System?

In [Date, 1990] it has been defined as a "basically computerized record-keeping system; that is a computerized system whose overall purpose is to maintain information and to make that information available on demand".

In a typical GIS environment database systems are used to store and administer non-graphical and - in some systems - geometrical data, too.

2. Components of a database system

Normally, a Database System is composed of four components: data, hardware, software and users.

Hardware:

Database systems are running now on small single user personal computers up to networked mainframe or massively parallel computers.

The hardware parts of a system are made of:
- The processor(s), main storage and I/O channels needed to run the database management software components and associated support programs;
- Secondary storage devices - typically magnetic disks - to hold the data permanently together with disk drives, magnetic tape drives, tape robots or other backup subsystems;
- Communication hardware - like terminals, communication controllers, LANs, WANs - needed to allow user access to distributed computer equipment.

Software:

A layer of software is located between the database and the users of the system. It is called the *Database Management System* or *DBMS*. The main task of the DBMS is to keep users from dealing with hardware level details of the database. All requests to access, change or delete data from the database are handled by the DBMS. It also allows to define a hardware independent view of the database, allowing the *Database Administrator* to change physical aspects of data storage and organization without disturbing users of the DBMS. High and low level programming interfaces and scripting languages are normally provided to allow *end users*, *programmers* and *administrators* to access the DBMS.

Users:

There are three distinct classes of database users:
- First, there is the *end user*, who is accessing the DBMS from a terminal device. End users can interact with the system using ready-to-use application programs or a built-in high level interface provided by the DBMS manufacturer. All systems provide some kind of an interactive query language for issuing high level commands to the DBMS.

- Second, there are *application programmers*. They are writing application programs accessing the DBMS for use by end users, typically in languages such as COBOL, FORTRAN, C or in the scripting language of the DBMS. The programs can be *batch* or *online* applications. Batch applications typically run offline, producing large reports, sorts or are used for bulk data backup, import or export, while online applications are operated by end users at data terminals.

- Third, there is the *database administrator* (*DBA*). The DBA is responsible for creating, maintaining and taking care of the database. The DBA has to implement security measures, optimize the performance of the DBMS and provide technical support for other users.

Data:
We can roughly classify database systems into *multi-user systems* and *single-user systems*. In a single-user system only *one* user can access the data at a given time, while in a multi-user system *several* users can access data simultaneously at the same time. Today the type of system does not depend on the hardware it is running on. Even on PC-class hardware multi-user systems are run successfully.

Data in a database system are *integrated* and *shared*. *Integrated* means, that for a given organization all data are consolidated and stored without redundancy in a well known location of a database system. *Shared* means, that data (or a subset) can be accessed concurrently by users with the appropriate acccss rights. As the data are shared and integrated, it is a requirement that the DBMS allows to define subset views into the database for different users.

3. Relational database systems

Nearly all database systems in production use today are based on the *Relational Model* ([Codd, 1970]), while the majority of older systems are based on inverted list, network or hierachic models. Research systems and some new production systems are using an object oriented approach (OODBMS) ([Kim, 1995]). We will not talk about these systems, because standards are not yet formally specified and current OODBMSes are quite vendor specific.

What is a relational system?

- Basically, it is a system where data is presented to the user as tables only and

- operators on tables (e.g. a data retrieval operator) generate always tables as results.

Most current DBMS are not only relational but SQL (Structured Query Language) systems as well. SQL (then called SEQUEL) has been published first by [Chamberlin, Boyce, 1974] and was used in the experimental IBM "System R". Later it has been refined and used in all of IBM's commercial DBMSes, like SQL/DS, DB/2 and relational systems of other vendors. SQL has been standardized by ISO as ISO/IEC 9085:1992 SQL-92 and also as DIN 66315 standard ([Demuth, Demuth, 1993]). Most current SQL systems do not yet fully implement the new SQL-92 language features but are based on SQL-87 (ISO 9075:1987) plus some SQL-92 and proprietary extensions. Full conformance has been announced to be available in the near future by the major DBMS vendors.

4. SQL - data definition language

SQL consists of two parts, the *Data Definition Language* (*DDL*) and the *Data Manipulation Language* (*DML*). The DDL part of the language allows to define tables, their names, data types and names of their columns. First, let us introduce some definitions:

A *table* consists of a row of column headings and zero or more rows of data values. It has at least one column heading which specifies the name and data type of the column. Each data row contains one data value for each specified column heading; all data values for one column heading are of the same type. Rows are unordered, while columns - at least in a SQL-system - are ordered from left to right. A *base table* is a special kind of table which exists from itself, is *persistent* and is *named*, in other words it is *autonomous* and carries a unique name. An example is printed in Fig.II-7.1:

Table COO

P-NO	EASTING	NORTHING	HEIGHT	COMMENT
17	1000,00	2000,00	20,00	No Comment
12	333,00	200,00	17,00	Unknown precision

Figure II.7-1: Example of a base table

The principal SQL DDL statements (Fig.II-7.2) are:

```
CREATE TABLE    CREATE VIEW    CREATE INDEX

ALTER TABLE

DROP TABLE      DROP VIEW      DROP INDEX
```

Figure II.7-2: SQL DDL statements

The general format of the CREATE TABLE statement is as follows:

Note: Statements in brackets ([]) are optional, keywords are written in all capitals, words in lower case have to be replaced with specific values by users, ellipses (...) mean, that immediately preceding entities may be repeated.

```
CREATE TABLE base-table
( column-definition [, column-definition ] ... );
```

column-definition means:

```
column data-type [ NOT NULL ]
```

here is the example of a CREATE statement which creates table COO above:

```
CREATE TABLE COO
    ( P-NO          INT        NOT NULL,
      EASTING       FLOAT      NOT NULL,
      NORTHING      FLOAT      NOT NULL,
      HEIGHT        FLOAT,
      COMMENT       CHAR(20)
    );
```

This statement creates a new empty base table named COO. Several entries describing the table are automatically entered into the catalog of the DBMS. The table has five columns with the defined data types which will be described below (Fig. II-7.3). Instead of full type designators like INTEGER or CHARACTER, abbreviations (INT, CHAR) can be used, too. For detailed information refer to [Date, Darwen, 1993].

In the real world sometimes data is missing or unknown. To deal with this, SQL systems allow to enter a special value called NULL. NULL means: This value is missing. In a SQL system any field can contain missing values (NULLS) unless it is specifically forbidden by using the NOT NULL attribute while specifying a column-definition.

INTEGER	binary integer data (normally 32 bit/word)
SMALLINT	halfword binary integer
DECIMAL(p,q)	packed BCD number, p digits and sign, decimal point q digits from right
FLOAT(p)	floating point number with p binary digits precision
CHARACTER(n)	string of exactly n 8-bit characters
CHARACTER VARYING(n)	string of up to n 8-bit characters
DATE	date in the form (yyyymmdd)
TIME	time (hhmmss)
TIMESTAMP	combination of date and time, optionally up to parts of a second

Figure II.7-3: Some scalar SQL data types

In our example table COO point numbers (P-NO), eastings and northings columns carry the attribute NOT NULL, because it makes no sense to enter coordinates into a database without these values. But heights may be unknown and comment texts are sometimes not available.

Like a new base table can be created at any time, an existing base table can be altered at any time by adding a new column at the right:

```
ALTER TABLE base-table ADD column data-type ;
```

For example:

```
ALTER TABLE COO ADD RMS-X FLOAT ;
```

adds column RMS-X to the right of the table by increasing the number of columns from five to six. If there are already rows with data in the table, the value of all added fields is set to NULL. It is also possible to destroy an existing base table at any time by:

```
DROP TABLE base-table ;
```

The table is removed from the DBMS and all catalog data as well.

5. SQL - data manipulation language

SQL offers four DML statements:

```
SELECT
UPDATE
DELETE
INSERT
```

these will be explained to some extent in this chapter. The syntax of the SELECT statement is:

```
SELECT [ DISTINCT ] item [, item, ... ]
FROM table [, table, ...]
[ WHERE condition ]
[ GROUP BY field [, field, ... ] ]
[ HAVING condition ]
[ ORDER BY field [, field, ... ]  ;
```

The simple statement to retrieve everything from COO:
```
SELECT *
FROM COO;
```

results in:

P-NO	EASTING	NORTHING	HEIGHT	COMMENT
17	1000.00	2000.00	20.00	No Comment
12	333.00	200.00	17.00	Unknown precision

The asterisk (*) is a wildcard character which means "all columns in the order of definition".
The SELECT statement shown above is hence equivalent to:
```
SELECT COO.P-NO,COO.EASTING,COO.NORTHING,COO.HEIGHT,
       COO.COMMENT
FROM COO ;
```

In a qualified retrieval, where we want to get specific numbers only from table COO we can use:
```
SELECT P-NO
FROM COO
WHERE P-NO >= 12
AND HEIGHT < 20,0 ;
```

which returns:

P-NO
12

The condition following the WHERE clause can include comparison operators:

=	equal
<>	not equal
>	greater than
<	less than
>=	greater equal
<=	less equal

and boolean operators AND, OR and NOT. Parentheses are used to indicate the order of evaluation
of expressions. It is possible to perform a retrieval with ordering by:
```
SELECT P-NO
FROM COO
WHERE P-NO >= 12
ORDER BY P-NO ASC ;
```

Result:

P-NO
12
17

the ORDER BY clause can be specified as:
```
column [ order ] [ , column [ order ] ]
```

where order means ASC for ascending and DESC for descending order, ASC is the default.
For further exercises let us define a second base-table as follows:

Table DIST

FROM	TO	DISTANCE	COMMENT
17	23	1000.00	No Comment
12	17	500.00	

which stores distance measurements between two points.

If we are interested to get all distance measurements from all known coordinate points we
have to ask for this by:

```
SELECT COO.P-NO, DIST.*
FROM COO, DIST
WHERE COO.P-NO = DIST.FROM ;
```

we get:

P-NO	FROM	TO	DISTANCE	COMMENT
17	17	23	1000.00	No Comment
12	12	17	500.00	

This operation is called an *equijoin*. The result is composed out of two input tables (COO, DIST)
which are connected via equal values in two columns (P-NO, FROM). The condition

```
COO.P-NO = DIST.FROM
```

is called a *join condition*. An equijoin always produces a result table with two identical
columns. If we eliminate one of these, the result is called a *natural join*:

Note: *The symbols /*   */ are used to enter comments into SQL scripts.*

```
SELECT DIST.*

/* or: COO.P-NO,DIST.TO,DIST.DISTANCE,DIST.COMMENT */

FROM COO, DIST
WHERE COO.P-NO = DIST.FROM ;
```

Result:

P-NO	TO	DISTANCE	COMMENT
17	23	1000.00	No Comment
12	17	500.00	

SQL DML includes, as already mentioned, three update keywords, the first is UPDATE with the
general format:

```
UPDATE table
SET   field = scalar-expression
      [, field = scalar-expression ] ...
[ WHERE condition ] ;
```

Example:
```
UPDATE DIST
SET DISTANCE = DISTANCE * 1,001,
SET COMMENT = 'Measured by Mekometer'
WHERE FROM = 12 ;
```

results in:

FROM	TO	DISTANCE	COMMENT
17	23	1000.00	No Comment
12	17	500.50	Measured by Mekometer

All records (rows) in table DIST matching the condition clause are changed according to the SET instructions. SET DISTANCE = DISTANCE * 1,001 is an example of an arithmetic expression. It has to be mentioned, that UPDATE operations can only be performed to just *one* table in a single update statement.

The second keyword is the DELETE operation:
```
DELETE
FROM table
[ WHERE condition ] ;
```

In a simple example, we delete the coordinate record containing point number 12:
```
DELETE
FROM COO
WHERE P-NO = 12 ;
```

After executing this statement, the row containing point number 12 is removed from table COO.

The INSERT keyword allows to enter new data rows into an existing table. Its general format is:
```
INSERT
INTO table [ ( field [ , field ] ... ) ]
VALUES ( literal [ , literal ] ... ) ;
```

or

```
INSERT
INTO table [ ( field [ , field ] ... ) ]
subquery ;
```

In the following example, we insert a new data record into table COO:
```
INSERT P-NO, EASTING, NORTHING, COMMENT
INTO COO
VALUES ( 23, 300.00, 200.00, 'marked with yellow colour');
```

If we omit the field names, the insert statement assumes all fields (column headings) of a table be enumerated, similar to the * notation in SELECT *.

Questions

1. What are the basic DML and DDL keywords of SQL?

2. What is the row order of tables in a relational DBMS?

3. How many tables can be changed simultaneously with a single UPDATE statement?

4. Is it possible to add new columns to an existing table with the NOT NULL clause?

5. Write a SQL script which retrieves all measurements from table DIST shorter than 900.

References

Achilles, A.: SQL. Fourth Edition, Oldenbourg-Verlag, 1994, 220p.

Berkel, T.: Die relationale Datenbanksprache SQL. Fernuniversität Hagen CBT, 2 Disk. mit Begleitbuch, Addison-Wesley, 1995

Chamberlin, D. D., Boyce, R. F.: SEQUEL: A Structured English Query Language. Proceedings 1974 ACM SIGMOD Workshop on Data Description, Access and Control, Ann Arbor, Mich., 1974

Codd, E. F.: A Relational Model of Data for Large Shared Data Bank. CACM 13, No. 6, June 1970

Date, C. J.: An Introduction to Database Systems. Volume I, Fifth Edition, Addison-Wesley, 1990, 854p.

Date, C. J., Darwen, H.: A Guide to the SQL Standard - Covers SQL-2. Third Edition, Addison-Wesley, 1993, 432p.

Demuth, B., Demuth, F.: Altbausanierung - SQL-92 - die Norm. IX Multiuser Multitasking Magazin, Heise-Verlag, 12/1993, pp. 136-142

Gulutzan, P., Pelzer, T.: Optimizing SQL. Including Disk., Prentice-Hall, 1994, 300p.

Hein, M. et al.: Oracle 7 - SQL. Addison-Wesley, 1995, 650p.

Kim, W. (Ed.): Modern Database Management: Object-Oriented and Extended Relational Database Systems. Addison-Wesley, 1995, 600p.

Marsch, J., Fritze, J.: SQL. Second Edition, Vieweg-Verlag, 1994, 252p.

Perry, J., Salyer, C.: ORACLE 7 Developer's Guide. Including Disk., Sams, 1994, 1000p.

Sauer, H.: Relationale Datenbanken: Theorie und Praxis inklusive SQL-2. Third Edition, Addison-Wesley, 1995, 304p.

Stolniki, J.: SQL-Programmierung. Tewi-Verlag, 1994, 600p.

Vossen, G.: Datenmodelle, Datenbanksprachen und Datenbank-Management-Systeme. Second Edition, Addison-Wesley, 1994, 706p.

II.8 Database Systems II

Joachim Wiesel, Karlsruhe

Goal:

This chapter introduces advanced concepts of SQL systems.

Summary:

Architectural concepts, more advanced SQL statements and organizational aspects of relational systems are explained.

Keywords:

Database, relational model, SQL, ODBC, integrity

1. Introduction

After having introduced basic principles of relational database systems, we will now focus on some advanced aspects. Namely topics concerning administration, concurrency, security and architecture are handled.

2. An Architectural View

Normally a database is organized into levels which are from bottom to top more and more independent from hardware or system software restrictions. In a SQL system, users communicate with the DBMS via *views* and *base tables*.

2.1 Views

Views are created and destroyed by DDL statements CREATE VIEW and DROP VIEW. They do not consist of their own physical data, but they are *named and derived, virtual* tables. Their

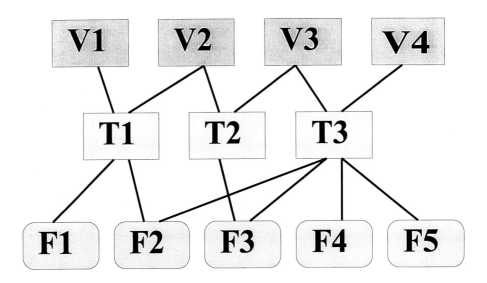

Figure II.8-1: View/Table/File relations in a DBMS

definition from existing base tables is stored in the catalog of the DBMS. Actually it is stored in a catalog table named SYSVIEWS. Similarly descriptions of all base tables are stored in a catalog table SYSTABLES, while all column definitions can be found in table SYSCOLUMNS. Fig.II.8-1 shows the basic architecture of views (V1-V4), tables (T1-T3) and files (F1-F5) and how they could be related to each other. Views are defined by this operation:

```
CREATE VIEW view [ ( column [ , column ] ... ) ]
AS subquery ;
```

For all query operations, views are identical to base tables, but this is not necessarily true for all classes of update operations.

Let us take our already known `DIST` table (see Chapter II.7) and extend it a little bit by adding some more rows of data:

FROM	TO	DISTANCE	COMMENT
16	23	1000.00	No Comment
12	17	500.00	
17	23	350.00	Elta-2
21	12	450.00	Elta-2

With
```
CREATE VIEW S-DIST (FROM, TO, DISTANCE)
AS    SELECT *
      FROM DIST
      WHERE DISTANCE <= 500.00 ;
```

we create a *view* `S-DIST` as the subset of base table `DIST` which is marked grey in the table above. The operation:
```
SELECT *
FROM S-DIST ;
```

results in:

FROM	TO	DISTANCE
12	17	500.00
17	23	350.00
21	12	450.00

A simple
```
DROP VIEW S-DIST ;
```

removes the *view* `S-DIST` from the system catalog - but it does remove neither the base table `DIST` nor any data from the base table from which `S-DIST` has been created.

2.2 The System Catalog

The system catalog contains some bookkeeping tables which can be queried to get more information what has been stored in the DBMS. Specifically we can retrieve data from `SYSTABLES` which contains a row for every *named table* (view or base table) in the DBMS. This row contains e.g. the table name (`NAME`), the creator of the table (`CREATOR`) and the number of columns in the table (`COLCOUNT`). The table `SYSCOLUMNS` contains a row for every *column* of all named tables in the system. It contains the name of the column (`NAME`), the table name (`TBNAME`), the data type of the column (`COLTYPE`) and more. Another important system table is `SYSINDEXES`. It specifies the names (`NAME`) of all indexes, the name of the table to which the index belongs (`TBNAME`) and the name of the creator of this very index (`CREATOR`). Other tables, normally very system specific, support user administration, file space layout, usage statistics and more.

Some examples show, how to use the system catalog to obtain useful information. The first question asks for the number of tables created by a certain user:

```
SELECT 'Number of tables =',COUNT(*)
FROM SYSTABLES
WHERE CREATOR = 'Wiesel' ;
```

which will result (as we have created just two tables up to now) in:

Number of tables =   2

Both columns of the resulting table do not carry a name, the first one is a constant field while the second one implements a *virtual* field (a computed field).

What does the special function COUNT(*) mean? SQL provides some special *aggregate* functions to help answering questions like "How many items?", "What is the average value of a column?" etc. The aggregate functions of SQL-87 are:

```
COUNT()            number of values in the column
SUM()              sum of values in column
AVG()              average of values in column
MIN()              smallest value in column
MAX()              largest value in column
```

The constant character string 'Number of tables =' is simply being printed on its very position (column 1) of the resulting table (which only has 1 row).

Another example: Which columns belong to table COO?

```
SELECT NAME
FROM    SYSCOLUMNS
WHERE TBNAME = 'COO' ;
```

results in a listing of all column names of table COO:

NAME
P-NO
EASTING
NORTHING
HEIGHT
COMMENT

2.3 Index

If we consider the situation, that we have to retrieve coordinates from table COO by using the key P-NO knowing that all tables in a relational (SQL) system are unordered sets of rows (or records) we can run into performance problems, if the number of rows is high. A table with N rows requires N/2 comparisons of a given value with column values to find a match if data are randomly distributed in the table. To shorten access times, auxiliary data structures (an *index* or *indices*) are created. An index is a special kind of a file with exactly two values (columns): a data value and a pointer to the record (row) of the table containing the appropriate value. Fig.II.8-2 illustrates this.

Now the search procedure can be performed a lot faster than before, because the index is sorted or otherwise organized to allow fast access, e.g. as a B-tree, by pointer chains or hashing ([Knuth, 1973], [Martin, 1977]). The SQL commands for creating indices on columns of tables are CREATE INDEX and (to remove them) DROP INDEX. Of course it is possible to define more than one index per table. Indices are optimization tools, most widely used by database administrators to reduce the CPU-load and response time when accessing data frequently. The disadvantage is the increased demand for space (disk or memory) and some added overhead for inserting or updating data.

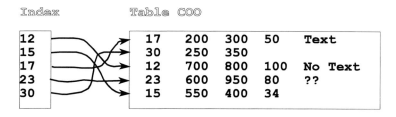

Figure II.8-2: Indexing table COO

2.4 Processing Architectures

As database systems are running on different machines - from PCs to mainframes to large networks - we will look into the basic structures of real world systems. Generally database systems can be considered to have a *frontend/backend* structure: the backend is the SQL engine, while applications, user interfaces and tools are forming the frontend (Fig. II.8-3).

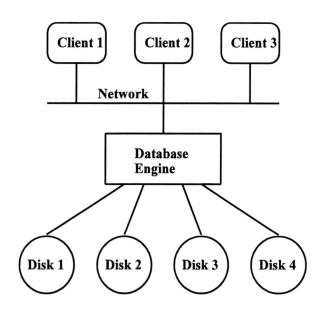

Figure II.8-3: Client/Server Architecture

As the SQL backend can be implemented as several concurrently running processes, multiple requests from clients can be processed simultaneously. System throughput can be improved by using a computer with multiple processors (CPUs) and by distributing table spaces over several disk drives.

Client systems nowadays are typically PCs running the MS-Windows operating system. Standard Office Automation and desktop mapping programs (e.g. MS-Excel, Corel Word Perfect, Arcview) are equipped with modules for accessing remote SQL databases. The computer industry has specified a standard for client programs to do this: ODBC (Open Database Connectivity).

To be able to access e.g. a remote Oracle System running on a Novell Netware server, a Unix server or a Windows NT server ODBC drivers are needed to be installed on the client workstation and an ODBC server module on the server side. Client drivers are available from nearly all database software companies and also from some independent software companies specialized in ODBC connectivity.

With this technique, a part of the computing burden can be moved from the server to the client workstations in exchange for increased network traffic created by transferring large result tables from the server to the client computer.

Questions

1. What is the purpose of creating an INDEX?

2. What are the tradeoffs of Client/Server Architectures?

3. What is the system catalog?

4. What is the purpose of ODBC?

References

Corrigan, P., Gurry, M.: Oracle Performance Tuning. O'Reilly and Associates, 1993, 638p.

Date, C. J.: An Introduction to Database Systems. Volume I, Fifth Edition, Addison-Wesley, 1990, 854p.

Date, C. J., Darwen, H.: A Guide to the SQL Standard - Covers SQL-2. Third Edition, Addison-Wesley, 1993, 432p.

Demuth, B., Demuth, F.: Altbausanierung - SQL-92 - die Norm. IX Multiuser Multitasking Magazin, Heise-Verlag, 12/1993, pp. 136-142

Demuth, B., Demuth, F.: Intergalaktische Kommunikation - SQL3 - Die Datenbanksprache für das Jahr 2000. IX Multiuser Multitasking Magazin 3/1994, Heise-Verlag, pp. 50-61

Gulutzan, P., Pelzer, T.: Optimizing SQL. Including Disk., Prentice-Hall, 1994, 300p.

Hein, M. et al.: Oracle 7 - SQL, Addison-Wesley, 1995, 650p.

Knuth, D.E.: The Art of Computer Programming. Vol. III: Sorting and Searching, Addison-Wesley, 1973, 730p.

Martin, J.: Computer Data-Base Organization. Second Edition, Prentice-Hall, 1977, 540p.

Marsch, J., Fritze, J.: SQL. Second Edition, Vieweg-Verlag, 1994, 252p.

Marx, S.: Überholspur - Parallele relationale Datenbanksysteme statt Mainframe-DBMS? IX Multiuser Multitasking Magazin 11/1994, Heise-Verlag, pp. 50-55

Niebling, R., Diercks, J.: Schwieriges Geschäft - Entscheidungskriterien zur Datenbankauswahl. IX Multiuser Multitasking Magazin 3/1994, Heise Verlag, pp. 32-42

Perry, J., Salyer, C.: ORACLE 7 Developer's Guide. Including Disk., Sams, 1994, 1000p.

Sauer, H.: Relationale Datenbanken: Theorie und Praxis inklusive SQL-2. Third Edition, Addison-Wesley, 1995, 304p.

Schmidt, W.: Fenster zur Datenbank - MS-Windows-Produkte als Front-Ends für Oracle7. IX Multiuser Multitasking Magazin 9/1994, Heise-Verlag, pp. 114-119

III.1 Cartographic Reasoning and Cartographic Principles

Colette Cauvin, Strasbourg

Goal:

To define the basic terms in Cartography and the role of the map.
To set out and describe the steps in thematic map creation.
To show the relationship between cartographic reasoning and scientific experimental approach.

Summary:

After definition of the field of cartography as well as the role of a map as medium of communication, the stages in cartographic reasoning - relevant to the scientific experimental approach - are explained. The acquisition of raw data and its transformation (pre-processing) into suitable data for further processing in a GIS are two fundamental steps. Commonly the last step of GIS procedures will be the visualization of the results. In the case of geo-related data the elaboration of a map - the very heart of cartographic logic - is necessary. This procedure can be devided into three parts: representation (which includes map construction and the correspondence to a 3D coordinate system (georeference system)), cartographic design (grammar and perception aspects), and the layout or map organization. The production of such a map (the analogous realization of the document) will be briefly introduced.

This chapter shows that all decisions are interdependent during the procedure of creating a map. Therefore, they cannnot be separated from the underlying logic and the general context in which a map or a data layer is created.

Keywords:

Anamorphosis, cartography, cartographic reasoning, cartographic transformation, communication, data transformation, geoinformation system, graphic design, scientific experimental approach, thematic map, visualization, visual variables.

Geographisches Institut
der Universität Kiel

1. Introduction

The purpose of this chapter[1] is to lay out the basic principles of thematic map compilation. In Chapter II.3, topographic map has been defined and presented independently: Many authors see this as a special type of map, precisely as a reference map without which no other map can be constructed. This position is sometimes debatable and will be considered here, within the structure of this work. In this chapter will thus be proposed a rigorous process of thematic map compilation which excludes topographic maps.

However, in order to understand this process, a certain number of terms need to be defined and the cartographic representations placed in the context of geoinformation system logic and in the wider context of communication logic.

We should indeed keep in mind that the purpose is not to give an exposition of the cartographic principles for their own sake, but to show that they are necessary in order to create the various data layers used in geoinformation systems. These data layers are nothing but maps, simple or complex thematic maps, but maps nonetheless. If the principles of map conception and construction are unknown it becomes impossible to build the necessary layers of information. This explains the first part of this chapter which is a review of definitions and of the primordial role of all maps. A brief presentation of the steps in cartographic reasoning logically leads to the next parts, from data capture and transformation to map compilation itself, with its material layout on a precise support.

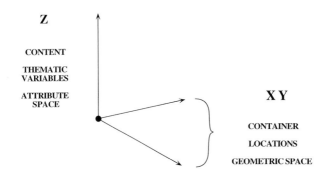

Figure III.1-1: The map and its components

2. Definitions and fundamental principles

In going over the various elements taken from geography or cartography books, a first definition of the map and of cartography can be retained.

Cartography is a set of concepts, methods and techniques used for representing part of the Earth surface (or surface of any other planet) on a plane and for communicating information to individuals through this representation called "map".

A map is a partial representation of the Earth surface to a given scale, made out of two components, the container or geometric space, characterized by coordinates [XY], and the content or attribute space, identified by the letter [Z] (Fig. III.1-1)[2].

These general definitions need to be clarified and refined in order to understand what thematic cartography is, and what it implies for map construction. This should be independant from the subject. An introduction to the cartographic fields thus becomes necessary.

2.1 Cartography: a multitude of fields

The definitions above underline two important points about cartography and its product, the map:

- the transposition from the earth surface to a plane which implies the necessity of precise locations where positions on a volume (or geoid, three-dimensional) correspond to homologuous positions on a plane (two-dimensional). The transposition, done through projection systems, cannot avoid some distortions. The operations necessary to determine the locations belong to *geometric cartography*, a field of specialists in astronomy, geodesy, topography and photogrammetry.

- the identification of the attributes to be placed in the reference plane, i.e., the Z content, described by valued observations of "thematic" space, which requires specialists from various disciplines, according to the subject.

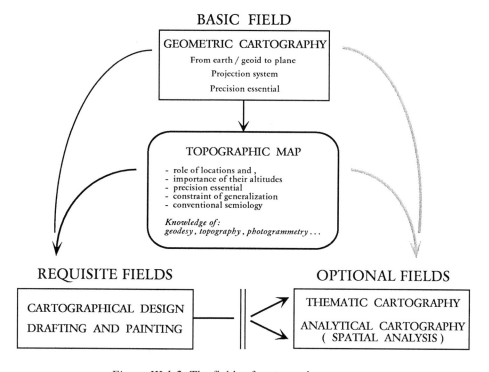

Figure III.1-2: The fields of cartography

The map - the cartographic space - is thus placed at the intersection of the two spaces: geometric space and thematic space.

It becomes necessary to call upon several disciplines for its elaboration; cartography covers many fields, some indispensable to the construction of all maps, others depending on selected options and objectives.This explains the reason why we find basic fields, requisite fields and optional fields in cartography (Fig. III.1-2). Basic fields mostly consist of geometric cartography, a set of concepts, methods and techniques with which a transposition from the Earth to a plane produces those well-named basic maps, or topographic maps. This type of cartography requires first and foremost great precision.

Requisite fields are those used by everyone wishing to construct a map, whatever his/her objectives, theme or specialization. They cover graphic semiology, or graphic expression, and technical production, i.e. drafting and printing. Indeed, how can there be a map without signs or symbols ? Why conceive a map if not to materialise it one day on some support or other and then print it ?

We have to distinguish between two other fields, which may be considered as optional as they depend on the overall purpose of the map: thematic cartography and analytical cartography. In the first case, it is the subject under study, or attributes Z, as well as the methods to represent it clearly which are important. In the second case the locations, linked in varying degrees to attributes Z, are paramount. It consists mainly of analysing surfaces and what is on them. These two fields can be more precisely characterized, in the same manner as the definitions given previously:

Thematic cartography is a set of concepts, methods, and techniques used to represent a selected phenomenon, localizable on a plane corresponding to a part of the Earth surface. In this type of cartography, the base map becomes a secondary although definitely non neutral data, where deformations may occur under the influence of the transferred phenomena. The final goal is to produce a map which communicates information to a precise public with a specific end in view.

Analytical cartography is a set of concepts, methods, and techniques used to determine the spatial characteristics of a selected phenomenon. Container is essential in this field as it is explicitly taken into account by the analyses. The base map characteristics, whether linked to the content of the phenomena or not, are to be put forward through coefficients, graphics, maps etc. Analytical cartography does not necessarily imply the production of a derived map.

In both cases, production can be turned towards research or applications, but the bases are the same. In fact, the distinction between those two optional fields is important because it refers directly to two fundamental stages in geoinformation systems. Thematic cartography corresponds mainly to the stage of data layer creation, while analytical cartography[3] corresponds to the spatial study of data set layers possibly leading, to the creation of derived data set layers. This second stage is too often left out; the last works on GIS, however, seem to stress the research aspects again, namely, the spatial analysis methods downstream from layer construction.

In this chapter, however, we will concentrate on the thematic cartography which can create data set layers and on the requisite fields of cartography, graphic semiology and technical production[4]. We must not forget that in every case the objective is to convey information, to communicate knowledge on space to one or more individuals. It becomes thus necessary to mention a few of the essential elements of the communication logic with which every map is laid out, before giving a general outline of the cartographic process.

2.2 The map: a medium of communication

Whatever the final goal, a map conveys a message, an information to a "receiver"; it is a medium of communication between author and reader, sometimes one and the same person. To ensure the effectiveness of a map - to bring new information to the reader - we must take into account that it depends on two persons working under different, if not outright conflicting, constraints :

- the thematician who knows the subject to be represented, knows what is important and how to find opposite information;
- the cartographer, who knows the methods and techniques to use in order to put localizable phenomena in evidence, and so always has to proceed towards thematic or spatial generalizations.

The wishes and desires of the above two persons are not always compatible; Laswell's[5] communication diagram adapted to cartography may be used to show this (Fig. III.1-3).

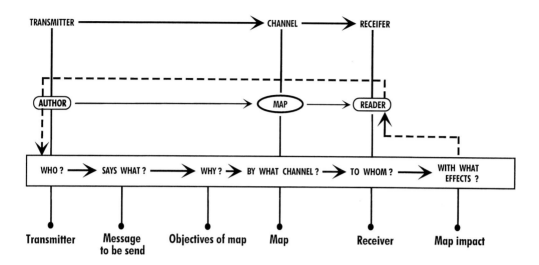

Figure III.1-3: Lasswell communication diagram adapted to cartography (adapted from [Lasswell, 1966])

This diagram is based on 6 interdependent questions.

Who ?   The transmitter, i.e. the one who communicates with the author. The author determines the goal of communication and map contents; proceeds to observe selectively, according to the overall purpose as well as to the receiver whose particularities must be well known.

Says what ?   The message; what the author extracts from data and displays, what he desires to convey to the receiver, according to the initial hypothesis and the overall purpose.

Why ?   The map objectives, determined beforehand; depending on the overall purpose. Maps with identical subject and reader can turn out to be very different.

By what channel ?
                  The map, with all its characteristics: the communication medium, that needs to be improved by preliminary processing, graphic design etc.

To whom ?   The receiver, or reader with his/her specific requirements and capacities. The reader has a personal "repertory" composed of his knowledge, his physiological and psychological characteristics. Transmitting a message is not enough; communication works only if the receiver is ready or prepared to receive the message.

With what effects ?
                  Assessed by the map impact on the reader: this implies finding out how well the map is received, evaluating the conformity between message sent and message received, i.e. is the map "efficient"? Answering this is very difficult as it requires that the readers be subjected to tests, in order to determine whether the intension of the author was realised. These visual perception tests are cumbersome to set up and rarely done[6]. Much remains to be done in this particular field.

Therefore, according to this diagram, one must determine whether or not the reader or user of the map can understand the message sent and further, whether or not he can receive the bare initial information or an improved version of the same. In the framework of geoinformation systems, a produced map is useless and largely unusable if it cannot permit combinations or support relation set-ups. That is why, even though we will attempt to discuss cartographic communication in some detail, especially with regard to design, it should be clear from the very onset that a map must alwaysbe conceived in terms of information communication.This is where a diagram can be most useful, provided it is integrated into cartographic reasoning, a subject that we wish to now examine.

2.3 Cartographic reasoning: a scientific approach

One cannot produce a map merely as an end in itself. A map must have a purpose, otherwise it is a waste of time for both, author and reader. A map is a reasoned construction which rests on a double logic, that of the subject and that of cartography. Disciplinary logic and cartographic logic interlock, along with geoinformation system logic of course. These logics are part of a scientific experimental approach, which allows each step to be checked, explained, and discussed. Let us be more precise.

"Scientific" refers to disciplinary logic, to the consensus on the scientific subject under process at a given period. "Experimental" means that the approach is transparent and can be checked by everyone.

Scientific experimental approach is composed of several stages, in which cartography appears systematically, once the hypotheses are set down. Thus we can also propose a cartographic reasoning completely tied to scientific experimental approach and associated to the problem underlying the purpose of the research, the purpose of the map production (Fig. III.1-4).

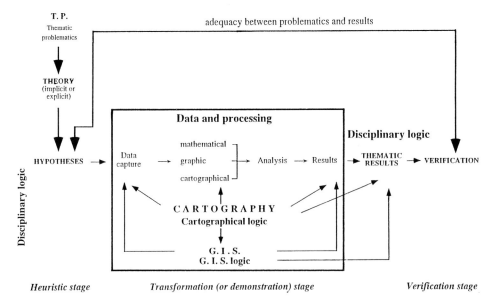

Figure III.1-4: Cartographic reasoning and scientific experimental approach

In this context, cartographic reasoning can be characterized by 5 fundamental stages, three of which command the whole of the map elaboration process (Fig. III.1-5). Let us consider these briefly before discussing them in more detail later on.

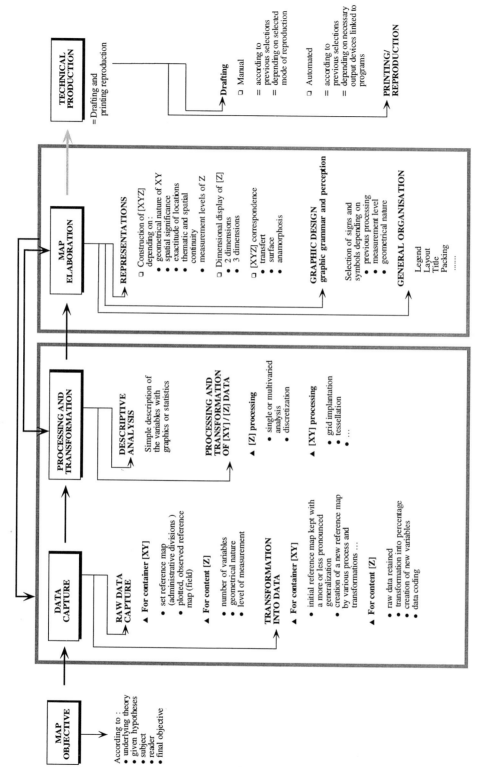

Figure III.1-5: Detailed steps of the cartographic approach

Step 1 : **Purpose of the map.**
This is tied to the theory and hypotheses to be confirmed or rejected, which are in turn, related to the general problem of the study. It corresponds to the work and is an integral part of the disciplinary logic.

Step 2 : **Data capture.**
Where located data, transformed and quantified according to the problem, are obtained. The researcher collects raw data from various sources, which are then transformed into data according to hypotheses.

Step 3 : **Preliminary processing and transformations.**
This step is necessary, as it is impossible to represent several variables simultaneously if a map is to remain legible, and this despite of new visualization technologies. It aims at structuring data and defining underlying fundamental elements.

Step 4 : **Map elaboration.**
This represents the basis of cartographic reasoning. It leads to the production of data layers adapted to the subject to be treated, to derived or resultant maps in the context of scientific experimental approach.

Step 5 : **Map production.**
Essentially technical, this step is different from the others; it will not be broached here. However, it is indispensable in order to present a document materially. It entails manual or automated design (CAD) and the possible reproduction of the document thus obtained.

By going through all these steps that it becomes possible to produce maps which are usable and useful, on their own or in a located data base context (or GIS). Outlining these steps is not enough though, and we are going to describe them in more details. However, the first step - the purpose of the map - cannot be independently presented, being an integral part of the thematics under study. It is obvious that the hypotheses to be verified through the map must be set out clearly and accurately. The goal must be set down without any ambiguity if an "efficient" map is to be the result. This particular step cannot be described generally, because of its thematic specificities. Therefore, we will start with the general rules of data capture, the second step in our cartographic reasoning.

3. Data capture

A map is the intersection of two spaces. Therefore, capturing associated data requires collecting two series of data corresponding to each space: locations [XY] and attributes [Z]. Independent of the information sources, the goal of this part is to bring out the general characteristic of located data capture; in order to do this, we must first clarify a few terms.

3.1 Raw data and data: a fundamental differentiation

A distinction was made between raw data and data several years ago, which was emphasized in [Beguin, Thisse, 1979] and taken up again by [Rimbert, 1979].
Raw data are a quantity or quality either surveyed, measured, or directly collected. They had not been transformed and, exceptional cases apart, cannot be processed without preliminary modifications related to the problem to be solved. Precipitation readings on a pluviometer are raw data, so is the surface covered by woods in a given space. Without a link to a precise subject, without more or less complex tranformations, these raw data do not allow a significant representation in relation to the problem to be solved.

Data are raw data quantified or qualified in agreement with the given problem. According to the announced hypotheses, the researcher, cartographer or thematician will transform the raw data to allow a verification of the hypotheses. For other authors, even more strictly, data is raw data transformed in order to integrate it into a model. In every case, data can not be disassociated from the person with the "thematic" knowledge of the problem to be solved. It can be expressed as follows :

Raw data + Researcher = Data

A generally extendable example from a geographical table will make it easier to understand the difference between those two terms and to appreciate the importance of the transformation of raw data into data.

Let us take a geographical information table, with places in rows (spatial units, [*XY*] locations) and attributes [*Z*] in columns. These attributes can be either information or data. If [*Z*] is the active population by professional sector for urban areas in France, for instance, it will be expressed in absolute values. Depending on the manner selected for transforming the initial table, data significance will change and the cartographic representation will be different.

By reading the table in rows (Fig. III.1-6), one finds the geographic object. A spatial unit is described by the number of people in each profession for example. By dividing each cell of the row by the sum of active population of the row $[Z_i / \sum_j Z_i]$, we obtain percentages for each considered unit, classical and well known percentages. One speaks then of internal ratio.

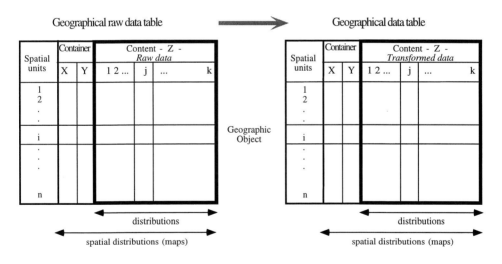

Figure III.1-6: Geographical table: raw data and data

By working with columns, without taking locations into account, one has a statistical distribution (Fig. III.1-6): every value of a raw data or data is known for every studied unit. Associated to the container, each of the "variable" allows the construction of a map and the representation of the phenomenon's spatial distribution.

If the value of a given column is divided by the sum of this column, or $[Z_j / \sum_i Z_j]$, we find the weight of one spatial unit in proportion to the whole of the considered spatial units. We are thus able to construct an external ratio whose significance is quite different from that of the internal ratio. If the problem is expressed in terms of spatial unit comparison, the internal ratios are the most interesting, but if the zone and its variations are relevant, then external ratios are best suited to grasp the respective weight of each spatial unit in the system. These are 2 examples of possible transformation of the same raw data in order to solve two different complementary problems.

Raw data collection is only a first step in data capture. With identical information, depending on the problem, data transformation varies; several data tables may stem from the same information table, but each will be associated with a particular hypothesis. Transformation or adaptation problems, however, are not identical for container and content data.

3.2 Container data capture (locations [XY])

The location of raw data collection and their transformation depend on the purpose of the representation and the phenomenon to be represented. Both elements influence the type of generalization to be adopted.

3.2.1 Container and purpose of the representation

Depending on the purposes of the representation two referential constraints are to be selected: the scale and the projection system, the two being closely linked.

The *scale* of a map is the ratio of the distance between 2 points on the map and the distance between the same 2 points on the Earth´s surface. The scale is large when the denominator is small and vice-versa. The smaller the zone in range (large scale), the more details can be seen, the finer the resolution. Varying the scale, objects can present several different aspects and change dimensions. A forest of 100 ha at a scale of 1:500.000 is a dot on the map (zero dimension), at 1:10.000 it is an area (dimension 2). Finally we must be conscious of the fact that some features of a landscape do not appear at every scale. A minor road will be recognizable on a 1:25.000 map, but not on a 1:1 Mio. We can choose any scale, provided the phenomenon under study appears clearly.

So, we are led to proceed to operations of aggregation (or disaggregation), depending on the scale selected with the general objective in mind. If a "department" (county) is studied, the "canton" (district) may be considered as the basic spatial unit; if the initial reference map is composed of "communes" (villages), which are the basic administrative unit in France, they must be aggregated into "cantons" (districts). The problem is simple as the units fit perfectly (Fig. III.1-7). However, this is not always the case; some of these new spatial units may impinge on units considered as basic for a given layer. At the reference map level, the construction is possible by intersection of the two layers, but there are difficulties with the valuation of these units created by these intersections, which we will not tackle here[7].

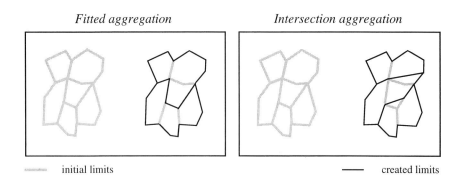

Fitted aggregation   *Intersection aggregation*

⸺⸺ initial limits   ⸺⸺ created limits

Figure III.1-7: Spatial aggregation

The *projection system* is closely related to the notion of scale. Here we will only mention it briefly, because it has been described in detail in Chapter II.1 ('Reference Systems'). As the selection is done at the level of container collecting, we cannot totally avoid the subject here: measurements can turn out to be completely absurd if the projection system is badly chosen. Their role is really important on a small scale; if the selected scale is large, the projection system generally plays a smaller part in simple thematic maps. But it is less obvious in the case of GIS where geometric corrections[8] are necessary when maps coming from different scales or satellite images have to be fitted in. Then it becomes essential to know the projection system characteristics in order to avoid serious mistakes. The constraints of projection systems are a lot stronger with GIS than with isolated maps, quite independent on the scale.

Both choices - scale and projection system - are made in connection with the map objective, but other decisions relevant to the spatial support selection are connected to the phenomenon that is to be represented.

3.2.2 Container and phenomenon to be represented

The phenomenon to be represented is located, but its location can vary in nature or dimensions on the geometrical level. The geometrical nature of a phenomenon is also called graphic position or graphic primitives. In this context, we will speak of "geometrical nature" with associated dimensions. A phenomenon, a cartographic object can be a point with dimension zero, a line with dimension one or an area with dimension two. Let us explain: A city on the scale of a country corresponds to a point on the map and is of dimension zero. Again, the same city but studied on its own becomes an area and is of dimension two. Cartographic point objects are of dimension zero, and can be simple or labelled points or topological nodes; whatever their characteristics may be, they keep the same spatial significance, the same nature (Fig. III.1-8a).

Figure III.1-8a: Cartographic point objects of dimension zero (adpt. from [Clarke, 1990])

Cartographic line objects (Fig. III.1-8b), such as roads or rivers, are of dimension one, but their thematic significance can differ from their geometrical nature. For instance, a set of roads corresponds to a set of lines, but the roads form a network irrigating a territory, or surface: a network can thus have a linear nature and an areal significance. These notions stand out very clearly in fractal geometry[9] to which we refer our readers.

Areal cartographic objects are of dimension two (Fig. III.1-8c) and can be land plots, or city blocks, or a plant cover. The pixel (picture element) is considered as the smallest indivisible area element.

The notion of continuity must be associated to the one of geometrical nature as it will play a role in choosing one type of map or another. A point phenomenon is by definition discrete, discontinuous: for instance, a set of points representing urban areas can be defined as *point pattern*. A line phenomenon can be continuous or discontinuous. In fact, it would be more accurate to speak of "sequenced continuous". A road is continuous between 2 crossroads but it can be interrupted and change its type (from national to district road for example); it is continuous in sections, in sequences.

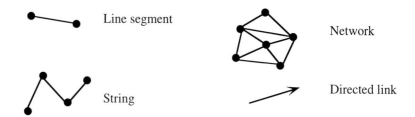

Figure III.1-8b: Cartographic line objects of dimension one (adpt. from [Clarke, 1990])

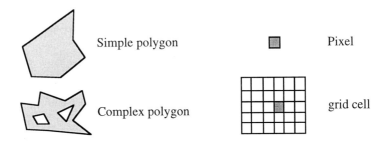

Figure III.1-8c: Cartographic area objects of dimension two (adpt. from [Clarke, 1990])

Finally, some areal phenomena are continuous - e.g. altitude - while others are discontinuous, such as administrative units. These differences are important for location data input in a database and for making ulterior choices (processing, graphic design etc.).

The collection of raw location data depends on the map objective and on the phenomenon to be represented. Their transformation into data mainly consists of a generalization which itself depends on those 2 previous elements. Generalization can be specific for an isolated map but not for a set. In the context of setting-up information layers for a system of located data, the problem is different and the generalization rules must be more rigorous and allow comparisons.

3.2.3 Container and generalization

A generalization is almost always essential if legibility and communication are to be maintained. A map is not a "glory-hole" where anything can be put in and then be expected to stay legible. According to [Tobler, 1974], "generalization is the application of a transformation which modifies raw data", thus initial information is changed. Generalization is not a neutral operation; a complete definition such as [Robinson, 1978] is essential to grasp the fact: "generalization is the modification of specific raw data to increase the communication process by neutralizing undesirable effects of reduction".

Selections of transformations depend on the map objective, the chosen scale, physical graphic limits and on data quality. Three main operations are necessary:
- simplification: the important characteristics of spatial data are determined, the essential ones retained and possibly emphasized and the not useful details reduced or eliminated;
- classification/clustering: data are rank-ordered, graded and grouped in clusters, in order to put forward the main characteristics of the phenomenon to be represented;
- symbolization: selected significant characteristics are graphically coded and the previous sections are made visible.

According to the degree of generalization, the plotting will be more or less close to the original; associated measurements will be of varying accuracy as shown in Fig. III.1-9, where generalization is done manually, as an example. It should be borne in mind that manual generalization can lead to results that may vary from one individual to another, as [Jenks, 1963] emphasized.

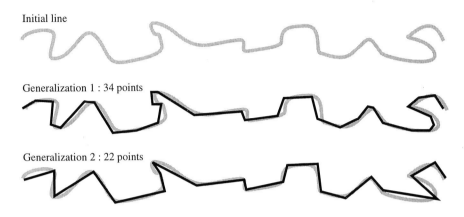

Initial line

Generalization 1 : 34 points

Generalization 2 : 22 points

Figure III.1-9: Manual generalization

To avoid individual bias, generalization methods have grown in number with the arrival of computers; starting from the first manual attempts and [Jenks, 1964] very important work, generalization algorithms have improved and decisions are not visual anymore, as the latest works on fractals can attest[10]. As this aspect is also discussed in other chapters, it will not be described here in detail, but we should not forget that generalization can greatly modify a map and thus its combinations with other maps. In no case it can be considered as a neutral operation used only to facilitate legibility. Inserting data into a GIS has put generalization conception right back on the agenda, e.g. which generalization best applies to content data.

3.3 Content data capture (thematic attributes [Z])

Three fundamental elements must be taken into account to study raw content data and their transformation into data: measurement levels, type of formalization and geometrical nature, the latter similar to container data geometrical nature.

3.3.1 Content and levels of measurement

The levels of measurement (or "scales" of measurement) express accurately the mathematical properties of the variables to be represented. There are 4 of them, the first two often called "qualitative" levels and the other two called "quantitative or superior" levels.

NOMINAL level of measurement: this "basic" level identifies objects with a number by assigning each object only one class. *Building a nominal scale is reducing n individual observations to k classes, by determining an exhaustive and exclusive rank-ordering procedure.* Mathematically, the nominal scale is characterized by an *equivalency relation*, determining the partitioning of an observation set and is reflexive, symmetrical and transitive. *At a nominal level, it is the differences between classes which are interesting and not the differences between individuals.* Land use can be expressed through a nominal scale by alloting a different code to each category: for instance 1 for woods, 2 for prairies, 3 for fields, etc.

A particular case of nominal scale is the *binary nominal scale*, where the observation set is divided into 2 disjointed parts. The solution for each element of the set is yes or no, presence or absence, 1 or 0. This measurement level is of great importance for several statistical and computer processings. Every nominal scale can be reduced to a binary nominal scale.

ORDINAL level of measurement: identifies objects and arranges them in relation to one another; it is reflexive if the order is complete, antisymmetric or if the order is partial and transitive. Ordinal scales are adapted to phenomena with hierarchical structures but, as previously, classes of individuals and not individuals are used. Thus roads can be characterized by an ordinal scale: let 1 be for local roads, 2 for district and 3 for national roads; this time, codes are rank-ordered and cannot be reversed without changing the scale significance.

INTERVAL measurement level identifies the interval or distance between 2 points, measured as an euclidian distance[11]. Only operations such as additions, substractions, can be done, but the origins of the value are not fixed, the scale zero is arbitrary. Calendar dates and temperatures are variables read at an interval level. The Celsius zero is not the same as the Fahrenheit zero and the christian era zero is not identical to the Hegira zero.

RATIO level of measurement has the same properties as the interval level but here the zero is fixed. "Each individual is given a decimal or a real number Calculating a ratio between observations is possible, if the name of ratio scale is given to this measurement" ([Marchand, 1972]). It is the most common level, in which every mathematical process is possible.

Mathematical processes vary from one measurement level to another and must be correctly identified in order to avoid serious mistakes in the preliminary processing. Fig. III.1-10 sums up the measurement level properties and processing possibilities.

Knowing if the thematic variables are continuous or not is also useful as this property is equally involved in the processing and the representation. Finally, measurement levels and continuity also play a role in the choice of symbols in graphic design. *These thematic variable characteristics are not gratuitous. Processing, representation and symbolism depend on them.* Variable formalization also has to be taken into account.

Types of Variables	Measurement Level	Relations	Mathematical Characteristics	Additional Information	Mathematical Properties	Descriptive Central Tendency	Parameters Dispersion
Cardinal	nominal nominal binary	equiva- lence	partition in n classes partition in 2 classes	Classes and not individuals are of interest	reflexivity symmetry transitiveness	mode	entropy
Ordinal	ordinal	equiva- lence order [≥ or ≤]	partition in n classes ordered	Countings are done	reflexivity (total order) asymmetry (partial order) transitiveness (partial order)	median	percentile
Real	interval	equivalence order ratio of intervals [≥,+,−,*,/]	arbitrary zero	restriction on direct ratio calculations		mean	variance
	ratio	equiva lence order ratio of intervals ratio of scale values [≥,+,−,*,/]	fixed zero	complete mathema- tical operations			standard deviation

Figure III.1-10: Levels of measurements and properties

3.3.2 Content and formalization

This distinction is not always highlighted, although it intervenes explicitly or implicitly at several levels. The form in which variables are expressed must be determined: are they absolute values, percentages, frequency, coded, simple or composite values, individual or aggregate ?

Absolute values, for quantitative variables, allow every possible transformations but cannot always be directly represented, particularly on surfaces, introducing in some processings a size factor altering the results (principal components in factorial analysis for instance). Relative values are easier to represent and process, provided one knows perfectly how those values have been calculated.

Simple values (percentages, absolute values etc.) can be directly represented but composite values (factor scores for example) often lead to difficulties with symbols and legends. Factor scores of factor analysis or standardized residuals of regressions cannot obey to the same graphic rules as percentages of the internal rate type. As for the coded values of a hierarchical classification, result interpretation is the only thing which determines the graphic representation. The formalization of variable values is not a neutral operation either. Lastly, we must still put these characteristics in relation to spatial support to which thematic phenomena are linked.

3.3.3 Content and geometrical significance

Thematic data can be linked to point, line or area type locations, on which their cartographic representations depend. Those different types of geographical positions and their importance have already been explained in Section 3.2.2. To complete this part on raw data collection and their transformation into data, it becomes indispensable to place all these components in relation to one another (Fig. III.1-11).

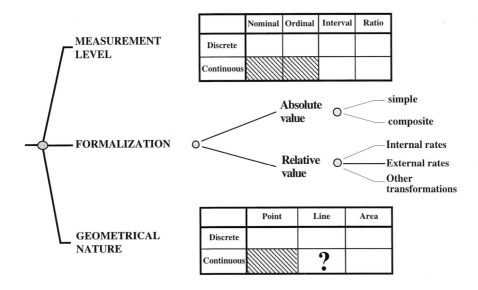

Figure III.1-11: Levels of measurement, formalization and geometrical nature

We can conclude that if raw data collection is not neutral, then transformation into data is even less so, as it is intimately linked to research objectives, hypotheses and the data itself. We also draw attention to the fact that the selection of a type of data has consequences on the cartographic development, the preliminary processing and the representation, as we will see now.

4. Preliminary analyses, processing and transformations

Seldom collected raw data transformed into data according to established hypotheses can be directly transcribed as a map, whatever the chosen process. It is essential to know the statistical and graphical characteristics of the variables perfectly, to be able to represent them cartographicly, i.e. to create data layers usable in a located data base. Moreover, data juxtaposition and superposition period is past even if still useful in certain cases. The combination of data by graphic process or through mathematical tools leads to a more understandable and more accurate representation, as much for communication as for the creation of data set layers. Without going into details for all the possible analyses or raw data processing, we will suggest a few directions in which to use at the best located raw data prepared at the collecting stage.

4.1 Statistical and graphic analysis of data

Whatever the measurement level at which a variable is surveyed, its statistical character must be known to avoid interpretation errors and faulty processing. Each variable must be studied separately. Propositions are more common and more numerous for variables surveyed at a "superior" measurement level, but possibilities exist in each case.

4.1.1 Qualitative data

The first graphic to set up is a distribution diagram, where variable modalities (ordered only if measurement level is ordinal) are on the abscissa, and in column a symbol indicating the presence of units with this modality. It is not really a graphic in the mathematical sense of the word but a graph preliminary to analysis which allows a first approach to distribution (Fig. III.1-12a).

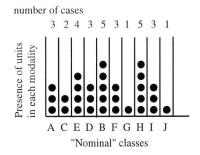 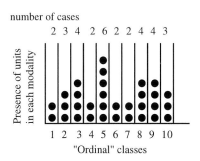

Figure III.1-12a: Qualitative variables: distribution diagrams

In the case of nominal variables, the letters A,B,C etc. merely represent codes; *because of this the diagram is not unique and could present quite another aspect according to the order in which the letters corresponding to modalities are placed.* This is different in the case of ordinal variables where order has a significance. Each person constructing this graphic will arrive at the same display. But in both cases, *as the classes are set and correspond to discrete phenomena, it is impossible to build a histogram or a variation curve.* The only simple graphs characterizing distribution are the classic column charts where the height of the plotting is proportional to the occurence frequency of the phenomena in the class. Each plotting must be seperated from the other by even intervals (Fig. III.1-12b).

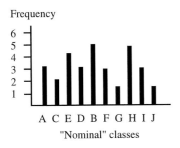 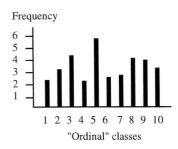

Figure III.1-12b: Qualitative variables: bar chart

Usable statistical parameters are indicated in Fig. III.1-10. For nominal variables, only mode can measure central tendency and only entropy can measure dispersion, or "disorder". For ordinal values, the median expresses central tendency and quantiles express dispersion. These parameters are often used in cartography representations but the possibilities for "quantitative" variables are much more numerous.

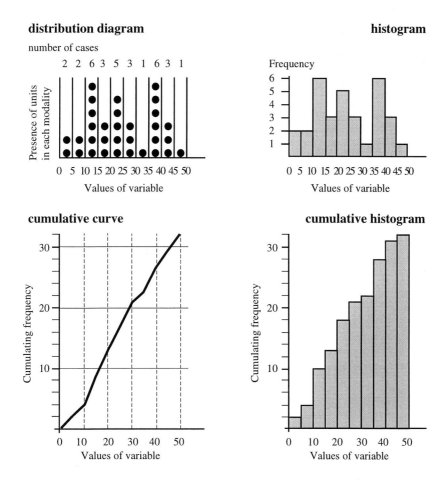

Figure III.1-13: Quantitative variables: graph representations

4.1.2 Quantitative data

For these variables, values can be discrete or continuous, but we can work directly with individuals: class analysis is a choice and not a necessity. The previous graphs are still acceptable; we even strongly recommend beginning with the construction of a distribution graphic where variable values (or class values selected and not set) under study are in abcissa and value frequency occurence in column. But from now on, the order of those values is always significant. It becomes thus possible to build a histogram or a cumulating histogram which will express statistical distribution of the variable (Fig. III.1-13); the rectangular surface of the bars is proportional to the occurence frequency.

Classic statistical parameters are mean (central tendency), standard deviation (dispersion) as well as kurtosis and skewness. One must also check if statistical distribution follows the normal distribution or not. Some of the ulterior processing are not applicable if the distribution is very different from a normal distribution. The use of functional graph papers can turn out to be interesting in this case.

A fairly good knowledge of the variables and their distribution allows pertinent selection of preliminary processing and of cartographic representation. It does help to find the methods that highlight the data structure and spatial organization methods. These will briefly be presented.

4.2 Processing and transformations

Data Z can rarely be represented directly on support $[XY]$. Most of the time preliminary processing of the thematic variable or transformation of the reference map are necessary.

4.2.1 Preliminary processing of content variable [Z].

For a long time, it was considered that only graphic design could propose and improve on cartographic representation and big efforts were made in this field. But the superimposition of symbols, the multiple associations, even if they are logical and well-thought-out, can never turn out to be easy reading. They mostly bring out - and badly at that - what is already known. The contribution of preliminary graphic or mathematical data processing is to bring out the underlying data structure before going on to cartographic processing. From here on, it is a two step process: first process data, at best according to set hypotheses, the try to cartography, to create data set layers expressing the open structure. The methods are numerous, and we are not going to present an exhaustive list here, but we will rather propose a few criteria for choosing amongst potential processing. Whatever the selected method, it all boils down to representing one unique variable Z, simple or composite, qualitative or quantitative, with discrete or continuous values. As it is almost impossible to obtain a legible map if every value of the phenomena is put down on the map, *we are led to "discretization," a mandatory step preliminary to any representation*. This is why after proposing a few decisional elements and listing several of the more usual methods, we will detail what is commonly called "class interval" or "discretization".

4.2.1.1 Selection criteria

Because the preliminary processings required are numerous, we will merely indicate the selection elements (Fig. III.1-14) and propose a general classification table of the methods (Fig. III.1-15).

Selecting a process depends essentially on three elements: variable characteristics, the nature of the process itself and the approach to the problem. *The principle characteristics of variable* are their number and measurement level. With only one single variable to be represented, the problems to be solved are relatively simple. Processing is tied to discretization, i.e. to cluster the thematic values of quantitative variables by class.

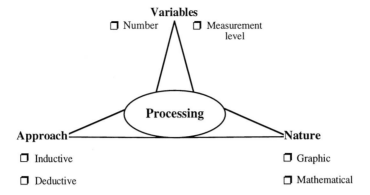

Figure III.1-14: Processing selection (elements of variation)

		Qualitative variables	Quantitative variables
Graph processing		Taxonomy dendrogram Scalogram Bertin matrix	Ternary plot Quadrant method Profile method Taxonomy dendrogram Scalogram Bertin matrix
Mathematical processing	**Inductive approach**	Multiple correspondence analysis Mutidimensional scaling	Principal components analysis Correspondence analysis Multiple correspondence analysis Hierarchical classification k-means clustering Mutidimensional scaling ....
	Deductive approach		Simple, multiple regressions Stepwise regressions Path analysis Trend surface analysis Spectral analysis Various models (diffusion, …) ....

Figure III.1-15: Pre-processing of Z: a few methods

But when the number of variables to be simultaneously represented increases, the possibilities in methods vary with up to 5, or even 10 variables, graphic processing such as the triangular graph or "the profile method" is possible. However, should the number increase again, mathematical processing will become indispensable and the measurement level of the variable will play a fundamental role. Qualitative and quantitative variables cannot be processed identically, one must apply different statistical fields.

The second element in selecting is the *nature of the processing*, graphic or mathematical. It is quite possible to structure data with graphs, as in the quadrant method or with the taxonomic dendrogram. These methods achieve good results if the variables are not too numerous and they are easier to explain to an uninformed public than the more sophisticated mathematical methods. However, if there are too many variables, they become difficult to use.

Lastly, *the approach to the problem*, is tied to the way in which the problem has been defined. An inductive approach does not allow generalization of the results without numerous repetitive experiments, while methods following a deductive approach facilitate comparisons. But the first approach is perfectly justified if one has but faint knowledge of the phenomenon and no preliminary theory. The second is more adapted to verifying an already established hypothesis.

These selection components are interdependent, and knowing them helps thought without imposing a decision in any way. They do, however, help to understand the general methods table proposed here (Fig. III.1-15); it is not exhaustive, merely indicative, separating the method families by the nature of the processing and the measurement level of the variables. It can be completed as new methods appear.

Whatever the method retained, a small number of variables or at least one has to be mapped. Therefore, we will develop this special processing, indeed necessary in cartography, of a unique variable.

4.2.1.2 Compulsory class intervals

When spatial units are numerous, the possible values of variables are also numerous. In order to communicate correctly, information must be simplified. Generalized: class intervals become imperative but the choices linked to the processing are far from obvious. Several questions arise: which elements play a role in deciding the type of class interval? Are there any basic rules? How many classes? And principally, with what limits? We will attempt to answer these questions while indicating the general rules and principal methods, although we refer the reader to specialized works for more details[12].

Fundamental elements and basic rules: map objectives and variable characteristics[13] are fundamental elements in choosing class intervals. Is the map made for a general public or a mature reader, will it be used as an illustration or become a basic map in a GIS, or be compared to another? For each objective, each type of reader only one or two methods will be acceptable. With the variables, distribution is important; you cannot apply the same statistical class interval to a variable following the normal distribution and to a bi- or tri-modal variable. Then again, one must know if all the possible values of a variable are significant, if the variable is simple or composite, if percentages or factor are scored. For each type of formalization, some methods apply, while others are excluded.

Basic rules are essentially logical. First of all, classes must cover the whole of the variable values, which means that the total range of classes corresponds to the variable range: there must be no missing interval. An individual, a spatial unit can belong to one class and one class only, as it is a partitioning; no class, exceptions apart, may be empty which means that each class has at least one individual. Finally, one must avoid false precision in class limit values which would include non-significant break.

Taking into account our remarks on objectives, variable characteristics and basic rules, the question arises on which number of classes to decide?

Number of classes: three criteria or constraints are necessary to select the number of classes. *The first is tied to logic. The number of classes determines the quantity of details to be read on a map and therefore the degree of thematic generalization.* As the number increases, the generalization becomes more correct, and one finds more initial information. Inversely, too few classes conceal variations and the map loses all interest. To find an average between those 2 extremes, 2 statistical equations can help, equations that "give a reasonable ratio between the precision of a map and the quantity of available data" ([Haggett, 1972]). The first is Brooks and Carruther's (1), the second Huntberger's (2):

(1) $k \leq 5 \log_{10} n$ (2) $k = 1 + 3.3 \log_{10} n$

with k: number of classes,   and n: number of observations (spatial units)

Huntberger's equation leads to fewer classes than Carruther´s. The number obtained compared to the two other criteria, however, is always a little higher.

The second criterium is of a technical order. Despite the technological progress emphasized by W. Tobler as early as 1973 and his propositions, it is difficult in ordinary practice to have as many grey tints as there are thematic values. If colour is not used, it is clear that 6 or 7 different shadings of grey for areal symbols is the maximum. With point symbols, the problem is different but we are then faced with the *third criterium linked to visual perception*, where communication rules apply. What is the use of a beautiful map with numerous grey shadings or as many sizes in circles as there are distinct values for a variable if the eye cannot distinguish those differences? According to the works of cartographers and psychologists, more than 6 or 7 shades are already difficult to differentiate. If a greater number of divisions are created, then the reader carries out regroupings, which may not necessarily be those intended by the author.

Now we have to define the limits and range of classes. The methods are numerous and can be divided into seven main families, arbitrary, exogenous, mathematic, statistic, probability, graphic and experimental. In most cases, either the variable itself, or the transformed variable (if the distribution form implies a normalization of the variable), can be worked over. To avoid encumbering the text, we will sum up those main methods in a table (Fig. III.1-16) where the name of each method is indicated as well as the computation techniques for the class range and limits, the advantages and drawbacks as well as the uses. A few methods will here be quoted as these require developments difficult to fit in this context. One point has not been tackled in this table, the mathematical transformation of variables. When variables do not follow a normal distribution, which then excludes several class interval methods, it is possible to work on them with transformations of the square root type, or root or log etc. which "normalizes" them. Most methods apply to these "normalized" variables, but generally the equal range class or normalized class methods are preferred. The only problem is the transformation selection, which is solved depending upon the distribution. This requires a knowledge of mathematics.

The abundance of methods becomes clear. The question is to determine the optimal classes of intervals in a given case. The first element of the answer is to think about the purposes of the representation and to know how the distribution looks: some techniques become then imperative while others are discarded. Various graphics and indexes also help to take a decision: error measurement (Jenks and Coulson index), class homogeneity index (tied to the variance), information measurement resting on entropy, spatial relation measurement based on spatial autocorrelation. For all these indexes, we refer you to their authors or to our own work already mentioned where a comparative analysis was done. We must stay conscious of the fact that to select intervals amounts to generalizing thematic data, which means losing information. What can be considered an acceptable loss for a general public communication map and for a GIS data set layer? There is no absolute answer. Permanent reflection is as essential at this stage as it was in the previous ones; each decision can considerably modify the map and determine whether it falls on the preliminary processing of Z or even the XY transformation, which is far from secondary even if less well known.

4.2.2 Preliminary transformation of container [XY]

Once the base map is generalized and selected according to the phenomenon to be represented, some transformation can turn out to be judicious in order to show the phenomenon under different aspects. Studying a point pattern, a point distribution on a surface, whether climate stations or wells around the countryside or shops in a city for example, amounts to having a map to start with, a data layer where places are found with the outline of the zone around them. But analysing methods can imply transformations of this simple base map.

Although these transformations are described elsewhere and are an integral part of GIS set-up, we will not dwell at length on the subject in the context of cartographic principles.

FAMILY	METHODS	CALCULATIONS		ADVANTAGES	DRAWBACKS
		Preliminary Calculations	Limits		
ARBITRARY	a priori	variables as depends on knowledge of phenomenon		sometimes responds satisfactorily to observations	no possible comparison
EXOGENOUS	External referential	variables as depends on external referential		facilitates comparisons general studies	can conceal regional or local variations
MATHEMATIC	Equal ranges	range calculation $e = \dfrac{M-m}{k}$ constant interval	C1: [m,(m+e)[C2: [(m+e), (m+2e)[········· Ck:[(m+(k-1)e), (m+ke)]	simple method justified if distribution is close to normal	leads to errors if distribution is not symmetrical comparisons difficult
	Arithmetical progression	X common difference M-m=X+2X+.....+kX $X = \dfrac{M-m}{(1+2+.....+k)}$ class range C1: X C2: 2X ... Ck: kX	C1: [m, (m+X)[C2: [(m+X), (m+3X)[········· Ck: [(m+{1+2+...+(k-1)}X) +((m+{1+2+...+(k-1)}X)+kX)]	spreading out of small values classes of increased amplitudes	do not use with left asymmetrical distribution
	Geometrical progression	X common difference Xk=(M/m) $\log 10 X = \dfrac{\log 10M - \log 10m}{k}$	C1:[m,mX[C2:[mX, mX2[········· Ck:[mXk-1, mXk]	pronounced spreading out of small values	can not be used with zero minimal values
STATISTIC AND PROBABILITY	Percentile based class	equal absolute frequency (n) in each class $n = \dfrac{\text{total absolute frequency N}}{k}$ in case of regression residuals, calculation of standard error replacing standard deviation	limits determined by counting the number of n individuals in an ordered distribution	interesting in the case of very high number of observations eliminates the weight of extreme values data are reduced to an ordinal scale	results unsatisfactory if variable values have lots of equal chosing limits is difficult if variable presents discontinuities
	Standard deviation and mean	mean and standard deviation calculation	1° solution C1: less than $(\mu-1{,}5\sigma)$ C2: [$(\mu-1{,}5\sigma)$, $(\mu-0{,}5\sigma)$[C3: [$(\mu-0{,}5\sigma)$, $(\mu+0{,}5\sigma)$[C4: [$(\mu+0{,}5\sigma)$, $(\mu+1{,}5\sigma)$[C5: more than $(\mu+1{,}5\sigma)$ 2° solution C1: less than $(\mu-2\sigma)$	allows comparisons, as standardizing values amount to giving them a common origin and an identical measurement unit the first solution is more satisfactory	normal or transformed into normal distribution necessary difficult for some public

		Calculation / Construction	Example	Properties	Comments
			C2: $[(\mu-2\sigma), (\mu-1\sigma)[$ C3: $[(\mu-1\sigma), \mu[$ C4: $(\mu, (\mu+1\sigma)[$ C5: $[(\mu+1\sigma), (\mu+2\sigma)[$ C6: more than $(\mu+2\sigma)$		
	Nested means	$\mu1$ mean calculation and partitioning into two classes in relation to $\mu1$ calculation of second class means $\mu2a$ and $\mu2b$ if necessary, calculation of third class means	example of four classes C1: $[m, \mu2a[$ C2: $[\mu2a, \mu1[$ C3: $\mu1, \mu2b[$ C4: $[\mu2b, M]$	two interesting properties: frequencies are evenly distributed in classes and class ranges are very close no class empty	set number of classes division is „final" the mean value is used significantly instead of as a trend
	Variance (Jenks´method)	Fischer´s algorithm based on the notion of variance Minimizes intra-class variance and maximizes inter-class variance		minimizes errors due to generalization maximizes class homogeneity	limits are not necessarily adjacent division is „final"
	Equiprobability	method presented by Armstrong (1969) and developed by Grimmeau (1977)		if normal distribution, combines the advantages of the quantile and of the mean/standard deviation method. Otherwise allows the peculiarities of sub-distributions to show	calculations rather complex
GRAPHIC	Natural breaks	construction of frequency diagram (or histogram or cumulate curve)	class limits correspond to breaks or discontinuities read on graphs	adapted if breaks really appear	results unsatisfactory if no visible break no comparisons possible
	Spatial breaks	construction of clinographic curve	class limits correspond to breaks read on graph	integrates the quantity of space covered by the clustering principle	results unsatisfactory if no visible break no comparisons possible
	Spatial quantiles	construction of clinographic curve determination of the quantity of equal intervals on abscissa (depending on number of classes)	class limits correspond to spatial breaks indicated on graph abscissa	integrates the quantity of space covered by the clustering principle „visual" balance of the map	
EXPERIMENTAL		Methods depending on treated subject, selection of interval class is done according to experimental breaks, various laws (rank-size law, k distribution......)			adapted to particular problems

M: maximum, m: minimum, N: number of individuals, n: number of individuals in a class, σ: standard deviation, μ: mean, k: number of classes, C1: class 1

Figure III.1-16: General chart of discretization methods

Thus at the end of the first three steps, the cartographer and the thematician - for they are not to be separated - have data, i.e. transformed raw data, prepared with the hypotheses verification in mind; now the map or maps must be elaborated, where located data and thematic data will be concretely put in relation. Many solutions exist, but also many constraints. This is what we are going to see in the fourth step of this cartographic approach.

5. Map elaboration

Having data is one thing, representing them spatially is another. Many elements have to be taken into account and studied separately in order to be able to make choices, again, in relation with the overall purpose. There are at least 3 specific stages in map compilation with prepared data (Fig. III.1-17). The first concerns representation families, i.e. the type of construction according to [*XYZ*] characteristics, the type of "visualization" or dimensional display (2D, 3D etc.), the type of "correspondence". The second is the choice of the symbols in the representation, namely graphic design which includes graphic grammar and graphic perception. The last deals with the general organization of the map, everything which "surrounds" it; this may seem to be secondary but in reality it is often by this "packing" that the reader comes into contact with the document. We will detail those stages in the following sections.

5.1 Representation families

This title regroups the possible representation of data on a plane; different components come into it and interact with one another. These are, in particular, the *XY* and *Z* characteristics which play a role in the map construction, the type of display or of visualization according to the "dimensions" and the *XY* and *Z* correspondence.

5.1.1 Map constructions (according to [XY/Z] characteristics)

Depending on *XY* and *Z* characteristics, construction types are not identical. One must know the types of maps which are possible with the constraints on both sets of associated data: locations and attributes. This amounts to placing geometrical nature information in relation to measurement levels (Fig. III.1-10) and possible construction types.

It might have been clearer to describe the main types of maps associated with the geometrical nature of the phenomenon under study, or its position in Bertin's sense of the word, but in many cases, it is impossible to make a total abstraction of the measurement levels of the thematic data (Fig. III.1-18). This is why this part's presentation rests on the [*XY*] and [*Z*] characteristics and will systematically take up the same points:

- continuity and spatial significance of [*XY*]
- measurement level and thematic significance of [*Z*]
- accuracy and simplification of container
- [*XY/Z*] observation or construction
- possible construction principle

5.1.1.1 Point type maps

In this case, representations are always discontinuous, and, in principle, can express only discontinuous phenomena. However, exceptions exist, and point maps can have areal significance. All measurement levels coexist with a point geometrical nature, but the thematic significance depends on what is expressed, an association or differenciation, an absence or presence, a hierarchy, a quantity or a density. The "points", in the location point sense, can correspond to the accurate, observable positions of a phenomenon or represent several elements or again express created places.

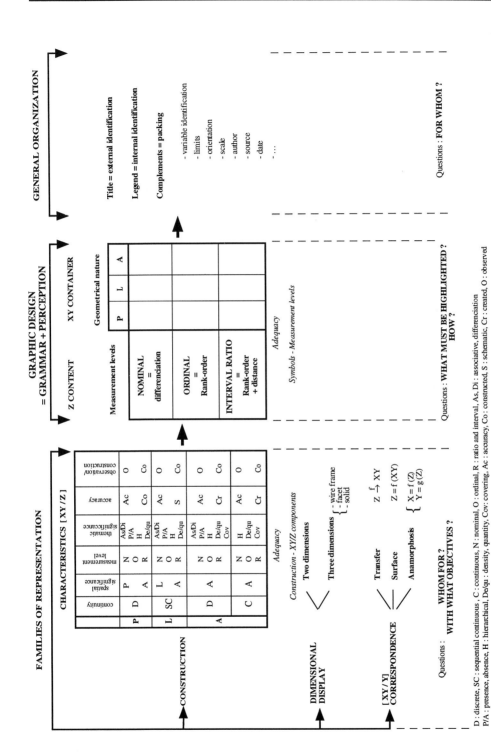

Figure III.1-17: Map elaboration: major steps

D : discrete, SC : sequential continuous , C : continuous, N : nominal, O : ordinal, R : ratio and interval, As, Di : associative, differenciation
P/A : presence, absence, H : hierarchical, De/qu : density, quantity, Cov : covering, Ac : accuracy, Co : constructed, S : schematic, Cr : created, O : observed

Five main types of point map constructions can be listed according to these various criteria:
- with pictorial or geometrical qualitative symbols
- with point patterns or distribution
- with rank-ordered point symbols
- with quantitative valued dot symbols, systematic or irregular in distribution
- with repetitive quantitative symbols
- with proportion or class graduated quantitative symbols

GEOMETRICAL NATURE	MAP TYPES

POINT	⇨ using pictorial or geometrical qualitative symbols
	
	⇨ point pattern map
	
	⇨ using point ordinal symbols
	
	⇨ dot maps (quantitative symbols)
	
	⇨ using quantitative point symbols
	  - repetitively
	  - graduated symbols
	  - proportionally graduated symbols

LINE	⇨ using differenciated linear symbols with accurate or schematic plotting ⇨ using binary nominal linear symbols ⇨ using ordinal linear symbols ⇨ using quantitative linear symbols ⇨ vector field

AREA	⇨ choropleth
	
	⇨ dasymetric
	
	⇨ with isolines
	  - isometric
	  - isopleth
	
	⇨ membership map
	  - no covering
	  - covering
	    • single membership (=partition)
	    • multiple membership (=overlay)

Figure III.1-18: Construction and types of maps

- maps with pictorial or geometrical qualitative symbols express the location of discontinuous phenomena, with no thematic differences except for their belonging to one class or another (Fig. III.1-19a). We are speaking of variables surveyed according to a nominal measurement level; the reader must easily associate or differenciate those symbols. In principle, symbols are laid-out on the accurate location of the phenomenon, but sometimes the size of the symbol can lead to shiftings and inaccuracies. In a communication map, this alteration may be considered as not too serious, but in a basic map for a GIS, it is fundamental; maps with qualitative symbols are critical, especially with pictorial symbols.

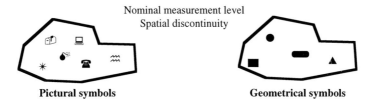

Figure III.1-19a: Point-related maps with pictorial or geometrical qualitative symbols

- *maps with point patterns or distribution* can express the presence or absence of a phenomenon, usually considered as discontinuous. A spatial distribution of "points" is fundamentally discontinuous, but if we consider the zone "attracted" to, dependent on this point, the significance can be areal. The variables can belong to any measurement level but they are always reduced to a nominal binary level. Only one distribution is represented on the same map, except for the aberration of finding two distributions on the same map, which makes it non-legible. GIS leads to combinations which avoid those serious mistakes if the initial maps are accurate, which is an essential characteristic of spatial distribution maps (Fig. III.1-19b).

- *maps with rank-ordered point symbols* correspond to discontinuous phenomena representations where the variable Z is surveyed according to an ordinal level and is thus hierachically ordered. Only graphic design can allow a choice (Fig. III.1-19c). Until today no particular construction that we know of has been invented.

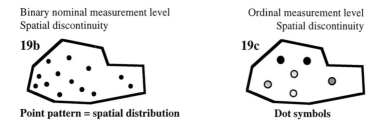

Figure III.1-19b,c: Point-related maps with binary nominal symbols and ordinal symbols

- *maps with quantitative valued point symbols,* systematic or irregular in spatial distribution, often called dot maps, represent a distribution with, generally, small one-sized dots which express quantities on a surface, a density more or less. Their spatial significance leans towards the areal. These maps are made when a population is statistically known in a given space, the farming population of a district for instance. Without going into construction details for this type of map (Fig. III.1-19d), the basic steps involved are as following:
- selection of a dot size to a quantity of population: 1 mm² for 100 inhabitants for instance
- calculation of the number of necessary dots for each spatial unit (district) by dividing the unit population by the quantity of population alloted to the dot; for 1500 inhabitants, 15 dots
- distribution of calculated dots on each unit: two possible solutions. Either distribute systematically (Fig. III.1-19d1) or take into account the unpopulated sectors (Fig. III.1-19d3). Usually, administrative boundaries are later taken away so that only the distribution shows up (Fig. III.1-19d2 and III.1-19d4). For a communication map, both solutions are acceptable; for a GIS data layer, the second is better. However, there are

other possiblities, to which we will return that should avoid this type of map in a GIS, insofar as the dots do not express a concretely observable phenomenon.

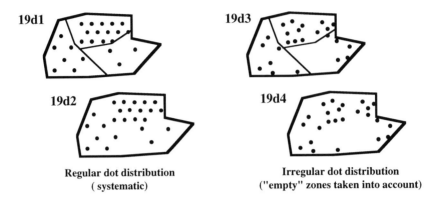

Regular dot distribution
(systematic)

Irregular dot distribution
("empty" zones taken into account)

Figure III.1-19d: Dot maps (equal value symbols)

- ***maps with proportion or class graduated quantitative symbols*** are the best known. They correspond to discontinuous point phenomena, of quantitative measurement level, where locations have to be accurate, even if sometimes the importance of the symbols brings about some shifting. Maps with repeated symbols are to be avoided for just this reason, as when the number of symbols to be drawn on the same spot becomes too high, inaccuracies inevitably occur. Maps with proportional circles, with varied sized squares, with cubes are frequently seen but they are essentially communication maps and not base maps (Fig. III.1-19e). Except from very exceptional cases, they must be conceived as GIS "output" and not as input maps.

The principal difference in map construction is the necessary choice to be made between class intervals and proportionality. This is a general problem but it is more crucial to point maps than to area maps. Author and reader often think that they are perfectly able to grasp the slightest nuance in the size variation of a circle or square, although both agree that only 6 or 7 shades of grey on a surface are really distinctly perceived. Whereas every graphic perception test shows that it is quite unwarranted. Reading quality is far from improved with point symbols as opposed to areal symbols. With proportional circles, one thinks one sees the differences but in fact creates ones own association: Each reader will then have his/her own reading of the map. The symbol proportionality option, if it satisfies the mind, remains difficult to accept on the communication level. Creating classes of symbol sizes, discretization of the data leads to the classic problems previously mentioned: how to choose the classes? But whether one works with class intervals or with proportionality, a difficulty, specific to point symbols, remains: should the symbol size be calculated with a radius, with a side proportionate to its thematic value or with a surface proportionate to this value?

Let us explain the problem in the case of a circle. Let P be the variable value used as reference (for example the minimum value of the variable) and R the reference radius associated to P if P_i is a value of this variable; we get radius r_i associated to p_i with the following equations:

- in the case of surface proportion $r_i = R \sqrt{\dfrac{p_i}{P}}$

- in the case of radius proportion $r_i = R \left[\dfrac{p_i}{P} \right]$

It is quite obvious that the difference will be emphasized in the radius case. Several authors, however, advocate the surface option which differentiates smaller values better than the larger ones. As it mainly concerns communication maps, the options allowing efficient reading should be retained.

The following aspects should be remembered:
- representations almost express spatial discontinuity
- only a few representations can go in a GIS data layer, particularly point patterns
- many of the representations are merely communication representations
- some choices are common to all positions, class intervals in particular
- a specific choice is the proportionality type in the case of quantitative thematic variables

We must note that from a basic point representation, if the hypothesis support transformations, areal maps can be constructed. Point locations then become bench marks. These issues will be discussed in the third section. As a preview, it is advisable to list the representations associated with linear positions; after all, a line may be considered as a succession of points linked up together.

	Proportional symbols	Class symbols
In proportion to radius		
In proportion to surface		
	5 dots = 5 different sizes	5 dots = 3 different sizes (= 3 classes)

Figure III.1-19e: Maps with graduated quantitative point symbols

5.1.1.2 Line type maps

Line representations correspond to phenomena continuous or sequentially continuous. They can also express thematically all types of measurement levels, without any of them being really original. More distinctly than point type maps, they exhibit location plotting, accurate or schematized and a linear or areal significance. These different criteria will direct the description of line type maps which will essentially be regrouped by Z measurement levels.

- nominal line maps are of two types: accurate plotting type and schematized plotting type, based on topology.
    The differenciated accurate-plotted line symbol maps have a locating objective and a purely linear significance: to identify the plan of a road and then distinguish between the various types of roads. Line plotting is usually accurate. The degree of accuracy depends on the map's final objective. The symbols are plottings which vary "differentially" and not quantitatively (Fig. III.1-20a). They can be continuous, discontinuous or double, but their width must be constant. The role of graphic design is essential. Whatever the phenomenon, the layout should be very fine in a data layer, in order to avoid introducing errors through the very thickness of the plotting.

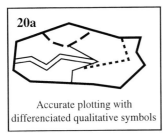
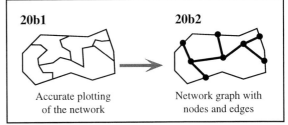

20a	20b1	20b2
Accurate plotting with differenciated qualitative symbols	Accurate plotting of the network	Network graph with nodes and edges

Figure III.1-20a,b: Linear representations with nominal symbols

The second type of nominal line map, the graph type, presents plotting variously schematized. The type of exchanges between railway stations (nominal variable) does not require an accurate line representation. It is the connection between these and its nature which are important, and thus is approximate plotting good enough. Network maps push this generalization to the utmost, they favour the connection between two points or the connection between a set of points. We are speaking of a binary nominal variable. Accurate plans are not adhered to, only the relative positions of points are maintained. In fact, these maps correspond to a topological construction where points, called nodes, are connected by edges. They are rather graphs (in the graph theory sense) than maps (Fig. III.1-20b). The summits or nodes are represented by point symbols, the edges by a line of constant thickness. These graphs are usually derived maps. They are not basic maps of a GIS, but they can be built directly from a data set layer. Their setting up is relative simple. The theory of the statistics linked to graphs can be applied to these and they lead to a fair knowledge of network characteristics. We find an intermediary map category, which belongs neither to the base map nor to the final illustration which is deduced from basic information and used in the final production. Finally if these graphs represent networks, their spatial significance is more areal than linear. It is difficult to conceive a network without a territory, i.e. a surface, that it irrigates. But network as well as lines can be hierachically ordered representations in which the express order must be given.

- ordinal line maps are built as previously, with an accurate plotting (hierarchically ordered roads) or in a topological manner with networks in which valuation is ordinal: But just as in

point representations, it seems that graphic symbols are the only ones able to translate the characteristics. And again, if grammatical solutions exist, the perception of the result is far from satisfying. For reasons explained in Section 5.2, the plotting must have identical width: only its content can express order, which is often badly perceived (Fig. III.1-20c). This problem disappears with quantitative line maps.

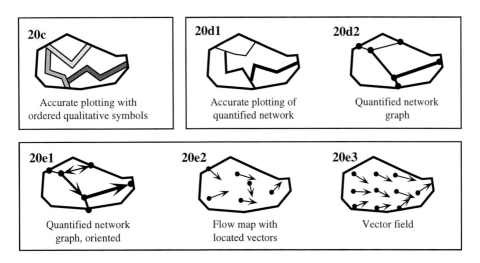

Figure III.1-20c,d,e: Linear representations for quantitative and ordinal variables

- *quantitative line maps* correspond to quantitative information which expresses the "importance" of the connection, independant of the width or capacity of the phenomenon, or flows or exchange ratios. The width of a road is quantitative data, the plotting can be proportionally thick, but in keeping with the scale of the map, exceeding a certain thickness in the plotting will damage the accuracy and legibility of the map. The exchange ratio between a pair of points is often represented by, what is partially incorrectly called, flow maps (Fig. III.1-20e). These are numerous and the possibilities for their realization depends on the nature, the direction of the flows and the number of exchange poles, which, however, are too complex to be detailed here. We must note that the spatial significance of flows can vary and so do their representations. If flows are considered as exchange with linear signification, arrows for the shape and size with varying plotting can be the answer. If one considers that flows correspond to the servicing of a territory, they take an areal significance and vector fields will be a more appropriate alternative (Fig. III.1-20e3). A phenomenon considered as linear as it corresponds to the connection between two places can thus be represented as a field or a surface, because its thematic significance is areal. Should we classify vector fields in areal representations because of that? The question is as yet unsolved in cartography.

5.1.1.3 Area type maps

For the map types already proposed, it was clear that the variable measurement levels were essential. The outline followed here stressed this scale predominance. With areal representations, the fundamental variation is linked to the notion of spatial continuity completed by the construction model used. The outline will thus be slightly different in order to avoid repetitions.

- the first distinction rests on the notion of continuity: are areas formed by distinct separate spatial units, such as administrative units (blocks in a city, districts, regions etc.), or by physical units identifiable in the field such as geological outcrops? Or are they continuous surfaces where markings express levels in graduation, a progression such as contour lines? Using this continuity criterium, a first separation between "discrete" maps such as choropleth and dasymetrical[14] maps, and "continuous" maps such as isoline maps, can be made.

Discrete spatial unit maps are used when spatial units are contiguous and when surface sectors under study are completely covered by the symbols of the selected phenomenon. But there is no continuity when going from one to the other. In *choropleth maps* (Fig. III.1-21a), values can be totally different between 2 neighbouring units without this absence of transition corresponding in reality to any observable difference. These maps are maps in which spatial units are discontinuous (units), not necessarily visible in the field when the limits are "artificial", set down, created by men, for instance administrative or political units. The variables "placed" in spatial units, usually very uneven and not always significant in size, can belong to any type of measurement level, any formalized category, which is why all of them cannot be used as criteria for characterization. This is also true for dasymetrical maps (Fig. III.1-21b), which are equally composed of discontinuous units. These units correspond to zones considered as homogeneous either through field observations (land use, soil zones) or through calculations such as Wright's works of 1936 (densities). Limits are either observed or calculated.

Both types of areal maps with areal significance, very frequent in GIS constitution, are characterized by the observed unit discontinuity. This discontinuity does not have the same meaning for each; they can be used for nominal, ordinal or quantitative variables, which is not the case for isoline maps.

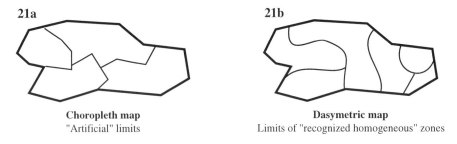

Figure III.1-21a,b: Discontinuous area representations

Continuous areal maps, isoline maps, are representations with variables necessarily quantitative, continuous, where the basic spatial constraint is spatial continuity as their construction supposes an interpolation, a differentiate calculus. Isoline maps have different names: isovale, isometric, isopleth. Isolines are lines on a surface, joining points of the same value to one another for a given phenomenon; they express nothing more than the passing of one value to another, on each side of the line; a gradient can be calculated from any point on the map. These maps are built from point location, already valued, from which the value in all points of the surface is calculated. Several components play a part and introduce variations in the elaboration of these surfaces; we will list the main ones, namely the control points, the interpolation function and the geometrical principle of value distribution (Fig. III.1-21c1).

Each component has its own importance, and can lead to very different maps; therefore, the use of isoline maps, even if very rich, should be cautiously practised. If the number of basic points is low, or if these points are unevenly distributed, one must have an interpolation procedure to take these "deficiencies" of the initial point pattern into account. The choice of the interpolation function is linked to the phenomena under study: is it spatially linear, quadratic, or cubic? This is far from obvious and few publications have attempted to establish an equivalence between thematic phenomenon and the interpolation function, even if some authors have presented syntheses on isoline maps[15]. A classical isopleth map is proposed in Fig. III.1-21c2.

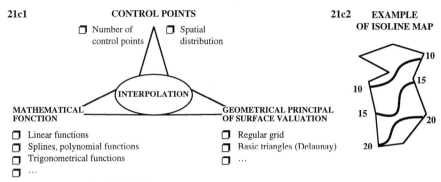

Figure III.1-21c1,2: Interploations: principal components of variation

Isoline maps are constructions expressing thematical phenomena as theoretical characteristics. They are useful as GIS basic maps or derived maps. They are unique, once their elaboration criteria are selected, and their correct selection is a point to study in more details. With this process, we are already touching the second distinctive element of areal maps, the construction model.

- the second distinctive element of areal maps rests on the use of models. A new family of maps appears with this criterium, the membership map[16]. It corresponds to representations resting on models leading to alloting place to one or more groups. They are derived maps, the set-up of which is greatly facilitated by the development of GIS. We can separate non-covering and covering membership maps, and in the latter distinguish between single membership (partition) and multiple membership (overlay) maps.

The *non-covering membership maps* may correspond to envelop maps. Referring to the mathematical set theory; envelopes are lines "approximate in their plotting" which regroup point or areal units with the same peculiarities (Fig. III.1-21d1). The property that characterizes this process is the membership relation. Closed plotting underlines the zone where locations with the same properties related to the same phenomenon are gathered; envelopes cannot intersect but do not have to interlock either like isoline maps. This representation combines a mathematical with a graphical logic. Since envelopes can regroup point or areal units, and can be contiguous or not, they do not necessarily overlay.

Covering membership maps have an areal significance, but a place can belong to a single zone or again to several sets simultaneously. The *first type expresses a partition*: a point belongs to one set and one set only. The tessellation (or paving) methods produce maps in which a partition of space is effectively built from the known points in space without taking the values alloted to the points into account.They lead to discrete representation, where limits have a precise significance that can be interpreted (Fig. III.1-21d2a and d2b).

The second type of map is the *multiple membership map*. If partition is apparently satisfying (cartographicly) as graphic solutions are geographically known, a place (in wider sense) can have multiple membership for one or more criteria. To conceive this motion, one has only to superimpose several maps of the same sector and see how new sets appear through intersections of the various underlying (class) intervals. A township can be attracted by city A for its high-level services and by city B for the more common ones, and so belong to both A and B.

The important aspect is that each point belongs or does not belong to a set, or belongs simultaneously to several sets. This type of map is not easy to construct, even less to express graphically. The fuzzy set theory proposes a mathematical solution which has largely being unexploited in cartography until now.

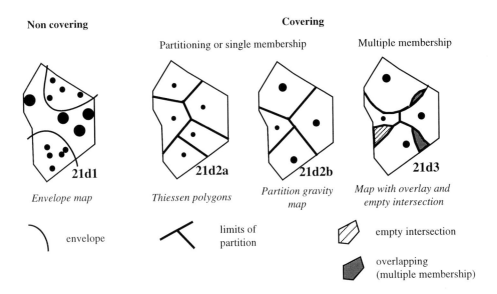

Figure III.1-21d: Membership maps

We will not elaborate this subject now, but it is clear that this type of representation corresponds more closely to what can be observed than the partition maps; GIS development also facilitates their construction and multiplication and thereby, our better understanding of the same.

We can thus see that map construction types are numerous and vary according to the geometrical nature of the phenomena to be represented, their measurement levels, the spatial continuity and the selected models. Some produce maps that can be used as data layers, such as distribution maps (point patterns), line plotted maps or certain area maps, whereas other representations should not be used as anything other than interpretable maps, illustrations or as communication maps. Other factors, however, play a role in map differentiation, the visualization mode or dimensional display in particular.

5.1.2 Visualization or dimensional display

Through the use of the above terminology we are able to establish whether a map is represented in two dimensions (flatly) or in three dimensions (volume, perspective). In the case of a two dimensional map (2D), the variable is projected on plane *XY*; the differences in thematic values are essentially shown through the graphic symbols. In a 3D map, the thematic variable rises over the plane, its height being linked to variable values (Fig. III.1-22). Choropleth maps then appear as contiguous blocks of uneven heights, isopleth maps as a more or less undulating surface. 3D maps belong rather to the communication map class than to the GIS basic map class. However, at the moment, a lot of research is being undertaken to investigate on how to use 3D maps as an analytical tool with specific constructions for cross sections or profiles[17]. In this context we will merely speak of communication 3D maps, often called "false 3D" or "2,5D".

These maps, attractive to the reader, are diversely interesting, but can only be constructed in areal positions with variables of a quantitative measurement level. They can bring out nuances not easily seen in 2D, as every variable value can be taken into account due to the proportionality between thematic values and graphic heights. But the choices are difficult to make, many parameters are involved which lead to very different maps.

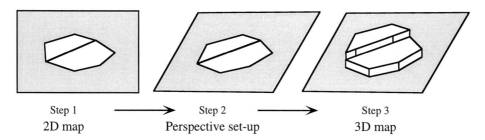

Step 1	Step 2	Step 3
2D map	Perspective set-up	3D map

Figure III.1-22: Choropleth maps 2D and 3D

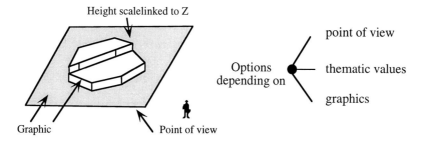

Figure III.1-23: Three-dimensional maps: principal parameters

The principal parameters, for 3D maps, are linked to the relations between the graphic scale of heights, the thematic values and the graphics which can have heavily variegated expressions (Fig. III.1-23). We must stress that the point of view selection is essential. It has three components, the azimuth angle, the elevation or site angle and the viewing distance.

The *azimuth*, angle going from 0° to 360°, corresponds to the observer's position (or point of view) on the plane. Its choice is tied to the locations of the elevation values of variable Z. It should permit a perception of the differences in height linked to value, the weaker variations should not be concealed behind a "wall" due to higher thematic values.

The *elevation over the horizon* or *site angle*, is an angle that varies from 0° to 90°, expressing the observer's position above the reference plane. It is particularly linked to the thematic value range. If the range is low, differences must be emphasized and a small site angle (10°-30°) has to be chosen, if the range is wide we have to pick an angle going towards 90° which will flatten the relief a little bit.

Finally, the *distance between observer and reference plane* is a length (with units varying according to software programs), difficult to manage as it has no extreme boundaries. A close point of view emphasizes differences while distance tones them down progressively.

These parameters are not independent; their combinations in order to realise the best effect is more a question of practice than of rigorous logic. Parameter value selection for a 3D representation does not follow strict rules that everyone can apply; because of being a communication map and not a map to be integrated in a GIS. This display mode has a very strong influence on understanding phenomena and can be very interesting as a GIS output, as other parameters - closely tied to graphic design, which will be explained later on - can strengthen the effects we wish to emphasize, if well chosen.

At this stage, it becomes obvious that the real choice is to know if we want to and if we can build a 3D map, which boils down to the objective of the map. A communication map in this final phase can advantageously be a 3D map. A basic map on the other hand cannot without going into very sophisticated methods realize this. Where *XYZ* correspondence is concerned, which is the third component for representation type selection, the choice becomes even less obvious.

5.1.3 [XY] and [Z] correspondence

This distinction in map construction was introduced by [Rimbert, 1979][18]; even if it is still in progress, it deserves to be mentioned as it allows a better understanding of the relations between the 2 principal components of a map: locations [*XY*] and attributes [*Z*]. Three main families of correspondence are recognized today:
- the transfer system
- the surface system
- the anamorphosis system

The transfer system belongs to an *f* application of the space attributes in the location space :

$$Z \quad \xrightarrow{\quad f \quad} \quad XY$$

Thematic values are projected, tacked on the surface containing the locations as if this support was neutral. Most maps - point maps, network maps, choropleth maps etc. - belong to this category.

The surface system expresses the fact that Z attributes are a mathematical function *f* of the locations *XY*, or

$$Z = f(X,Y).$$

The phenomenon under study is fundamentally tied to the locations: spatial support is no longer considered as secondary, or neutral but as an intervening element in the understanding of this phenomenon. Thematic variable values are calculated due to the locations. The function *f* can vary: polynomial function of varying degrees, trigonometrical function etc.

Methods for obtaining these maps usually belong to the regression model family. They are always isoline maps, expressing a spatial trend in quantitative variables[19]. Maps built according to this system can be integrated into a GIS database. They also could be used as interpretation or communication maps.

In *the anamorphosis system*, which is by far the least known, though the most used without map authors being conscious this: [*XY*] are functions of [*Z*]

$$X = f(Z)$$
$$Y = g(Z)$$

This is the converse to the surface system. Locations will be modified according to thematic values of the variable and not always identically in both directions at that. This is quite usual, no map can be made without some distortions since they are -as we have already mentioned- the representation of a volume on a plane. One cannot flatten out a balloon without tearing (distorting) it. This also applies for the earth. The projection systems allow a correspondence between the points on the Earth geoid surface and the points on a plane, according to a mathematical system which is set out like the equation system above (see Chapter II.1). So all basic maps are already built according to the anamorphosis system.

We can modify the basic maps still further by having the thematic data intervene as space factors and by introducing distortions in these spaces. These are called *derived maps* which usually express spaces really used, such as railway or road travel spaces. But there are many types of anamorphosis maps. According to the function chosen or the variables used or even the construction methods, several transformation families tied to the anamorphosis system have been identified which we will list here:

- geometrical tranformations which correspond to projection systems and which are elsewhere;
- spatio-thematic transformations, where the thematic variable is considered as a weight or connection which distorts, transforms and creates space ("piezopleth" maps, configurations etc.)
- spatio-differential transformations which express trends or differences in homologuous spaces, these differences being shown through transformed, distorted images (azimuth transformation, bidimensional regression etc.)

We will not dwell on this third system, which except for projection systems is still not well developed. Its application is difficult, as one must choose transformations which best express the underlying spatial structures. Equivalences between mathematical models and spatial phenomena are necessary. This field though still under study is rich as far as its understanding of the subjects under discussion and its communication possibilities are concerned.

In contrast to the dimensional display type, correspondence systems intervene as much in basic as in final map construction. Choosing a representation type depends on many components which must all be well known, if we do not want to obtain a document which has nothing to do with the overall objective. Whatever its place in GIS; these selections can still be modified, modulated, or distorted depending on whether graphic design is applied in agreement with or in contradiction to the previous decisions.

5.2 Graphic design

Graphic design is a language, a sign language, which can translate every decision (made in the previous steps) to graphic signs and it is not neutral. It is advisable to know both its grammar and its perception. If part of the French school has turned with J. Bertin in the direction of a deeper understanding of graphic grammar, anglo-saxons with J. Olson and numerous others are greatly interested in graphic perception. The one does not work without the other, in any language, one can write without a grammatical mistake and, alas, without any style. Graphic language follows the same constraints: a map must be correctly built on the grammatical level but if it doesn't attract the reader, for communication maps at least, it has not achieved its goal. Graphic design must lead to a precise and attractive communication of the message stemming from the whole of the reasoning. It is an integral part of it and its grammar and perception should be studied a bit deeper.

5.2.1 Graphic grammar

Graphic grammar sets down correspondence between signs and their signification. We find here F. de Saussures' works in linguistics, with the notions of signifier and signified which J. Bertin[20] has most interestingly studied, even if some positions can be discussed or refined according to recent knowledge. We will not repeat here all the elements of J. Bertin's works; we will mention only the essential notions according to the objective of this section i.e. six visual variables which are elementary variations of a spot or a symbol: size, value, grain, colour, shape, and orientation. Each visual variable possesses properties, can bring out associations, differences, order, proportionality. There is a direct relation between sign systems and measurement level of thematic variables[21]. It would be absurd to keep a sign expressing order if the thematic variable is read according to a nominal level, expressing thus only a membership relation. Each visual variable must be defined in a precise manner[22].

Size is the variation in surface of a sign (Fig. III.1-24a). It is quantitative and so quite ideal for quantitative thematic variables, and *for every point or linear phenomenon, but it cannot be selected in the case of a variable of an areal nature*: A change in size implies surface variation and so shape variation, and consequently an anamorphosis. Quantitative variable and size variation are incompatible and so one must choose between the two, for example: does one prefer the spatial property and lose quality on measurement level or vice-versa.

Point position	Linear position	Area position

Figure III.1-24a: Visual variable "size"

Value variation is "the continuous progression perceived by the eye in a series of greys on a scale from black to white". In a given surface, it is expressed by the black and white quantity rate and thus gives an order. It is quite indicated for ordinal measurement levels, whatever the geometrical nature of the phenomenon, but gives better results on surfaces. It can be used for quantitative level of measurements, either by redundancy or by replacing size in areal position, knowing that the quantitative measurement level is then lost.

Point position	Linear position	Area position

Figure III.1-24b: Visual variable "value"

Grain is the quantity of discernable points on a unitary surface for a given value. Null grain is a grain where points are so fine as to escape detection. Like the value, grain expresses an ordering and can be used as a redundancy or as a replacement for thematic variables of quantitative measurement levels. It can be used with every geometrical positions, but - analogous to the value - differences on surfaces will be more obvious to the reader.

Point position	Linear position	Area position

Figure III.1-24c: Visual variable "grain"

Colour is a physical "decomposition of light into its spectral components"[23]. By regrouping two or more of these components, we obtain colours through additive synthesis (Fig. III.1-24d) of the three primary colours: red, green and blue (RGB). By doing the converse operation, i.e. by subtracting synthesis, we also obtain new colours: cyan, magenta and yellow (CMY). Colours are distinguished according to three criteria (IHS system): intensity, hue (the colour itself) and saturation (degree of purity).

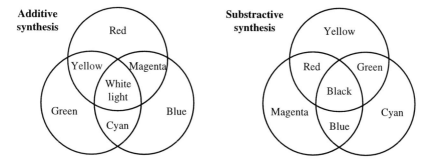

Figure: III.1-24d: Visual variable "colour"

Taking this brief reminder into account, we have to point out that colour is neither ordered nor quantitative. It should be used only to represent nominal thematic variables. Yet, by respecting certain rules, it is sometimes used for expressing an order. It can be applied easily to all types of cartographic objects, and facilitates differentiation. Colour is an attractive visual variable, rarely used until recently because of its expensive printing; nowadays this constraint is quickly disappearing, but the variable is often used mistakenly, not being well understood. Satellite images in particular, have forced cartographers to seriously think about this component and to use it correctly.

Orientation can take an infinity of values from 0° to 360° (Fig. III.1-24e). It is strictly nominal and so is used only to show up associations or differences. A certain surface is necessary to grasp all its variations, and so it gives the best results on an areal position. What is more, the differences in angles must be sufficiently marked so that the eye can perceive them. Finally, vertical and horizontal plotting on a surface are not read in the same way, and so distortions, inexistant in reality, can erroneously be perceived.

Point position	Linear position	Area position
⊘ ⊗ ⦶ ⊖	▱▱ ▱▱ ▱ ▱	◸◸◸◿

Figure III.1-24e: Visual variable "orientation"

Shape is given by the outline of the symbols, and its possibilities of variation are infinite. It corresponds to a nominal measurement level and so it can equally express associations or differentiations. Differences in shapes can only be perceived if the symbol has a minimum size. Point position often gives disastrous results: the reader is bewildered by diversity and cannot organize his/her visual comprehension.

Point position	Linear position	Area position
▲ ❑ ■ ○ ❖ ◗ ✳	▱▱▱ ▱ ▱ ▱	◸◸◸◿

Figure III.1-24f: Visual variable "shape"

These simple definitions help to understand why visual variables cannot be studied without taking the geometrical position and the measurement level into account; their interrelations, purely grammatical, are shown in Fig. III.1-25. We must still study the rules of perception.

5.2.2 Graphic perception

Before broaching the subject, we must complete what has been said on communication rules in our first part, especially by detailing some of the peculiarities of symbol legibility and by analysing reader behavior and reading habits.

5.2.2.1 Legibility rules

Again, the three main rules of legibility rest on the works of J. Bertin. The first is *graphic density* or sign density per cm^2. If the image is overloaded, the eye will not perceive anything;on the other hand, too few symbols on a map will bring too few information, and this will not justify its construction. One must be able to decide on the optimal quantity of signs so as to have a rich and legible map, but there is no recognized threshold value.

Measurement levels	Geometrical position		
	point	line	area
Nominal = **Differenciation** or **association**	SHAPE + COLOUR + ORIENTATION (?) Size Value Grain NO	SHAPE + COLOUR + ORIENTATION (?) Size Value Grain NO	SHAPE + COLOUR + ORIENTATION (?) Size Value Grain NO
Ordinal = **Order hierarchy**	In a given shape, orientation or colour VALUE + GRAIN SIZE NO	In a given shape, orientation or colour VALUE + GRAIN SIZE NO	In a given shape, orientation or colour VALUE + GRAIN SIZE NO
Interval and ratio = **order** + **distance**	In a given shape, orientation, colour, grain or value SIZE **Value and grain** possibly redondant	In a given shape, orientation, colour, grain or value SIZE **Value and grain** possibly redondant	In a given shape, orientation, colour, grain or value VALUE + GRAIN = loss of information **Size impossible**

Figure III.1-25: Visual variables, measurement levels and geometrical implanting

The second rule is *angular separation*, and corresponds to the eye capacity at distinguishing between symbols. Signs must have a minimal size to be perceived, a point for instance must have at least a 0.2 mm diameter. A separation/differentiation threshold must equally be accepted, so that the differences between symbols can be recognized. Small equilateral triangles and isosceles triangles are hard to distinguish.

The third concerns *retinal legibility* which is the degree of contrast of the map. If the support is visualized in light grey and the drawing in dark grey, the contrast is weak and the map not much legible. Conversely, a white background with thick black plotting has too much contrast and will incommodate the reader. Here too, one must find a compromise between a too sober and a too aggressive map.

These three rules are absolutely general and must be completed with the knowledge of reader's behavior and particularities of graphic readings.

5.2.2.2 Reading characteristics and reader's behavior

The first general remark is that graphic reading leads to information losses; it classifies and separates but it is not accurate, as authors such as Mac Carthy, Salisbury or J.C. Muller[24] have shown through comparisons. Only a part of the transmitted information is received. Looking at a map, one does proceed to various tasks such as classification, inventory and so on, as J. Morrison and many other cartographers and psychologists have pointed out, but reading is first of all global; the map is seen as a whole. A map is a bidimensional display and not a linear display like a text. Bidimensional reading of a map or painting is such that no order imposes itself spontaneously. This characteristic is at the same time an advantage (one "imagines") and a disadvantage (the message received can be totally different from the one sent). Consequently, one must know the reader's behavior before using such a language.

The reader constitutes blocks, sets of homogeneous zones, which he/she orders hierarchically and around which the reading is organized. The reader goes to the essential, and this may surprise some, i.e. the title, the legend. According to several tests, is seems that 84% of his/her time is spent looking at what is around the map, which explains the importance attached to the packing of the map. It is furthermore obvious that the reader does not always have much time to read the map. The reader has a certain rate of implication, and according to it, will see the map with a greater or lesser interest. He/she must not be put off. Finally the attention span is limited, and the reader always looks unconsciously to reducing his/her mental effort.

Taking these characteristics into account, it is obvious that the reader's habit must be followed and the legibility rules respected so as to avoid useless efforts, if a map is to be efficient. Complexity leading to injurious dispersion must be eliminated by reducing the number of signs. Easily assembled blocks must be constituted, symbols clearly differentiated, with enough surface to be legible and to allow clustering. Let us remember that the eye perceives only certain contrasts and certain differences.

This knowledge in graphic perception is essential at the communication map level, for final maps. Basic maps in a GIS rely more on preciseness and accuracy. Symbol selection must be guided by these 2 principles, no complexity must appear at this stage. On the other hand, for communication maps, the rules and principles set out must be respected while taking into account interrelations between grammatical rules, perception rules and every decision reached during the previous phases. The last element which plays a role in map producing is the packaging or - more accurately - the general organization of the map.

5.3 General organization of the map

This aspect of map construction is often neglected. Concretely, it is the organization of the space and text surrounding the map, whatever the final objective of the document. We just saw that according to tests done by psychologists, the reader spends a lot of time looking around the map; this must be taken into account to have efficient maps. "Cartographic environment" must be simplified so that the reader concentrates on the map itself, but he/she must also have sufficient information to understand the map. The entire document must be balanced, not overloaded. Further, it must be without any unjustified empty spaces and not too misleadingly attractive.

Independent from technical components map organization on screen or on paper is composed of 3 fundamental "elements": title, legend, and packing. According to J. Bertin, the title is the *external identifier*, the one through which the reader "meets" the map. A good title is concise, indicative, without article or preposition, easily noticeable, "winking" at the reader. Whatever the map objective, the title is under one form or another indispensable.

The *legend* or *internal identifier* allows a recognition of all the elements on the map itself, with the same characteristics, dimensions and orientations. Their signification must easily be understood, the legend must not be too long. It is adapted to the public: the same map for two different publics will bear two different legends. In presentation, symbols come before text, as the reader begins with symbols on the map and goes on to the legend to identify the signs, first of all by looking for the symbols. For data layers as well as for communication maps, the legend must be complete and precise, only the presentation can differ. Texts must be easy to read.

Finally other information must also appear on the map, gathered under the term *map "packing"*; this is the minimum list:

- sources for the reference map and attributes;
- author(s) and place where the map was created;
- the date, essential, especially for a GIS data layer;
- the scale, graphically presented, extremely accurate for a GIS;
- the orientation or partial plotting of the parallels and meridians. Always accompanied by the projection system used and the reference meridian in the case of a GIS, without which no registration of maps or satellite images is possible, for instance;
- the methods used, for map construction and various Z variable processing;
- authors of methods, software programs, or programs used for map elaboration.

These are not superfluous information but indications for the better understanding of the document, without which there might be hazardous interpretations. Packing may appear as a complement to map elaboration, but its absence can be injurious. Every information that may help must be present even in a simplified shape.

The layout of all this information depends on the map objective. For a basic plan which will be used several times and which is mostly used for ulterior processing, essentially, the document must be clear and without ambiguity. For a communication map, aesthetics play an important role and vary with the author. This third stage in map elaboration is partially free in its final organization, but its contents must always be present.

However, representation families, graphic design and general organization are interdependent to varying degrees. It is certain that for a basic map, a data layer, the first point is essential; for a communication map, the whole of the steps are for various reasons almost equally important, since a mistake on one level can annihilate the entire efforts made. A well-conceived isopleth map with symbols representing differences instead of progressions will be a failure: the reader will not understand. At each step of the map elaboration, choices exist, but some decisions are fixed.

It is obvious that a map is not built anyhow: the choice of representation type, the symbol selection, the general organization of the map depends on the purpose of the map. Here we find what was stated in the first part: a map with no precise end in view will not, besides exceptions, give satisfying results. Each decision has to be materialized on paper or on screen but we will not go into the technical aspects of manual or automated drawing, or printing.

6. Conclusion

At the end of this - certainly incomplete - presentation of cartographic reasoning and cartographic principles linked to research problems, some conclusions come imperatively to mind:

No map can be established if its objective is not clearly defined. A map used as a data layer in a GIS will not follow the same constraints as a map for the general public or land planners. Every decision we mentioned that has to be taken, at every stage in the process, must be explained and justified; if the author hesitates, he/she must refer to the objective before making a decision. Cartographic decisions are integrated in a scientific approach, in a precise problem; one must get rid of the usually simple mental approach to map construction. Even apparently secondary decisions can be harmful to the final production: all the stages are interdependent.

It also became clear through the discussions in this chapter that not all cartographic problems can be solved. Not all maps are real constructions, sometimes only graphic design, greatly subjective after all, can propose a solution. Principles exist; a rigorous and logical approach is proposed, but cartography is a developing science. It is the very foundation of GIS, its integration with computer science led to the development of this new field. So cartographic and GIS principles must not be set yet. They must improve and transform each other, regularly taking new technologies into account. But technical aspects must never be allowed to dominate; they help to rethink certain problems and to think up a few new ones; they must always be integrated in a logical framework if an efficient document, answering to the initial needs, is to be produced.

Questions

1. Give two examples of raw data and data in view of constructing a map of the annual precipitations of a small district, first for a land planner, then for a "general public".

2. Express the land use of a village (6 categories) according to:
    - a nominal scale
    - a binary nominal scale
    - a ratio scale

3. To study the distribution of unhealthy trees in a forest (large scale), what cartographic constructions are possible? In a GIS context, is a grid preferable to a Thiessen polygon type of tessellation?

4. What cartographic construction would you use to represent the pollution degree in a river: on a mean scale in view of drawing public opinion?

5. Can we associate a thematic variable of the
    - nominal point type to the size visual variable?
    - quantitative line type to the orientation visual variable?
    - ordinal area type to the grain visual variable?
    - quantitative area type to the size visual variable?

6. A zone intended for a new waste dump is to be set up near a city of 50.000 inhabitants through which runs a river from east to west. The mayor asks a research consultant to draw up a/several map(s) so as to be able to decide on the location, knowing that agricultural lands north, south and west of the town are productive and that the principal winds blow from the north-east. The city and district are serviced by a road network of good practicability.

    Propose a map of potential sites intended for the town mayor and council, describing and explaining each step of its compilation.

References

Beguin, H., Thiesse, J.: An axiomatic approach to geographical space. Geographical analysis, N°4, Vol. II, 1979, pp. 325-341

Bertin, J.: Sémiologie graphique - Les diagrammes, les réseaux et les cartes. Mouton, Gauthier-Villars, Paris, 1967, 431 p.

Bonin, S.: Initiation à la graphique. Epi, 1975, 171 p.

Burrough, P.A.: Principles of geographical information systems for land resources assessment - Monographs on soil and resources survey. Oxford science publications, N°12, Clarendon Press, Oxford, 1986, 194 p.

Cambray, B. (de): Modélisation 3D : état de l'art. MASI, UA 818 CNRS, Paris, 1992, 98 p. annexes

Cauvin, C., Rimbert, S.: La lecture numérique des cartes thématiques. Ed. Universitaires de Fribourg, 1976, 172 p.

Cauvin, C., Reymond, H., Serradj, A.: Discrétisation et représentation cartographique. Collection Reclus modes d'emploi, Montpellier, 1987, 116 p.

Clarke, K.C.: Analytical and computer cartography. Prentice Hall, London, 1990, 290 p.

Coombs, C.H.: A theory of data. Wiley & Sons, New York, 1964, 585 p.

Cuff, D.J., Mattson, M.K.: Thematic maps - Their design and production. Methuen, New York, 1982, 169 p.

Davis, P.: Data description and presentation. Science in geography, Oxford University Press, N°3, 1974, 119 p.

Dickinson, G.C.: Statistical mapping and the presentation of statistics. London, Ed. Arnold, 1967, 195 p.

Evans, I.S.: The selection of class intervals. Transactions of Institute of British geographers, Vol. 2, N°1, 1977, pp. 98-123

Flowerdew, R., Green, M.: Statistical methods for inference between incompatible zonal systems. In: Goodchild, Gopal (eds.): Accuracy of Spatial databases. Taylor and Francis, London, 1989, pp. 239-248

Foley, J., et al.: Computer graphics - Principles and pratice. Addison-Wesley, New York, 1992, 1174 p.

Frankhauser P.: La fractalité des structures urbaines. Anthropos-Economica, Collection Villes, 1994, 291 p.

Gold, J.R.: Communicating images of the environment with case studies of the use of the mediatii. West Midland overspill schemes, Occasional papers 39, Birmingham, University of Birmingham, 1974

Grimmeau, J.P.: Cartographie par plages et discontinuités spatiales - Application à la mobilité de la main-d'oeuvre en Belgique. L'espace géographique, N°1, 1977, pp. 49-58

Jenks, G.F.: Generalization in statistical mapping. Annals of Association of American Geographers, Vol. 53, 1963, pp. 15-26

Jenks, G.F.: Thoughts on line generalization. Proceedings of the international symposium on cartography and computing AutoCarto IV, ACSM, ASP. Vol. 1, 1980, pp. 209-220

Joly, F.: La cartographie. Magellan, PUF, 1976, 276 p.

Lam, N.S.N: Spatial interpolation methods : a review. The American Cartographer, Vol. 10, N°2, 1983, pp. 129-150

Lasswell, H.: The structure and function of communication. In: Berelson, Jonowitz (eds.): Reader in public opinion and communication. New York, Free Press, 1966, pp. 178-190

Lewis, P.: Maps and statistics. Methuen & Co., 1971, 318 p.

Mac Carthy, H.H., Salisbury, N.E.: Visual comparison of isopleth maps as a means of determining correlations between spatially distributed phenomena. Department of Geography, State University of Iowa, Vol. 3, 1961, 81 p.

Mandelbrot, B.: The fractal geometry of nature. W.H. Freeman & Co., New York, 1982, 468p.

Marchand, B.: L'usage des statistiques en géographie. L'espace géographique, N°2, 1972, pp. 79-81.

Meine, K.H.: Thematic mapping : present and future capabilities. World Cartography, 1979, pp. 1-16

Monmonnier, M.S.: Maps, distortion and meaning. Association of American Geographers, Resource paper N°75-4, Washington, 1977, 51 p.

Morrisson, J.L.: Towards of functional definition of the science of cartography with emphasis on map reading. The American Cartographer, Vol. 5, N°2, 1978, pp. 97-110

Muercke, P.C.: Thematic cartography. Association of American Geographers, Resource paper N°19, Washington, 1972, 66 p.

Muller, J.C.: Fractal and automated line generalization. Department of Geography, University of Alberta, Edmonton, 1986, 16 p.

Raper, J.: Three dimensional applications in geographical information systems. Taylor & Francis, London, 1989, 189 p.

Rimbert, et al.: Cartographie informatisée et géographie humaine. Rapport final, Expérimentations en cartographie transformationnelle, Tome 2, Fascicule A, 1979, 77p.

Robinson, A., Sale, R., Morrison, J.: Elements of Cartography. Wiley & Son, 4 ed., New York, 1978, 448 p.

Saussure, F.: Cours de linguistique générale. Payothèque, Paris, 1979, 509 p.

Spiess, E.: Some graphic means to establish visual levels in maps designs. IX International Conference on Cartography, Maryland, 1978, 16 p.

Taylor, D.R.F.: Graphic communication and design in contemporary cartography. Progress in contemporary Cartography, Vol. 2, Wiley & Sons, New York, 1983, 314 p.

Tobler, W.R.: Numerical map generalization. Ann Arbor, Michigan, 1974, 26 p.

Tobler, W.R.: Analytical Cartography. The American Cartographer, Vol. 3, N°1, 1976, pp. 21-31

Wright, J.K.: A method of mapping densities of population with Cape Cod as an example. The Geographical Review, Vol. XXVI, 1936, pp. 102-110

1 We are very grateful to A. Reymond who translated the whole chapter with many technical words, and to H. Haniotou who kindly read over once again.

2 All our thanks to N. Barthélemy who accepted to complete these graphics and adapted them for the english version and to H. Haniotou who proof-read the final version.

3 The expression of "analytical cartography" was first used by W.Tobler in 1976 in a paper bearing that same title. Many authors consider analytical cartography to be the equivalent of spatial analysis. We simply indicate its double designation.

4 We will barely tackle the latter here, as a special chapter is given over to it further on.

5 [Lasswell , 1966], quoted by [Gold, 1974]. The diagram was adapted by C. Cauvin.

6 Few research has been done on the subject, where anglo-saxons are to the fore, with the works of H. MaC Carthy and N. Salisbury as early as 1962, then those of J. Olson, and many others. Let us quote for Europe, however, the works of [Jouhaud, 1979] on planning and development in Alsatia and those of [Dias, 1988] on scolarity maps in Portugal.

7 We will not develop the problem here, but we again refer to some specialized works such as [Flowerdew, Green, 1989].

8 Geometric correction or registration: operation where origins and axes of map coordinates (or parallels and meridians) or of scale images or/and of different projection systems are put in correspondence.

9 On the subject, see [Mandelbrot, 1982]; [Frankhauser, 1994].

10 [Muller, 1986].

11 Distance definition: "let there be a set of objects called points; to each pair of point (i,j) of the set [E X E] let there be associated a single-valued element of R, number symbolized by d(i,j), called distance between point i and j" [Coombs, 1964]. It posesses three properties, unicity, symmetry and triangular inequality.

12 A clear synthesis is done in the paper of [Evans, 1977]: another with associated program program was realized by [Cauvin, Reymond, Serradj, 1987].

13 Let us remember that these problems concern quantitative variables only, as for qualitative variables, classes are fixed.

14 Term introduced by Wright J.K. in 1936 for homogenuous zone population maps.

15 [Lam, 1983].

16 The expression "membership map" is a personal proposition which seems to fit this map family; a more appropriate term may appear, however, so the above mentioned one should be considered as temporary. Its meaning is clear, the term itself is to be confirmed.

17 See [Cambray (de), 1992]

18 [Rimbert et al., 1979]

19 This is an affirmation to be qualified, as several methods are now being developed which allow nominal data processing.

20 [Saussures (de), 1979]; [Bertin, 1967].

21 At this stage, it becomes advisable to be very cautious with the term "variable"; in practise it is used to speak of processed thematic data, J. Bertin uses it to express the components of a signs system.

22 We are going to dip heavily into J. Bertin's works and those of his pupils for this section.

23 [Zwimpfer, 1985].

24 Quite a number of papers were consulted for this section; only the main authors will be mentioned in the bibliography.

III.2 Geometrical Models for Imaging Systems

Hans-Peter Bähr, Karlsruhe

Goal:

The necessary steps during geometric image processing should be learnt. This chapter concentrates on analytical models and gives an insight for rigorous and approximative processing.

Summary:

Rigorous geometrical models for conventional sensors (photogrammetric cameras on airborne platforms) are extended for scanner systems and satellite platforms. Errors for approximative solutions are presented (panoramic distortion, impact from height differences).

For both rigorous and approximative models results are given as well as the conditions to obtain them.

Keywords:

Opto-mechanical, electro-optical scanner; image coordinates; collinearity equations; polynomials; digital terrain model; orbital parameters; residuals

1. Introduction

Images are a very important source for a GIS Data Base (DB). It is essential for such a DB to show exactly congruent geometry in order to allow common processing. The necessary steps for transferring an original image into a "rectified" one for a GIS DB are:

    1. development of appropriate geometric models for the respective image types

    2. determination of parameters for the models related to the individual images

    3. rectification of the images

The three steps are basically different in nature: while (3) is a matter of image processing, (2) requires adjustment procedures and (1) physics and analytical geometry. (1) is the topic treated in this chapter.

2. Coordinate systems for imagery

In Photogrammetry, the geometrical model used for conventional cameras is the central projection

$$r' = c \tan \tau \qquad\qquad\qquad (\text{III.2-1})$$

according to Fig. III.2-1.

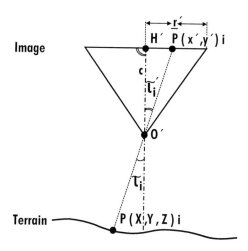

Figure III.2-1: Aerial photograph and central projection

In stereophotography the image coordinates measured are $(\bar{x}', \bar{y}')_i$ and $(\bar{x}'', \bar{y}'')_i$ for the left and right image, respectively (see Fig. III.2-2).

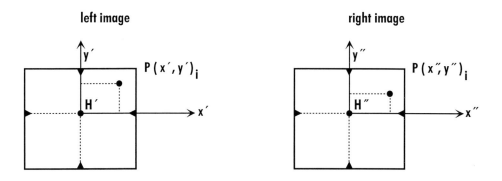

Figure III.2-2: A photogrammetric stereo pair; image coordinates

For 3 dimensions, image coordinates may be written as $(x', y', -c)_i$ according to Fig. III.2-3.

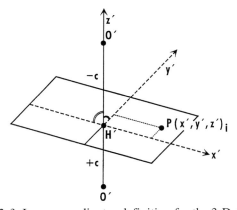

Figure III.2-3: Image coordinates definition for the 3-D domain

Here $0'$ means the centre of perspective which may be defined above or below the image plane. For simplification we will take the "positive case", i.e., the centre of perspective above the image plane. Though contrary to reality, this yields the same results as for the opposite case and typically shows $-c$ for z' :

$$p_i \begin{pmatrix} x' \\ y' \\ z' \end{pmatrix} = \begin{pmatrix} x' \\ y' \\ -c \end{pmatrix}_i \tag{III.2-2}$$

For many reasons the central projection represents an idealized model which has to be completed in order to describe the physical "reality" better, and to allow more precise geometrical processing. Consequently, the image coordinates in Fig. III.2-2 have to be changed into

$$x_i' = (\bar{x}' + \Delta x')_i$$

$$y_i' = (\bar{y}' + \Delta y')_i \tag{III.2-3}$$

where Δ represents the differences between the central projection and the revised model. The revision affects the "interior orientation", i.e.

c =   focal length, camera constant

H =   location of the "principal point" (see Fig. III.2-2)

$(\tau_i - \tau'_i) =$ image distortion (see Fig. III.2-1)

The parameters of interior orientation for calibrated photogrammetric cameras are delivered by an individual certificate for each camera. For further calibration, for instance during data processing, additional parameters are applied.

It is necessary to point out that the limiting factor for geometrical image processing is the precision of measured image coordinates.

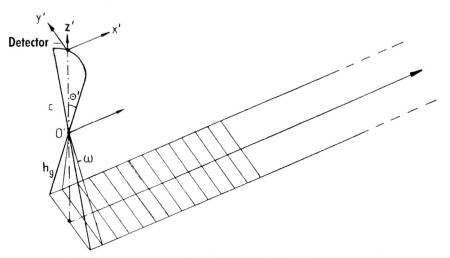

Figure III.2-4: Principle of the opto-mechanical scanner

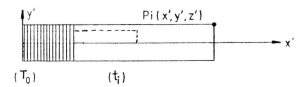

Figure III.2-5: Scanner systems: the x-coordinate is a function of time

Fig. III.2-4 shows the principle of **opto-mechanical** scanners. The 2D-image is acquired by the continuously advancing platform (airplane, satellite) moving in the *x*-direction whereas a rotating mirror or prism scans in the *y*-direction line by line. Note that there is no shutter, and that continuous strips are produced.

Geometrically, scanner systems differ completely from the central projection and have to be modelled by taking into account their specific design. In Fig. III.2-4,

h_g =   flying height
ω =   instantaneous field of view
θ =   scan angle

The condition for having the scanned lines put together one after the other without gaps or overlap may be expressed by the equation

$$\frac{v}{h_g} \overset{!}{\leq} \omega f \qquad\qquad (\text{III.2-4})$$

where v is the velocity of the platform and f, the scanning frequency. However, Eq. (III.2-4) neglects the varying scale along the lines and refers to the image axis. In practise, the tendency is to allow minor overlap in order to avoid gaps. This results in affine deformations of the scanner imagery, and produces strips "which are too long".

The **x-coordinate** of the imagery is a function of time as shown in Fig. III.2-5; though y is time-dependent too, this may be neglected in most practical cases. For x, however, we simply write, according to Fig. III.2-5 :

$$x_i' \quad \sim \quad \Delta t_i \ k$$

$$\Delta t_i \quad = \quad (t_i - T_o) \qquad\qquad (\text{III.2-5})$$

$$k \quad = \quad \frac{v \ c}{h_g}$$

Imaging systems of that type, i.e. systems with time-dependent x-axis, are called "dynamic imagery".

The **y-coordinate** of the imagery is varried by rotating a mirror or prism as displayed in Fig. III.2-6 where y_z' corresponds to the central projection and y_a', which is a linear function of the scanning angle θ', is acquired by scanning.

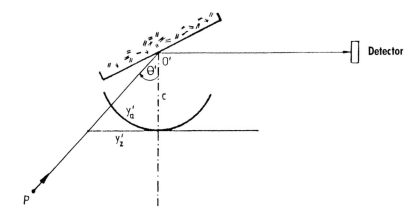

Figure III.2-6: y'-coordinate for opto-mechanical scanners compared to conventional imagery

The respective equations are

$$y_z' \ = \ c \ \tan \theta' \ > \ y_a' \ = \ c \ \theta \qquad\qquad (\text{III.2-6})$$

The difference

$$d_v \;=\; c \;(\tan \theta' - \theta') \tag{III.2-7}$$

is called "panoramic distortion" and represents the geometric difference between "conventional photographic imagery" and "scanner imagery" in y'. In practice, for "small" scanning angles, no difference is visible (e.g. LANDSAT imagery, where $\theta_{max} \sim 6°$) but for "large" scanning angles information at the margins of the image strips is compressed.

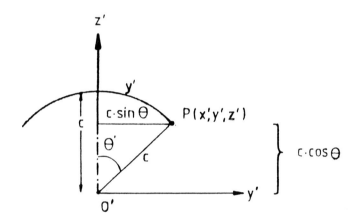

Figure III.2-7: **y-** and **z**-components for opto-mechanical scanners

Fig. III.2-7 shows the interrelation in y'- and z'-directions for the y-coordinate:

$$\begin{aligned} y' &= c \;\sin \theta' \\ z' &= c \;\cos \theta' \end{aligned} \tag{III.2-8}$$

$$P \begin{pmatrix} x' \\ y' \\ z' \end{pmatrix} = \begin{pmatrix} 0 \\ c \;\sin \theta' \\ -c \;\cos \theta' \end{pmatrix} = \begin{pmatrix} 1 & 0 & 0 \\ 0 & \cos \theta' & -\sin \theta' \\ 0 & \sin \theta' & \cos \theta' \end{pmatrix} \begin{pmatrix} 0 \\ 0 \\ -c \end{pmatrix} = A_\theta \begin{pmatrix} 0 \\ 0 \\ -c \end{pmatrix} \tag{III.2-9}$$

or Eq. (III.2-8) and (III.2-9) do not give any value for x' which was defined by Eq. (III.2-5) and has to be considered separately.

With **electro-optical scanners** where the rotating components are substituted by a linear array of semiconductors, the y'-coordinate is geometrically identical to the central projection, i.e. no panoramic distortion occurs:

$$P \begin{pmatrix} x' \\ y' \\ z' \end{pmatrix} = \begin{pmatrix} 0 \\ y' \\ -c \end{pmatrix} \tag{III.2-10}$$

Eq. (III.2-10) stands instead of Eq. (III.2-9). As far as the x'-coordinate for electro-optical scanners is concerned, it is identical to opto-mechanical instruments.

3. Generalized collinearity equations

3.1 Derivation of basic functions

The rigorous model in Fig. III.2-8 shows the analytical connection between the image coordinates for the central projection p_i' (see Fig. III.2-3) and the ground coordinates P_i.

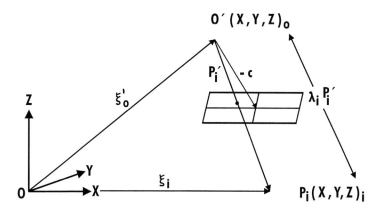

Figure III.2-8: Analytical conditions for conventional photogrammetry

The vectors in Fig. III.2-8 have to form a closed triangle

$$\xi_i = \xi_o' + \lambda_i \, p_i' \qquad\qquad (III.2\text{-}11)$$

where λ_i is a scale factor and 0 is the centre of perspective located at $(X,Y,Z)_0$. The transformation of the measured image coordinates $(x', y', - c)$ into p_i' is performed by

$$p_i' = A^T \begin{pmatrix} x' \\ y' \\ -c \end{pmatrix}_i \qquad\qquad (III.2\text{-}12)$$

A^T represents the orientation of the image composed of the rotations φ, ω, κ in $0'$. Combination of Eq. (III.2-11) and (III.2-12) leads to

$$\begin{pmatrix} x_i' \\ y_i' \\ -c \end{pmatrix} = \frac{1}{\lambda_i} \; A \begin{pmatrix} X_i - X_0' \\ Y_i - Y_0' \\ Z_i - Z_0' \end{pmatrix} \qquad\qquad (III.2\text{-}13)$$

The scale factor λ_i may be eliminated. This results in "collinearity equations" for the central perspective of the following form:

image coordinates $= f$ (ground coordinates, orientation parameters)

The unknown orientation parameters (φ, ω, κ, X'_0, Y'_0, Z'_0) may be determined by an adjustment procedure, provided that more than 3 ground control points and their respective image coordinates are available.

The step from central perspective geometry of conventional cameras to scanner geometry is very simple: we just have to substitute the image coordinates in Eq. (III.2-13), i.e. the left side of that equation, whereas the right side may remain, in principle, as it is. We have to take into consideration, however, that the orientation of dynamic scanner systems is time-dependent, which is expressed by the index j. Thus we have:

$$\begin{pmatrix} 0 \\ c \, \sin \, \theta'_i \\ -c \, \cos \, \theta'_i \end{pmatrix} = \frac{1}{\lambda_i} \, A_j \begin{pmatrix} X_i - X'_{oj} \\ Y_i - Y'_{oj} \\ Z_i - Z'_{oj} \end{pmatrix} \qquad \text{(III.2-14)}$$

for opto-mechanical scanners and

$$\begin{pmatrix} 0 \\ y'_i \\ -c \end{pmatrix} = \frac{1}{\lambda_i} \, A_j \begin{pmatrix} X_i - X'_{oj} \\ Y_i - Y'_{oj} \\ Z_i - Z'_{oj} \end{pmatrix} \qquad \text{(III.2-15)}$$

for electro-optical scanners, according to Eq. (III.2-9) and (III.2-10). The x'-coordinate remains a function of time as expressed by Eq. (III.2-5).

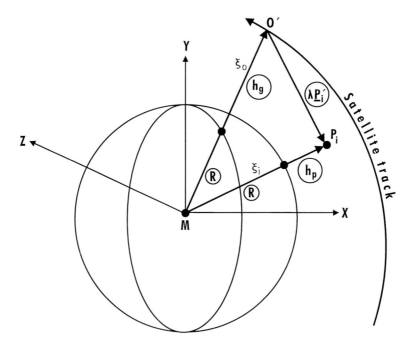

Figure III.2-9: Analytical conditions for imagery aquired from satellite platforms

Although the analytical derivation was done in close relation to conventional analytical photogrammetry there exist some special conditions for remote sensing.

3.2 Satellite Platforms

The vectors in Fig. III.2-8 may be specified for satellite platforms as shown in Fig. III.2-9. Here, the Earth has been approximated by a sphere of radius R and centre M, but the conditions can be expressed more rigorously, if necessary.

3.3 Orbital Parameters

In contrast to airborne platforms, the centre of perspective 0 of satellite platforms is predetermined by the orbital parameters of *Kepler's laws*. Fig. III.2-10 shows the corresponding conditions for a geocentric sphere:

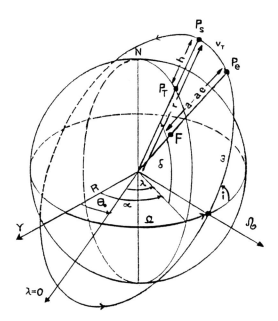

Figure III.2-10: Orbital parameters of a satellite P_S

- P_S is the position of the satellite
- P_T is the subsatellite point on Earth
- The major semi-axis a and the excentricity e describe dimension and shape of the orbit
- Argument of perigaeum ω and mean anomaly v_T define the position of the satellite within its orbit
- Inclination i and location of the ascending node Ω give the mutual relationship between the orbital and the equatorial plane

The 6 orbital parameters v_T, e, ω, i, Ω and γ are provided by astronomical observations and satellite tracking whereas the terrestrial dimensions (λ, δ, h) are provided by geodesy.

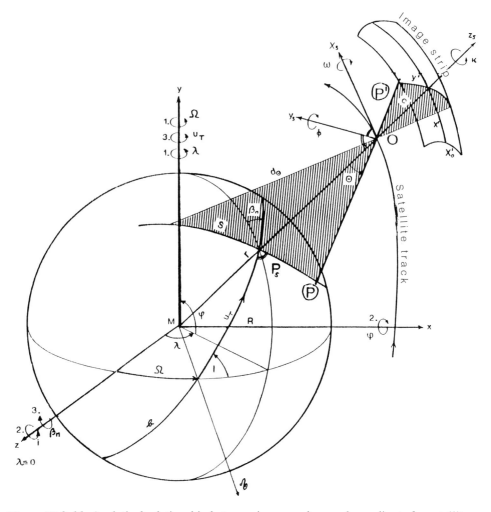

Figure III.2-11: Analytical relationship between image and ground coordinate for satellite-borne opto-mechanical scanner imagery

3.4 Collinearity equations for satellite imagery

Fig. III.2-11 applies collinearity equations to an opto-mechanical satellite system including orientation φ, ω, k. The terrestrial geocentric coordinates X, Y, Z for P may be linked to the corresponding image point P' $(x', y', - c)$ by 3 rotation matrices :

$A_{u, i, \Omega}$ in the geocentre M (u is composed of v_T plus ω)

$A_{\varphi, \omega, \varkappa}$ in the centre of perspective $0'$

A_{θ} performing the scanning (see Eq. (III.2-9)), correlated with A_{ω}

and by 2 translations where r points to the centre of perspective $0'$ and d_θ to an arbitrary point on the ground, P

$$d_\theta = r \cos \theta - \sqrt{R^2 - r^2 \sin^2 \theta} \qquad \text{(III.2-16)}$$

The final generalized collinearity equation derived from Fig. III.2-11 leads to

$$\begin{pmatrix} X \\ Y \\ Z \end{pmatrix} = A^T_{u,i,\Omega} \left[A^T_{\varphi,\omega,\varkappa} \; A^T_\theta \begin{pmatrix} 0 \\ 0 \\ -d_\theta \end{pmatrix} + \begin{pmatrix} 0 \\ 0 \\ r \end{pmatrix} \right] \qquad \text{(III.2-17)}$$

The image coordinates x', y' are "hidden" in the scanning angle θ (which - according to Eq. (III.2-6) - corresponds to y') and in time-dependent parameters (which - according to Eq. (III.2-6) - correspond to x') such as u.

3.5 Introduction of a Digital Terrain Model (DTM)

To limit geometric distortions, a DTM may have to be introduced into a geometrical model. Tab. III.2-1 gives the "critical" height difference of the terrain, a threshold from where a DTM starts to be necessary if geometric distortion has to be kept below 1 pixel. Tab. III.2-1 shows that a DTM has to be taken if topographic variations exceed 230 m within a LANDSAT TM scene. However, this limitations Δh are computed for the "worst case" at marginal points within the image strip. The table shows that for satellite imagery a DTM is not generally compulsory. For airborne systems, however, it is really necessary. For instance, for flying heights and strip widths of 1000 m, the critical height difference for the terrain is twice the pixel size on the ground!

Table III.2-1: The "critical" height difference, Δh

	Pixel on ground [m]	h_g [km]	swath width s [km]	θ max [°]	Δ h [m]
MSS	80	920	185	11,5	780
TM	30	705	185	15	230
SPOT	10	822	60	4.3	266
MOMS/83	20	299	140	26	85
RMK/83	20	250	180	42	55

3.6 Modelling the Variations of Orientation Parameters

Different from conventional photography, dynamic scanner imagery must consider variations of the orientation parameters as a function of time. A simple model would take the directions φ, ω, $\varkappa$ and the parameter R (see Fig.III.2-11) which corresponds to the flying height. This model is useful for satellite imagery, φ and ω being highly correlated to the X and Y components of the centre of perspective, which consequently are not taken for unknown parameters.

The variation of φ, ω, $\varkappa$ and R with time can be expressed by polynomials:

$$R = R_T + dR = R_T + a_0 + a_1 T + a_2 T^2 + \dots$$
$$\varphi, \omega, \varkappa = \dots$$

(III.2-18)

or more sophisticated models like Fourier Series or stochastic processes. It turns out that no model is significantly superior to others: most important is that the time variation is actually included in the geometrical model.

3.7 Conclusions for rigorous processing

We may summarize the experience with rigorous models by the following facts :

- Rigorous models have rarely been applied due to ground control deficiencies and because satisfactory results from approximate processing of satellite imagery, exception made of SPOT imagery, can be obtained.

- Generally, DTM integration does not improve the geometry of LANDSAT-TM scenes, though it is theoretically necessary. DTMs in TM imagery seem to be more important for radiometric modelling than for improving geometry.

- For airborne platforms, modelling time-dependent orientation parameters is crucial due to the necessarily dense population of ground control points. Accuracy depends primarily on

     - flight conditions (turbulences?)
     - terrain (topography?)
     - strip length

- Although the geometrical accuracy is best presented by visualizing error components ("residuals"), the root mean square error (rms) computed from the residuals is usually taken as a standard to compare results.

- For airborne scanner imagery rigorous models may reduce the rms to ± 1 pixel if the above mentioned conditions are favourable.

- For satellite scanner imagery rigorous models yield errors as low as ± 1/3 pixel.

- The residuals after adjustment tend to show effects introduced by image acquisition techniques, i. e. for scanner imagery, affine distortions are still dominant if data were not thoroughly processed.

4. Approximative solutions

4.1 Algorithms

Approximations take polynomials as mathematical models (see Chapter II.1), i.e., the physical reality may not be fully considered. Nevertheless, for satellite scanner imagery, results generally turn out to be quite good. Well-established commercial software programs frequently offer only polynomials as algorithms for rectification. One of the biggest disadvantages of using polynomials - compared to rigorous collinearity equations - is that they only give 2 dimensions instead of 3.

The simplest polynomial is the *similarity transformation*, i.e., the so-called "Helmert transformation":

$$X = a_0 + a_1 x' - a_2 y'$$
$$Y = b_0 + a_2 x' + a_1 y'$$

(III.2-19)

For determination of the unknown parameters a_i in a unique solution 2 ground control points are necessary. This algorithm corresponds to a transformation as indicated by Fig. III.2-12. Normally, more than 2 control points are used to obtain a higher reliability. In this case the unknown transformation parameters a_i have to be determined by an adjustment procedure.

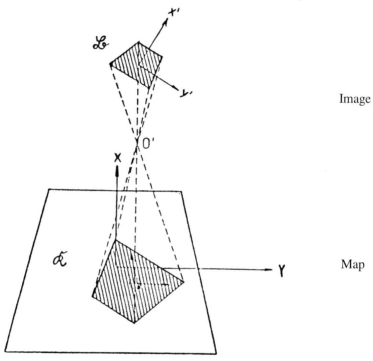

Figure III.2-12: Similarity transformation

As the internal geometry of the image is not affected by this transformation, the residuals (resulting from the adjustment procedure) show the geometric quality of the **original** image. The 4 unknowns require approx. 4 to 6 ground control points located at the image corners.

Another linear algorithm is the *affine transformation*, where 2 different scale factors for X and Y are introduced :

$$X = a_0 + a_1 x' + a_2 y'$$
$$Y = b_0 + b_1 x' + b_2 y'$$

(III.2-20)

As mentioned before, the affine distortion is a typical deformation of scanner imagery resulting from the imaging process. Therefore, Eq. (III.2-20) is generally very efficient. Here, approx. 6 ground control points should be introduced in order to determine the unknown parameters by certain reliability.

The next step after the affine transformation leads to the *second-order polynomial*:

$$X = a_0 + a_1 x' + a_2 y' + a_3 x'^2 + a_4 y'^2 + a_5 x' y'$$
$$Y = b_0 + b_1 x' + b_2 y' + b_3 x'^2 + b_4 y'^2 + b_5 x' y'$$

$$(\text{III.2-21})$$

which has lost any physical significance. This applies even better for third-order polynomials:

$$X = a_0 + a_1 x' + a_2 y' + a_3 x'^2 + a_4 y'^2 + a_5 x' y' + a_6 x'^3 + a_7 y'^3 + a_8 x'^2 y' + a_9 x' y'^2$$

$$(\text{III.2-22})$$

$$Y = b_0 + b_1 x' + b_2 y' + b_3 x'^2 + b_4 y'^2 + b_5 x' y' + b_6 x'^3 + b_7 y'^3 + b_8 x'^2 y' + b_9 x' y'^2$$

For Eq. (III.2-21) 9 well-distributed ground control points may be sufficient, whereas for Eq. (III.2-22) approx. 16 may be necessary. This may lead to problems since it is frequently difficult to find enough ground control points in both maps and images.

As far as the degree of the polynomials is concerned, one should try to choose the order as low as possible and only as high as necessary to model the geometrical displacements. This experience is caused by a specific characteristic of polynomials of higher order: these algorithms rectify the image parts around the control points very well, but tend to introduce additional displacements at the image parts between the control points.

4.2 Results for approximative processing

The overall quality of a geometrical model may be checked very well visually taking residuals. Residuals are error vectors at control points. A visual test is reasonable and reliable in case that sufficant points are available. This may be accepted, if the point number corresponds to the number of unknowns, which yields a redundancy by a factor of 2.

Fig. III.2-13 shows residuals at 234 control points in a LANDSAT-MSS-image from Bavaria ([Bähr, 1976]). The similarity transformation (Fig. III.2-13a) very clearly demonstrates the original geometry of the image, i.e. strong affinity (different scales in x and y). The systematic behavior completely vanishes after polynomials of 2nd order (Fig. III.2-13b). This of course may be shown numerically, too, but the visualization gives in addition a good impression to a human operator. Fig. III.2-13b clearly demonstrates that 2nd order polynomials are sufficient for the processed image.

We may summarize the experience of taking approximative models as follows:
- For processing LANDSAT-TM and -MSS imagery, polynomials are sufficient and yield up to $\pm 1/3$ pixel rms errors.
- 2nd-order polynomials (see Eq. (III.2-21)) for a full scene are "standard".
- The quality of pre-processing varies as a function of time and the place where the data were processed. Bad pre-processing quality, i.e. original images of rms error $> \pm 5$ pixel require 3rd order polynomials to reach the threshold of $\pm 1/3$ pixel.
- For small areas, i.e. for subsections of a full scene, linear models may be sufficient.
- The attention given to ground control point selection pays off, because control point quality is the limiting factor for precision of image rectification.
- Geometrical discontinuities may, however, occur during pre-processing: if duplication or omission of lines are present, ground control points will not eliminate this effect.
- Discontinuities of that type are crucial in GIS, when correlating different data layers.

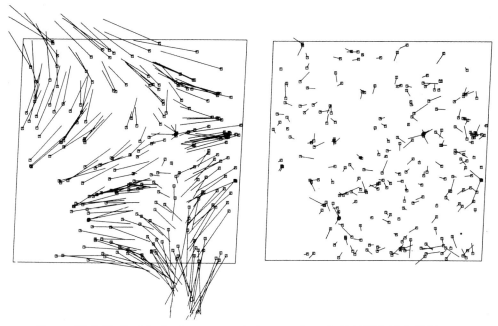

Figure III.2-13: Residuals after geometrical correction of a LANDSAT-MSS image.
a) after projective transformation.   b) after 2nd order polynomial
+-----+ : approx. 25 km on the ground, 250 m on vector

Questions

1.   What are the principal steps to produce a rectified image?

2.   What restrictions are introduced for satellite platforms by the individual orbital parameters?

3.   How do you get pass points?

4.   How to check geometric quality of an original/ a processed image?

5.   Why do we need equally distributed control points for geometric rectification?

6.   How many control points do you consider a minimum for rectification of a LANDSAT-TM scene in the Netherlands?

7.   What measures can be taken to overcome the weak geometric stability of airborne scanner imagery?

8.   How do houses appear geometrically in images taken by an opto-mechanical and an electro-optical airborne scanner system?

References

Bähr, H.-P.: Analyse der Geometrie auf Photodetektoren abgetasteter Aufnahmen von Erderkundungssatelliten. Wissenschaftliche Arbeiten der Lehrstühle für Photogrammetrie und Kartographie an der Technischen Universität Hannover, Hannover, 1976

Kraus, K.: Photogrammetry - Volume 1, Fundamentals and Standard Processes. Ferd. Dümmler Verlag, Bonn, 1993

Konecny, G., Lehmann, G.: Photogrammetrie. Walter de Gruyter, Berlin New York, 1984

Jacobs, H., Bähr, H.-P.: Impact of Landuse Classification by Geometrical Parameters. Publica do Instituto Português de Cartografia e Cadastro, 1995

III.3 Image Classification (Theory)

Hans-Peter Bähr, Karlsruhe

Goal:

Understanding of computer-controlled image interpretation.

Decisions based on statistics, using the Bayesian Estimator and the Gaussian distribution of samples.

This leads to a formalised mechanism of decision making and involves necessarily error generation.

Summary:

The classification procedure is developed through a graphically displayed histogram approach. Examples are given for one and two dimensional histograms. For the more rigorous, statistically controlled analysis the same procedures and drawbacks are included. The full theory may be simplified by modifying the covariance function of the multidimensional normal distribution.

Keywords:

Histogram, unsupervised classification, supervised classification, Bayesian Estimator, nearest neighbourhood classification, Maximum Likelihood classification

1. Introduction

Image Classification is a very general term and means assigning image pixels to well defined classes which may be very different in nature.

The present chapter will be restricted to *spectral* classifications. Therefore, the so-called "spectral signatures" are the dominating parameters for class definition. Spectral signatures are composed of grey values associated to every pixel of an image :

$$\textbf{Pixel} \qquad g_{i,k} = x_{i,k} = (x_1, \ x_2, \ x_3 \dots x_n)_{i,k} \qquad\qquad \text{(III.3-1)}$$

$$\text{where} \quad \textbf{\textit{n}} = \text{number of spectral bands.}$$

Eq. (III.3-1) gives a grey value $g_{i,k}$ as composed of n spectral components. The values $(x_i \dots x_n)$ correspond to the respective grey values.

If we take spectral signatures for classes, we neglect neighbouring effects between pixels, i.e. geometric information which could be taken for class description.

There is still another observation to be made for this chapter: the classes to be looked for in the imagery shall be restricted to "land use classes".

2. Monospectral classification

Images may be classified using only one spectral band (channel). In this case we assign the grey level values of one image channel to the respective classes. This procedure uses just one threshold but there are already advantages and disadvantages of multispectral classification worthy to be discussed. Therefore, we shall start by introducing monospectral classification and we shall do this for just one image.

2.1 Introduction of histograms

A histogram is the display of the grey value distribution in an image. Fig. III.3-1a, III.3-b and III.3-1c give a simulated example. Fig. III.3-1a represents an analogous image containing three classes: O_1, O_2 and O_3. Beside the three classes which are characterized by different grey values, there is the background of the image which shows a certain contrast to them. In Fig. III.3-1b, this analogous image is transformed into a digital representation. The human observer does not adapt very quickly to this type of representation. Numbers 0, 1 and 2 correspond to the three above mentioned classes whereas number 3 corresponds to the background. The *histogram* of this digital image is shown in Fig. III.3-1c. All grey values are on the abscissa and the number (frequency) of the respective grey values are shown on the ordinate.

A histogram immediately tells an expert the main characteristics of an image. If the scale of grey values ranges from black to white, the distribution of certain specific values shows whether we are dealing with a very light or a very dark image, or if there is a dominant area, which could indicate low contrast.

Fig. III.3-2 shows an original satellite image with very low contrast. The human eye readily detects three distinct clusters of grey values. The grey value distribution in the histogram confirms this impression (Fig. III.3-3). Here again we have three clusters: the darkest one corresponds to water, the lightest to land and the values in between to sandy marshland. The three types of surface are "classes" in the sense explained above. The variance of the distribution is not equal for all classes. We see quite distinctly, for instance, that the class "land" has a larger variance than the class "water" or "sandy marshland". The typical variances for the different classes are easy to understand.

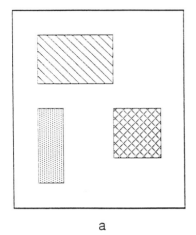

3	3	3	3	3	3	3
3	0	0	0	3	3	3
3	0	0	0	3	3	3
3	3	3	3	3	3	3
3	2	3	3	1	1	3
3	2	3	3	1	1	3
3	2	3	3	3	3	3
3	3	3	3	3	3	3

a   b

c

Figure III.3-1: 1-dimensional histogram for a simulated image

2.2 Class separation by histogram manipulation

Digital image processing allows very simple histogram manipulations. For instance, in Fig. III.3-3 we may eliminate those grey values which we consider to belong to the class "land". The remaining grey values will be equally distributed over the whole scale, from black to white.

The result can be appreciated in Fig. III.3-4 where the information in the land area has been almost totally eliminated. The contrast between water and marshland, which in the original image was not too good, has now been strongly enhanced. There are, however, some structures in the land area, which still remain visible; this means that some pixels in the land area obviously fall into a grey value which corresponds either to the black area of water or to marshland. This again can be easily understood by the human observer.

But we may as well do exactly the opposite: Fig. III.3-5 shows the effect attained when the classes "water" and "marshland" have been eliminated and the class "land" has been distributed over the full range of grey values from black to white. The result, of course, is a considerable gain of contrast within the "land" area, but marshland can no longer be identified.

Thus, a very simple example already shows the basic principles of image classification for land use:

1.   Some a-priori knowledge about classes in imagery is essential.

2.   Separation of classes is an arbitrary process. The thresholds are defined by the human observer.

3.   We may not expect 100 % accuracy by this procedure.

4.   The three classes show very different statistical behaviour. This means that the spectral signatures display larger or smaller variances according to the classes they have been assigned to.

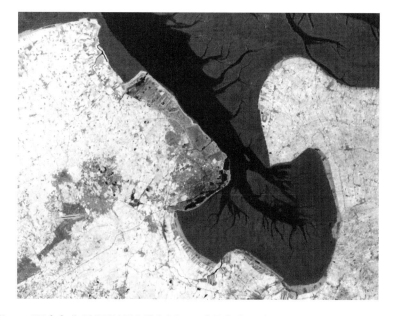

Figure III.3-2: LANDSAT-MSS (channel 7=infrared), showing Jade estuary / Wilhelmshaven North Sea |---------| = approx. 5 km

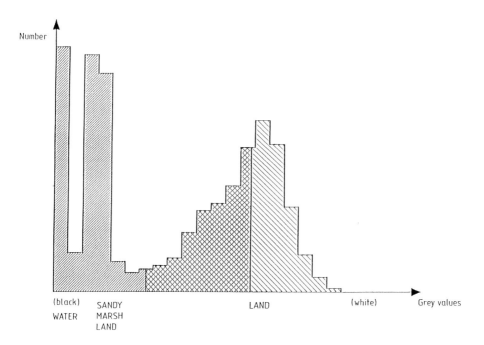

Figure III.3-3: Histogram (generalized) taken from Fig. III.3-2

Figure III.3-4: Result after elimination of the grey values for "land" (see Fig. III.3-3) and stretch for the whole range of 256 grey values

Figure III.3-5: Result after elimination of the grey values for "water" and "marshland" and stretching to the whole range of 256 grey values

3. Introduction of 2-dimensional histograms

In the preceding section, classes were defined on the basis of just one image channel. Modern imaging systems, however, offer a great number of different channels, i.e. different spectral bands which obviously provide more information to be used in classification procedures.

This may be very well visualized when taking two parameters, i.e. *two spectral channels.* Since classification, in general, is not necessarily a function of *spectral* bands, we shall take Fig. III.3-6 as an example. Here we analyse the attributes "age" and "salary" of two classes. Taking age and salary as axes of a coordinate system, every person will find his well defined position in it. Plotting a number of samples from the classes "student" and "professor", two clusters will evolve, one cluster corresponding to higher age, very much correlated to higher income, and the other class to lower age and lower income. There is no doubt that one cluster can be assigned to the class "professor" and the other to the class "student".

Here again we must point out that exceptions may occur. For instance, students are not necessarily of low age and income but, if not so, it is really an outstanding event. The expected case, the highest probability, corresponds to high age and high income for the class "professor". *All properties are relative, of course.*

Fig. III.3-7 gives a similar example. Here the attributes on the axes are height and weight. These two parameters are available for every person. The distribution in Fig. III.3-7 shows two clusters for adult persons. Analysing these clusters we discover that the distribution of men and women is a function of height and weight: the probability for men to belong to the class of high weight and height is higher than to be included in that of low weight and height. For low weight and height we expect women. We all know that it is not necessarily so, but the probability is like that. Clearly the class boundary cannot be unmistakably be defined. Nevertheless, with a certain probability, we may have two sets of some 70 % correct values.

It is obvious that classes like "professor", "student", "man" or "woman" cannot fully be described by just two parameters but this separation already yields certain results with a certain probability. We have to stress here that in multispectral classification, the results are of similar nature: *we get results with the highest statistical probability but not necessarily correct ones.*

The third example for 2-dimensional histograms as shown in Fig. III.3-6 and III.3-7 is given by Fig. III.3-8. Here the properties on the axes are the visible spectral range and the infrared range. Taking samples from "vegetation", we expect clusters of healthy vegetation with high infrared reflection and low reflection in the visible range. On the other hand, dry vegetation or harvested fields should show up in the area corresponding to low infrared and, possibly, high reflection in the visible spectral range. The example in Fig. III.3-8 shows a clear distinction between healthy vegetation and sand.

The central question is how to separate analytically the clusters in the *feature space*.

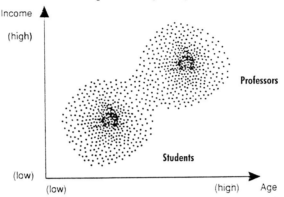

Figure III.3-6: Feature space "age" and "income" for the classes "students" and "professors"

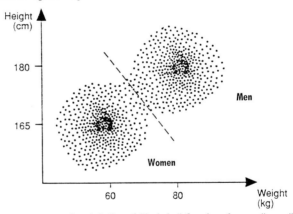

Figure III.3-7: Feature space "weight" and "height" for the classes "men" and "women"

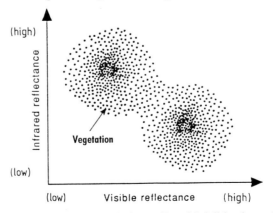

Figure III.3-8: Feature space "infrared channel" and "visible channel" for the classes "vegetation" and "sand"

4. Unsupervised classification

The above examples already disclose the specific nature of multispectral classification. Attributes of classes, boundary of classes and probability of assigning pixels to correct classes are problems inherent of all classification procedures. Fig. III.3-9 represents a real cluster obtained from a LANDSAT-MSS image. As a function of channel 1 and channel 2, we readily detect four different populations which may be assigned to four different classes. These four classes are "aprioristically" unknown. Moreover, we may ask ourselves whether the biggest class in the centre is not eventually composed of two separate classes instead of being only one. Class boundary means separation of the clusters in the diagram. While this separation is easy for the two classes on the right and below the centre, it is not so easy for the class on the left side, and it is particularly complex if we wanted to define two classes from the central cluster.

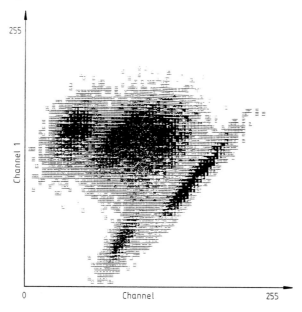

Figure III.3-9: Real cluster from a LANDSAT-MSS image ([Dennert-Möller, 1983])

This procedure (separation of classes and identification of classes by this method) is called *Unsupervised Classification*. By doing this we learn another very important rule for general classification: the human operator has to bring into the procedure his a-priori knowledge. Without knowing the real data, it would be impossible to assign real classes to the clusters. This means that for unsupervised classification we first will get a distribution of grey values in the area of the 2-D diagram and afterwards, the human operator has to relate such clusters to classes which are the most probable ones, compared with the real scene (in this case the LANDSAT scene) on the screen or on the hard copy image.

In general terms, unsupervised classification does not yield very satisfactory results. It may be very useful, however, to determine possible classes. Therefore, it is often used as a pre-processing procedure before going on with the rigorous *Maximum-Likelihood-Classification*.

5. Theory of supervised classification

Supervised classification is based on statistical algorithms. The a-priori knowledge of the operator, indispensable for any classification, is introduced at the beginning of the procedure, not at the end as is the case with unsupervised classification.

5.1 Statistical parameters

$$x = (x_1, x_2 \dots x_n)^T \qquad\qquad \text{grey value vector} \qquad\qquad\qquad\qquad \text{(III.3-2)}$$

$$\Omega \equiv (\omega_1, \omega_2 \dots \omega_m) \qquad\qquad \text{vector of } m \text{ classes} \qquad\qquad\qquad\quad \text{(III.3-3)}$$

$$p(x/\omega_j) \qquad\qquad\qquad \text{distribution function of } x \text{ in class } \omega_j \qquad \text{(III.3-4)}$$

$$p(\omega_j) \qquad\qquad\qquad\quad \text{a-priori probability of class } \omega_j \qquad\quad\; \text{(III.3-5)}$$

$$p(\omega_j/x) \qquad\qquad\qquad \text{a-posteriori probability of } x \text{ in class } \omega_j \quad \text{(III.3-6)}$$

The above list shows and defines the necessary statistical parameters for supervised classification. Eq. (III.3-2) corresponds to Eq. (III.3-1), showing the vector of grey values for 1 pixel, where n is the number of channels involved. The vector of grey values x for n channels is an element of the feature space. Ω is a vector of m classes where each class is any of the classes $\omega_{1,2\dots}$. The distribution function of a grey values x within a class ω_j is written like Eq. (III.3-4) which is the algorithmic form of the clusters shown in the example of Section 4. The distribution function is a-priori unknown. In Fig. III.3-9 we see that the respective classes show different distributions and, consequently, have different distribution functions. What is necessary for our statistical estimation is the a-priori probability of class ω_j which is Eq. (III.3-5). In general, the a-priori probability is also unknown. It may, however, be derived from "experience". For instance, it is extremely improbable (though not impossible) to find an urban settlement right in the middle of a large water body. And, to give a second example, it is not very probable to find large areas of green vegetation in a desert. The probability of the existence of a class in a certain environment (i.e., in a certain scene) is a very important information for statistical estimation. *Precise* numbers, however, are not required.

Finally, Eq. (III.3-6) is the result we are looking for, i.e. the a-posteriori probability for the estimation that the grey value vector x is an element of a certain class ω_j. This means that we look for a procedure by which we assign an arbitrary grey value vector x to a well-defined class ω_j.

5.2 The Bayesian estimator and the law of classification

A basic law in statistics which is frequently used for classification is the *Bayesian Estimator*

$$p(\omega_j/x) \quad = \quad p(x/\omega_j)\,p(\omega_j)/p(x) \qquad\qquad\qquad \text{(III.3-7)}$$

where

$$p(x) \quad = \quad \sum_{i=1}^{m} p(x/\omega_i)\,p(\omega_i) \qquad\qquad\qquad \text{(III.3-8)}$$

$$m \quad = \quad \text{the number of classes involved.}$$

The other components of the equations have been given in the preceding section.

The *Bayesian Estimator* is a method to determine the probability that a random pixel x belongs to a class ω_j. It is determined by the distribution functions for x within a class ω_j multiplied by the a-priori probability of the real existence of the class ω_j. Those two components are normalized by dividing them by $p(x)$. $p(x)$ gives the same multiplication as shown in the numerator but it is the sum of all classes from $i=1$ to m. The computed a-posteriori probability is high if the position of a pixel x is close to the expected central value of the distribution function.

It seems very logical to attribute a random pixel x to the class which shows the highest probability. This simple and very logical rule is expressed by the *Classification Law* :

$$x \in \omega_i \ , \quad \text{if} \quad p(x/\omega_i) \, p(\omega_i) \ > \ p(x/\omega_j) \, p(\omega_j) \qquad\qquad \text{(III.3-9)}$$

Separating functions between two classes will be determined when both members of Eq. (III.3-9) are set equal. Mathematically, these boundaries are surfaces of higher order (hyper-surfaces). The order of the boundaries is a function of the number of channels and the number of classes involved. The most important parameter in the Law of Classification, the distribution function $p(x/\omega)$ is still undetermined. It is mostly expressed:

$$p(x/\omega_i) \ = \ \frac{1}{\sqrt{2\pi^n \det(C_i)}} \ \exp\left(-1/2 \ (x-\mu_i)^T \ C_i^{-1}(x-\mu_i)\right) \qquad\qquad \text{(III.3-10)}$$

This is the well known *normal distribution* for i-dimensions. Statistical populations are frequently described by a normal distribution because their mathematical manipulation is not too complex and they often fit into reality (in Nature). The normal distribution for one dimension yields the well-known *Gaussian Functions* (Fig. III.3-10).

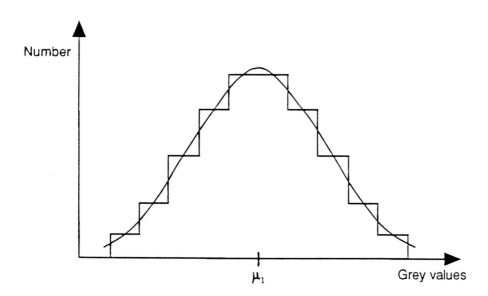

Figure III.3-10: Normal distribution (Gaussian function)

As stated above, each class has a specific distribution. Thus, the Gaussian functions for such classes have different mean values μ and variances σ^2. For two dimensions in the 2-D plane, the distributions form ellipses. These ellipses for real values are shown in Fig. III.3-9 where we see that mean values (central position of the clusters) and variance (form and orientation of the ellipses) differ according to classes. The big central clusters may be the result of two overlapping ellipses. For multispectral distribution, instead of the simple variance σ^2, we get the covariance matrix C (Eq. (III.3-10)).

The n-dimensional covariance matrix is composed of the variances and covariances between the n-channels :

$$C_i = \begin{pmatrix} \sigma_{11} & \sigma_{12} & \cdots & \sigma_{1n} \\ & \ddots & & \\ \sigma_{n1} & \sigma_{n2} & \cdots & \sigma_{nn} \end{pmatrix}_i \qquad\qquad\text{(III.3-11)}$$

$$\sigma_{ii} = E\left((x-\mu_i)^2/\omega_i\right)$$

Eq. (III.3-10) and (III.3-11) give the mathematical algorithm for what we call multidimensional normal distribution. For multispectral classification, we need the real values for the respective classes, i.e. the *unknown parameters* of the normal distribution. These are the mean value, μ, and the covariance matrix. These unknowns are computed for samples called "training fields" during the multispectral classification procedure. The unknowns are then calculated by least squares adjustment.

We should underline that if a normal distribution is taken as the distribution function, our final results will very much depend on how closely the real distribution of an arbitrary class in a multispectral image corresponds to its mathematical representation (Eq. (III.3-10)). Going back again to Fig. III.3-9, we may define the whole big central cluster as a single class or we may, as an alternative, split this big cluster into two normal distributions, i.e. two classes. On the other hand, there is a basic question: how correctly may the *real world* be separated into "classes which follow normal distribution"? *This must be kept in mind when doing multispectral classification.*

6. Derivation of three algorithms for supervised classification

Taking the basic Eq. (III.3-7), (III.3-9) and (III.3-10), we may derive three algorithms which are used for supervised multispectral classification. The complete equation is used for the most complex situations and to obtain the most "reliable" result. Simplification yields poorer results which are applied for didactic purposes and pre-processing or when limited computer power requires the use of "fast algorithms". For all three algorithms, there are no restrictions with regard to number of classes and number of bands, the *only difference lying in the design of the covariance matrix.*

6.1 Nearest-Neighbourhood-classification ($C_i = \sigma^2 E$)

Simplifying the rigorous Eq. (III.3-10) by substitution of the covariance matrix by the diagonal matrix E_1 , the significant term in Eq. (III.3-11) is transformed into $\sigma^2 E$

The real conditions may be explained by Fig. III.3-11. The distributions of the classes i and j are given as concentric circles where the central point is the mean value, μ. For any random pixel, $p(x_1/x_2)$, the distance to the central value of each of the classes is computed. By means of the classification Law, the pixel is finally assigned to that class which is at the minimum distance. Consequently, this method is also known as the *Minimum Distance Method.*

Class boundary of only two classes like those given in the example, is simply a straight line which intersects at right angles the line connecting the two class centres, at a point equidistant to both centres. The distance for a particular pixel is to be computed with of the location of the grey values in the 2-dimensional feature space and the position of the classes.

Although this method seems to be very clear and logic, it does not yield good results. The reason is that the classes are not carefully designed according to their real behaviour: in Fig. III.3-9, the clusters are not circles but ellipses.

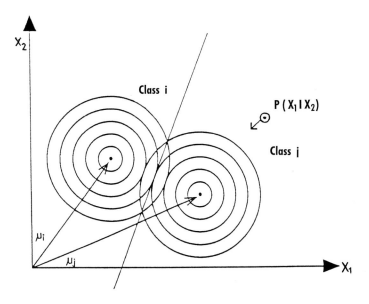

Figure III.3-11: Principle governing the Nearest-Neighbourhood-classification

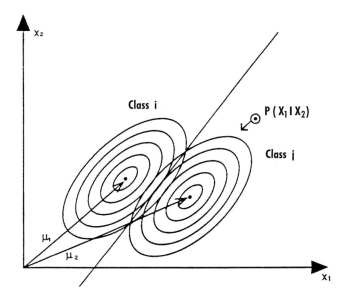

Figure III.3-12: Principle governing the Mahalanobis-classification

6.2 Mahalanobis-classificator   ($C_i = C$)

When introducing an ellipse instead of a circle into the 2-dimensional space, the covariance matrix is fully populated. If only *one* covariance matrix is taken instead of different matrices for any of the i-classes, we call this classificator the *Mahalanobis classificator*. It again uses the distance as explained by Fig. III.3-12.

This distance, however, is no longer the line between the centre of the class and the location of the pixel: it is interfered by the form of the covariance matrix. Since the distribution of all classes is identical, the separating line is again the straight line as shown in Fig. III.3-12.

6.3 The Maximum-Likelihood-classification (C_i , C_j for all $i \neq j$)

The use of different covariance matrices for each class leads to the *Maximum Likelihood classificator* (ML). This method involves the full theory developed in Section 4. From this point of view, the ML is *rigorous*. Compared to the two preceding methods, the only disadvantage is the considerably higher computing time required. Each covariance matrix of i-classes must be inverted.

Since the covariance matrix is different for every class, the separation is no longer given by a straight line but by a curve when more than two channels are involved, the order of the separation function is given by the number of channels and leads to surfaces and hyper-surfaces of higher order.

7. Concluding remarks

The theory of multispectral classification for imagery has been presented and the most important observations are :

1. A-priori knowledge of the operator is required.

2. Unsupervised classification is based on histogram evaluation and is mostly employed as a pre-processing method.

3. The most sensitive factor is the statistical distribution into "classes" and their mathematical formulation: Matching (Gaussian) distributions and *real world* manifested only by spectral signatures, involves many handicaps.

4. Nevertheless, the method yields good results and may be very effective when it is applied with care (see Chapter III.9).

Questions

1. Thresholding for 1-dimensional histograms and Maximum-Likelihood procedures: both do image segmentation. What is in common, what is different from the classification viewpoint?

2. The Nearest-Neighbourhood approach seems intuitively convincing. Why do we get relatively bad results?

3. The Bayesian Estimator requires a-priori probability of the classes to be determined. It is, however, not known in general. Discuss this issue!

4. What is the impact of sensor resolution on classification? Can we expect "better" results from higher resolution? Compare Landsat-MSS, Landsat-TM and MOMS-2!

5. What procedures are available to control classification results?

6. Discuss error sources in classification! Try to group them according to geometry, radiometry, algorithms, and semantics!

7. Do you see any possibility to automate the classification procedure completely?

References

Bähr, H.-P.: Abschätzung einiger geometrischer Fehlerkomponenten bei der multispektralen Klassifizierung. Bildmessung und Luftbildwesen, 1984, pp. 23

Jacobs, H., Bähr, H.-P.: Impact of Landuse Classification by Geometrical Parameters. Proc. ISPRS, Commission VII, Rio de Janeiro, 1994

Dennert-Möller, E.: Untersuchungen zur digitalen multispektralen Klassifizierung von Fernerkundungsaufnahmen mit Beispielen aus den Wattgebieten der deutschen Nordseeküste. Wissenschaftliche Arbeiten der Fachrichtung Vermessungswesen der Universität Hannover, Nr. 127, Hannover, 1983

Lillestrand, T.M., Kiefer, R.W.: Remote Sensing and Image Interpretation. Third Edition, New York, 1994

Niemann, H.: Klassifikation von Mustern. Springer Verlag, Berlin, 1983

Swain, P.H., Davis, S.M. (Hrsg.): Remote Sensing - The Quantitative Approach. New York, 1978

III.4 Geostatistics

M.J. van Dijk, T.H.M. Rientjes, R.H. Boekelman, Delft

Goal:

The goal of this model is to analyse the spatial characteristics of geostatistical data. These characteristics can be used to estimate values of the phenomenon under study at unmeasured locations.

Summary:

This chapter gives a brief description of the two most important procedures in a geostatistical analysis of spatial data: structural analysis and linear estimation. Structural analysis consists in choosing the most appropriate semivariogram model to characterize the spatial variability of a phenomenon. This semivariogram model can then be used in the geostatistical estimation procedure called "Kriging".

Keywords:

Random functions, regionalized variables, stationarity, semivariogram, interpolation, ordinary Kriging, co-Kriging

1. Introduction

In the field of geostatistics the value of a given variable, such as rainfall depth, at any point of a specified area is estimated when measured values of that variable are available at a finite number of points in that area. Techniques used for these estimations are called "interpolation schemes". The mathematical basis behind geostatistics is the theory of Regionalized Variables. This theory was initially based on the practical methods used in South African gold mines, notably by D.G. Krige, but was given a substantial theoretical basis by G. Matheron and others at the Centre de Morphologie Mathématique in Fontainebleau, France. This body of theoretical belief and the applied methods are known as "Geostatistics"; the practical estimation techniques that have been developed are collectively named "Kriging" techniques, after the pioneer D.G. Krige.

2. The regionalized variable theory

The regionalized variable theory is the statistics of a particular type of variable, the so-called *regionalized variable*. This regionalized variable differs from an ordinary scalar random variable because besides its usual distribution parameters it has a defined spatial correlation. Two realizations of a regionalized variable which differ in spatial location display in general a non-zero correlation, as contrasted with an ordinary scalar random variable in which successive realizations are uncorrelated.

A regionalized variable can be described by a function $Z(x)$, which gives the unique value of a property Z at any point x. Faced with the problem of discrete observations of a spatial phenomenon, the theory of random functions is increasingly applied, taking $Z(x)$ as a random process in a one-, two-, or three-dimensional real space.

The phenomenon characterized by the regionalized variable is in some way spatially stationary. The level of stationarity is very important for modelling the phenomenon; the magnitude of the parameters derived from the sample population depends on this level of stationarity. Since only one realization of Z is known (only a finite number of $Z(x_i)$ measurements is available), the ergodic hypothesis is introduced to estimate distributions of phenomena from this realization. It says that there is some form of stationarity; if different sequences of numerical values are taken, then the sequence of spatial averages will converge to their expectation. So any inferences about parameters of Z are based upon spatial averages of the single available realization of Z.

3. Stationarity of the phenomenon

Geostatistics is based on the first two moments of the random function. In practice, this means that the process Z, fluctuates in space around a given mean m, without a definite trend in any direction, the mean and variance being the same everywhere. The covariance C between the process at points x_i and x_j is independent on the individual locations x_i and x_j, and only dependent on the lag vector $h=x_i$-x_j :

$$E[Z(x)] = m$$
$$E\left\{[Z(x) - m][Z(x + h) - m]\right\} = C(h)$$

(III.4-1)

The stationarity of the covariance presumes the existence of the variance, which is finite and independent of x. If h tends to zero then the covariance approaches the variance of the random variable Z:

$$E\left[(Z(x) - m)^2\right] = Var\left[Z(x)\right] = C(0)$$

(III.4-2)

Also the semivariance has to be stationary:

$$\gamma(x,x+h) = \gamma(h) = \frac{1}{2} E\left[\{Z(x) - Z(x+h)\}^2\right] \qquad \text{(III.4-3)}$$

The semivariance is a measure of the similarity, on average, between points at a given mutual distance h.

Under second-order stationary[1] conditions there is a relationship between the variance, covariance and semivariance which makes one of these parameters unnecessary:

$$\gamma(h) = Var(Z) - Cov(h)$$
$$= C(0) - C(h) \qquad \text{(III.4-4)}$$

The semivariance is equal to the variance minus the covariance (Fig. III.4-1). This, expressed in a function, gives the semivariogram and the covariance function. These two functions express the spatial correlation of a phenomenon.

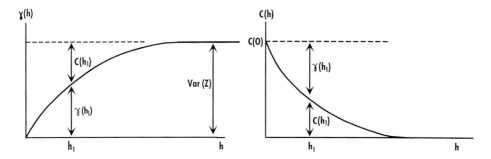

Figure III.4-1: Relation between the semivariance and the covariance

In many applications, however, the assumption of second-order stationarity may be too restrictive in terms of the constant spatial average over the domain. Therefore, a weaker hypothesis will be introduced, called by Matheron "the intrinsic hypothesis". This intrinsic hypothesis requires stationarity for the increments of $Z(x)$ only. Thus, the random process $Z(x)$, is said to satisfy the intrinsic hypothesis if its first-order differences, $Z(x_i) - Z(x_j)$, are stationary in the mean and variance:

$$E\left[Z(x) - Z(x+h)\right] = m(h) \qquad \text{(III.4-5)}$$

$$E\left[\{Z(x) - Z(x+h)\}^2\right] = Var\left[Z(x) - Z(x+h)\right] = 2\gamma(h) \qquad \text{(III.4-6)}$$

4. Semivariogram

The function γ was defined by Matheron as the semivariogram. It is used in graphic presentation and is one-half of the spatial variance, $Var[Z(x)-Z(x+h)]$. An estimation of the semivariance can be made from the available n Z_i measurements, at locations x_i and x_j. Each pair can thus be associated with a lag, or distance vector $h=x_i-x_j$. The pairs are then grouped in a limited number of lag classes in order to have a significant number of pairs in each class. The mean value of $\gamma(h_1)$, the semivariance of all pairs in a class with mean distance h_1, may be estimated from half

---

[1] Second-order stationarity requires only stationarity in the first two moments of the distribution: the mean and the variance.

the spatial variance $Var\ [Z(x)-Z(x+h)]$ divided by the n_1 possible pairs of observations in that class. Finally, the mean semivariance of the different distance lag classes is plotted against the mean distance of that class. The traditional non-parametric estimator is:

$$\gamma(h) = \frac{1}{2\ n(h)} \sum_{i=1}^{n(h)} \{Z(x_i) - Z(x_i + h)\}^2 \tag{III.4-7}$$

where

$$
\begin{array}{rl}
\gamma(h) & = \quad \text{the estimated semivariance for the distance classes } h, \\
Z(x_i), Z(x_i+h) & = \quad \text{the measured values within a distance class } h, \\
n(h) & = \quad \text{the amount of pairs in the distance class } h.
\end{array}
$$

The resulting plot is called an experimental semivariogram. The semivariogram shows the expected difference in value of two points against its distance.

Any experimental semivariogram found is only a reflection of the true semivariogram. [Kitanidis, 1993] gave some useful guidelines to obtain a reasonable semivariogram:
- use three to six intervals,
- make sure you have at least ten pairs in each interval,
- include more points at distances where the differences between calculated semivariances are larger. Especially for large values of h, there may be significant differences between semivariances computed from different subsets of data.

In practice, the experimental semivariogram must be fitted into an idealized model of the semivariogram, a theoretical semivariogram. The idealized curves are defined as simple mathematical functions which relate γ to h. The distance at which the maximum semivariance C - also called the "threshold" - is reached is known as the "range" a of the phenomenon. The range characterizes the zone of spatial dependency of the data. Almost all experimental semivariograms show an apparent discontinuity at the origin. This intercept C_0 is called the "nugget variance". Appropriate semivariogram models can be based on a linear fit, a spherical fit (Fig. III.4-2), an exponential fit, a Gaussian fit (Fig. III.4-3), a cubic fit or even a mathematical formula taking anisotropy into account.

Spherical model (Fig. III.4-2):

$$\gamma(h) = C_0 + C \left\{ \frac{3}{2} \cdot \frac{h}{a} - \frac{1}{2} \left(\frac{h}{a} \right)^3 \right\} \qquad \text{for } 0 < h < a$$

$$\gamma(h) = C_0 + C \qquad\qquad\qquad\qquad\qquad \text{for } h > a \tag{III.4-8}$$

Gaussian model (Fig. III.4-3):

$$\gamma(h) = C_0 + C \left[1 - \exp\left(-\frac{3h^2}{a^2} \right) \right] \tag{III.4-9}$$

The next step, the fitting of a smooth curve through the calculated values in order to express the semivariance by a mathematical formula, is crucial. By iterative changing of the lag distance, in order to optimize the compilation of the semivariogram expressed in C_0, C and a, an optimum fit will be achieved. It should be noticed that there must be sufficient lag classes to obtain a reasonable estimate of the semivariogram.

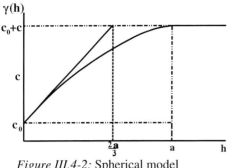

Figure III.4-2: Spherical model

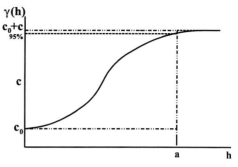

Figure III.4-3: Gaussian model

5. Kriging

By "Kriging Technique" we denote a class of interpolators based on a stochastic approach. There is a large array of Kriging techniques, sometimes obscured by cumbersome notations and mathematics. All these different Kriging techniques have a very simple and common line: they are all regression techniques and differ only by the particular set of functions of the data which are combined into an estimator. A unique advantage of the Kriging interpolation method is its ability to quantify the reliability of prediction, to provide an estimation plus a confidence interval.

5.1 Basic idea of Kriging

Estimation:
The basic idea of the different Kriging methods is the same. If x denotes a point in space and $Z(x)$ is a function of x which is known for the n observation points $x_1, x_2, .., x_n$, we look for an estimate $Z^*(x_0)$ of $Z(x_0)$ in a non-measured location of the form:

$$Z^*(x_0) = \sum_{i=1}^{n} \lambda_i \cdot Z(x_i) \qquad \text{(III.4-10)}$$

a linear combination of the observed values, supplied with a weight λ_i for each observation point.

Estimation error:
The difference between an estimated value $Z^*(x_0)$ and the true value $Z(x_0)$ for any set of weights is called the estimation error ϵ_{est}. If there is no trend (i.e., if the stationary hypothesis holds), $Z^*(x_0)$ is an unbiased estimation of $Z(x_0)$ and the expected value of the difference between the estimated value and the true value will be zero:

$$E\left[Z^*(x_0) - Z(x_0)\right] = 0 \quad \Longleftrightarrow \quad E\left[Z^*(x_0)\right] = E\left[Z(x_0)\right] \qquad \text{(III.4-11)}$$

Error variance:
The error variance ϵ_{est} (also called the estimation variance) of this difference is not zero: it is the mean of the squared difference. In the general case, with the estimated point being a linear combination of values 1, 2, ... , n, Eq. (III.4-10), the error variance, is as follows:

$$\sigma^2 = E\left[Z^*(x_0) - Z(x_0)\right]^2$$

$$= E\left[\left(\sum_{i=1}^{n} \lambda_i\, Z(x_i)\right)\left(\sum_{j=1}^{n} \lambda_j Z(x_j)\right)\right] - 2E\left[\sum_{i=1}^{n} \lambda_i Z(x_i) Z(x_0)\right] + E\left[Z(x_0)^2\right] \qquad \text{(III.4-12)}$$

$$= \sum_{i=1}^{n}\sum_{j=1}^{n} \lambda_i \lambda_j\, E\left[Z(x_i)\, Z(x_j)\right] - 2\sum_{i=1}^{n} \lambda_i\, E\left[Z(x_i) Z(x_0)\right] + E\left[Z(x_0)^2\right]$$

Each observation will contribute a different proportion to the total estimation variance of $Z(x_0)$. There are an infinite number of ways in which the weights can be allocated, and each will produce a different estimation variance. Among these at least one combination of weights must produce a minimum estimation variance. It is this combination which Kriging looks for. If the covariance function, the semivariance or the covariance is known, the weights λ_i can be calculated, as will be explained in 5.2.

5.2 Ordinary Kriging

Ordinary Kriging is the most utilized type of Kriging. It can be used when the variable is stationary and the covariance is known, in contrary the mean is unknown. Ordinary Kriging refers to the following model assumption:

$$Z(x) = m + \alpha(x) \qquad \text{(III.4-13)}$$

where m is the unknown stationary mean. The expected value of the error at any particular location is often referred to as the bias. Implementing the unbiasedness condition (Eq. (III.4-11)) on the linear estimation function (Eq. (III.4-10)) we can write:

$$E\left[\sum_{i=1}^{n} \lambda_i \cdot Z(x_i)\right] = E\left[Z(x_0)\right] = m \qquad \text{(III.4-14)}$$

This yields the first constraint:

$$\sum_{i=1}^{n} \lambda_i = 1 \qquad \text{(III.4-15)}$$

Unbiasedness is guaranteed because the coefficients sum up to 1.

The second constraint is that the weights λ_i have to be solved with a minimal error variance (Eq. (III.4-12)). To find the set of weights that will give the minimum mean square error, the technique of Lagrangian multipliers is used.

The optimal parameters satisfy:

$$\sum_{j=1}^{n} \lambda_j \gamma(x_i, x_j) + \mu = \gamma(x_0, x_i) \qquad i = 1,2,....,n$$

$$\qquad \text{(III.4-16)}$$

$$\sum_{i=1}^{n} \lambda_i = 1$$

where

$\gamma(x_i, x_j)$ = the semivariance of Z between the sampling points,
$\gamma(x_0, x_i)$ = the semivariance between the sampling points x_i and the estimated point x_0.

The reliability of interpolation (e.g. error variance) is found with the solutions of the weights:

$$\sigma_k^2(x_0) = E\left[(Z^*(x_0)-Z(x_0))^2\right] = \sum_{j=1}^{n} \lambda_j\, \gamma(x_0,x_j) + \mu \qquad\qquad\qquad\text{(III.4-17)}$$

The estimation error variance σ_k^2 can be regarded as depending exclusively on the number and the localization of the measurement locations. Therefore, σ_k^2 is an efficient tool for solving network optimization problems such as the optimal choice of measurement locations. It must be emphasized that σ_k^2 is not the variance of the real spatial estimation error but a modelled error. σ_k^2 provides a theoretical measure of the relative accuracies of the various estimates.

5.3 Co-Kriging

A more recent theory derived from the Kriging method is co-Kriging. This method uses one or more secondary variables which are usually spatially cross-correlated with the primary variable and thus contain useful information about the primary variable. The spatial cross-correlation between the variables is described by a cross-variogram which is almost similar to the semivariogram. A great advantage of co-Kriging is the fact that the good cross-correlation between the variables can be used to estimate less intensively measured variables with the use of one or more "cheaper" intensively measured variables.

6. Example

In the centre of the island of Cebu, the Philippines, a rain gauge network has been set up comprising 17 stations in an area of about 35 km × 30 km. For the period 1981-1990 all stations have a continuous time series. The mean annual precipitation values during that decade for the 17 stations are given in Tab. III.4-1.

Next, the covariation is analysed using semivariances calculated with Eq. (III.4-7). Through the experimental semivariogram a mathematical model must be fitted. A spherical semi-variogram model has been used, the parameters of which are given in Tab. III.4-2. The parameters are estimated by fitting the semivariogram model graphically through the experimental semi-variogram (Fig. III.4-4).

Table III.4-1: Mean yearly precipitation data

Station	Mean [mm]	Station	Mean [mm]
Adlaon	1666	Lahug Airport	1196
Biga	1329	Lusaran	1146
Bonbon	1514	Mactan Airport	1296
Bucaue	1695	Maribago	1300
Cambinocot	1412	RCPI	1589
Camp-7 BFD	1534	Sinsin	1636
Carmen	1399	Tabunan	1478
Das/UG	1345	Talamban	1371
Estancia	1217		

With the estimated model for spatial variation and the observed yearly data values Kriging interpolations can now be made of Central Cebu. To that end, 625 estimates were obtained on a 25 × 25 grid (2046 m × 1311 m). Fig. III.4-5 is a contour diagram of the 625 estimates generated by Ordinary Kriging.

Together with the Kriging interpolations a variance plot can be made; this consists of the Kriging standard deviations, a accuracy measure of the estimates. It shows where the Kriging map is reliable and where it must be interpreted with caution. Of course the standard deviation at the measuring points is equal to zero, the observed data are assumed to be the true values without measuring errors.

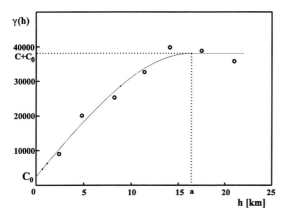

Table III.4-2: Semivariogram

	Constants
C_0	2500 mm^2
C	35500 mm^2
a	16300 m
h	3200 m

Figure III.4-4: Semivariogram

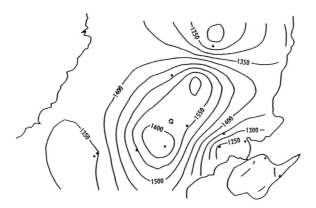

Figure III.4-5: Kriging interpolations, precipitation in [mm].

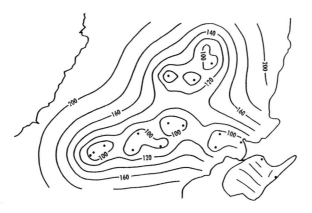

Figure III.4-6: Standard deviation of the estimated values in [mm].

Fig. III.4-6 shows the contours of the Kriging variance. The western part of the area, without data aquisition stations, is less trustworthy than the well-sampled Central and Eastern part, resulting in an estimation standard deviation of more than 200 mm.

Questions

1. What is the difference between the semivariance, the variance and the covariance?

2. Name 5 phenomena which can be treated as random functions.

3. When should a Gaussian semivariogram model be used (in terms of spatial scales)?

4. Name some advantages of the Kriging interpolation techniques over simpler interpolation techniques like Inverse Distance or Thiessen.

5. What is the role of variogram analysis in the Kriging interpolation techniques?

References

Bras, R.L., Rodriguez-Iturbe, I.: Random Functions and Hydrology. Addison-Wesley Publishing Company, 1985

Chua, S.H., Bras, R.L.: Optimal estimators of mean areal precipitation in regions of orographic influence. Journal of Hydrology, Vol. 57, No 1/2, May 1982

Clark, I.: Practical Geostatistics. Applied Science Publication, Essex, England, 169, 1979

Cressie, N.A.C.: Statistics for spatial data.

Delhomme, J.P.: Kriging in the hydrosciences. Advances in Water Resources, Vol. 1, No. 5, 1978

Dijk, M.J. van, Kappel, R.R. van: Optimalization of the Cebu rainfall network. M.Sc. Thesis, Hydrology Section, Delft University of Technology, 1992

Dijk, M.J. van, Rientjes, T.H.M.: Geostatistics and hydrology: part 1, Spatial and temporal variability; part 2, Estimation techniques. Publication of the Hydrology Section, Delft University of Technology, 1994

Guarascio, M., David, M., Huigbregts, C.: Advanced geostatistics in the Mining Industry. Nato Adv. Stud. Inst. Ser., D. Reidel, Dordrecht-Holland, 1976

Isaaks, E.H., Srivastava, R.M.: An introduction to applied Geostatistics. Oxford University Press, 1989

Javier Samper Calvete, F., Carrera, J.: Geoestadistica - Aplicaciones a la hidrología subterránea. Centro Internacional de Métodos Numéricos en Ingeniería, Barcelona, 1990

Kitanidis, P.K.: Geostatistics. In: Maidment (ed.): Handbook of Hydrology. McGraw-Hill, 1992

III.5 Interpolation Schemes

M.J. van Dijk, T.H.M. Rientjes, R.H. Boekelman, Delft

Goal:

This chapter shows how to estimate values of a given variable at any point in a specified area when measured values of that variable are available at a finite number of points in that area.

Summary:

In spatial sciences a linear estimator is often used when values at unmeasured points are estimated. This type of estimator is based on the assumption that data measured near the location are likely to be in better correspondence with the estimated value than data obtained further away. Measured data close to the estimated location receive a higher weight than measured data at a greater distance. An overview of often applied methods is given.

Keywords:

Interpolation, nearest neighbour method, Thiessen, distance weighting schemes, spline, fitting, Kriging, trend surface analysis

1. Introduction

For many spatial studies, distributions of phenomena have to be defined in real space, very often at (ir)regular space and time intervals. To define values at unmeasured locations, estimation techniques of interpolation and/or extrapolation are used.

An estimation technique can be used where a sample value is expected to be affected by its position and its physiographic relationship with its neighbours. It may be helpful to have a smoothed map to indicate the broad features of the data as well as an interpolated map for prediction. Techniques used for such estimations are called "interpolation schemes". In other words, interpolation is the process of transferring information from measured points into space. A large variety of (sophisticated) methods are available when interpolating randomly scattered data points in an area onto a regular grid. One must realize that all interpolation schemes make assumptions with regard to the possible behaviour of the phenomenon in question.

The most simple fitting procedure is the visual one: measured values of the phenomenon are plotted on a graph and a curve is drawn by hand through, or close to, the plotted points. Before computers became available this was the most widely used method. Although the method is somewhat subjective, the approximation will be reasonable when carried out by experienced scientists. The main advantage is that certain specific and local phenomena, which cannot be expressed mathematically, can easily be taken into account. Therefore, this method will always continue to be practised. In the following, more objective types of interpolation schemes will be discussed.

There are basically two types of interpolation schemes: local estimation techniques where only the known points in a restricted neighbourhood are used; and global estimation techniques where the entire set of points is used, usually by means of least-squares regression. Both techniques will be discussed in Sections 2 and 3, respectively.

2. Local methods

Local estimation techniques estimate the value at an unmeasured location, based on the known values within the small neighbourhood of this location. In Fig. III.5-1, A is the unmeasured location.

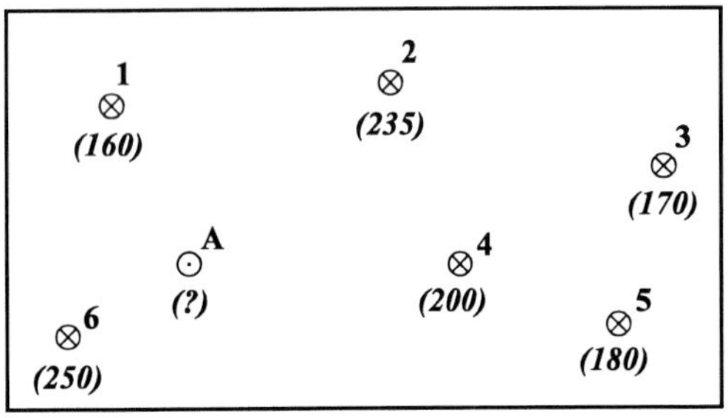

Figure III.5-1: Sampling situation

Table III.5-1: Values

Nr	X-coord	Y-coord	Z(x)
1	4	12	160
2	17	13	235
3	29	10	170
4	20	5	200
5	27	3	180
6	2	2	250
A	7	5	?

In a general case, suppose there are n measurements with values $Z(x_1)$, $Z(x_2)$, $Z(x_3)$, ... ,$Z(x_n)$. From these measurements we can form a "linear" type of estimator, that is a weighted average. This estimator is denoted by $Z^*(x_0)$ and is equal to:

$$Z^*(x_0) = \lambda_1 \cdot Z(x_1) + \lambda_2 \cdot Z(x_2) + \lambda_3 \cdot Z(x_3) + ... + \lambda_n \cdot Z(x_n) = \sum_{i=1}^{n} \lambda_i \cdot Z(x_i) \qquad \text{(III.5-1)}$$

where $\lambda_1, \lambda_2, \lambda_3, \ldots, \lambda_n$ = weights assigned to each measuring point.

A whole range of methods have been developed to decide on:
   - the method by which the "weight" has to be composed
   - the effect each weight should have on the interpolation scheme.

Such methods are mostly based on the distance of the sample to the point being estimated. The underlying assumption of these techniques is that data measured near the point of interpolation are likely to be in better correspondence with the value to be estimated than with other data. The various methods may be separated into simple and sophisticated ones, some of which will be reviewed.

2.1 Simple local estimation methods

The nearest neighbour method:
One of the first estimation methods was published by Thiessen in 1911. The idea behind this method is that the "best" information about an unvisited point can be gathered from the data point nearest to it. This method represents the simplest possible interpolation scheme and, due to its simplicity, is still widely used in hydrology for average rainfall estimation. The boundaries in a Thiessen polygon network are formed by the perpendicular bisectors of the lines joining adjacent measurement points. The region is divided in a way that is totally determined by the configuration of the data points. This method may be thought of as a weighted average with the closest point set to one, and all other weights, to zero. If used as a point value interpolation method, it gives a stepped and not a continuous varying surface.
   Let us consider the set-up of Fig. III.5-1, a Thiessen polygon can be drawn (Fig. III.5-2). The value of A can easily be estimated: it is given the same value as point number 6 (250).

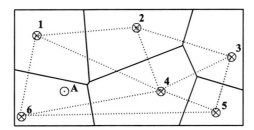

Figure III.5-2: Thiessen polygon

Distance weighting schemes:
The values at the unmeasured points are estimated by a weighted mean based on measured values of different points within a certain area. The weights assigned to the measured data points depend on a distance function. Delfiner and Delhomme made the following comment with regard to distance weighting schemes: "Clearly, no general rule can be derived from experiment on particular data and point configurations. Consequently, the choice of a distance weighting function is more or less a matter of personal belief, of tradition or of confidence in the device of influential authorities".
   Distance weighting schemes suffer from a certain arbitrariness in the selection of interpolation parameters and do not cope well with clustered data. An example of such a method is the inverse distance method, where the weights are inversely proportional to the distances. The weights may be raised to a power, not necessarily an integer value, to increase the effect of the weighting function. Inverse distance squared is most commonly used:

$$Z^*(x_0) = \sum_{i=1}^{n} \lambda_i \cdot Z(x_i) \qquad \lambda_i = \frac{\dfrac{1}{D_i^p}}{\displaystyle\sum_{j=1}^{n} \dfrac{1}{D_j^p}} \qquad i = 1,2,\ldots,n \qquad\qquad \text{(III.5-2)}$$

where D_i, (D_j) = distance between point x_i (x_j) and the point to be estimated
              p = power.

Let us consider again the example given in Fig. III.5-1. The estimated value $Z^*(x_0)$ can be calculated with an inverse distance function (Eq. (III.5-2)) if the distances between the measuring points 1 to 6 and point A are known (Fig. III.5-3).

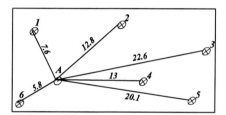

Figure III.5-3: Distances between points

The weights for the measuring points applying different power functions, together with the estimated value, are shown in Tab. III.5-2.

Table III.5-2: Inverse distance weights λ_i for different distance power functions

number	value	p=1	p=2	p=3
1	210	0.238	0.273	0.267
2	225	0.141	0.096	0.056
3	170	0.080	0.031	0.010
4	200	0.139	0.093	0.053
5	180	0.090	0.039	0.014
6	250	0.312	0.468	0.600
A	estimated value	217.3	227.8	233.5

This table shows that the higher the power of the distance weight, the faster the decline in weight (influence) will be and therefore, the less the effect of points further away will be on the estimation. For distance weighting powers higher than 3, the estimate of A will be in closer correspondence with the value of the nearest point.

2.2 Sophisticated local interpolation methods

The second group, the sophisticated methods, may all be considered to be contouring methods, among which optimal interpolation techniques are to be found. These techniques are optimal in the sense that they minimize the variance of the interpolation error. This class of interpolation techniques includes the various forms of Kriging. Generally, these techniques do outperform most of the other interpolation techniques.

Spline fitting:
Spline fitting aims to find the best polynomials through the observed points, to satisfy an optimal "'smoothness'" criterion. The coupled polynomials have to be of the same order to draw a smooth line through the points. This means that with splines it is possible to modify a part of the curve without having to recompute the whole.

Kriging technique:
The value at the ungauged point is estimated as a linear combination of n surrounding observed values, with a weight for each station. The weights are determined by minimizing the estimation variance. The technique assumes that the phenomenon under study has a certain local structure, i.e. relation, in the sampling domain. The measurements are used to represent this relationship in a model, which will be used to improve the accuracy of the estimated value. The Kriging technique is discussed in detail in Chapter III.4.

From a scientific point of view, one must be extremely cautious about likely accuracy of any interpolation scheme, including those of great mathematical complexity. This is because the validity of any interpolation scheme has to be seen against the extreme variability caused by the non-linearity of the phenomenon (i.e. the variable) under study.

3. Global methods

Global estimation methods try to find a universal spatial model which incorporates all the measurements. They principally take the long distance aspects of the spatial differences into account. This contrasts with local methods which mainly use local information to calculate estimates at non-measured locations. Although there are some special global methods like "choropleth mapping" and "trend surface analysis", almost all the local methods can be applied globally, i.e. to the entire domain.

3.1 Trend surface analysis

The idea of trend surface analysis is to fit a polynomial surface through the data points by least squares to describe the gradual long-range variations. The polynomials can be of the first, second or even higher order. Those polynomials represent all surfaces of the form:

$$f(X,Y) = \sum_{r+s \leq p} b_{rs} X^r Y^s \qquad \text{(III.5-3)}$$

of which the first three orders are:

$$b_0 \qquad \qquad \textit{flat}$$
$$b_0+b_1 X+b_2 Y \qquad \qquad \textit{linear}$$
$$b_0+b_1 X+b_2 Y+b_3 X^2+b_4 XY+b_5 Y^2 \qquad \textit{quadratic}$$

The integer p is the order of the trend surface; there are $P=(p+1)(p+2)/2$ coefficients of b_i. Trend surfaces can be displayed by estimating the value of Z at all points on a grid. The linear and quadratic fit of the example (Fig. III.5-1 and Tab. III.5-1) are shown in Fig. III.5-4 and III.5-5.

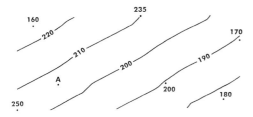
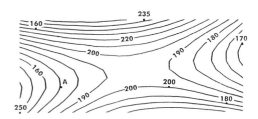

Figure III.5-4: Linear trend surface   Figure III.5-5: Quadratic trend surface

The coefficients b_i are found by minimizing:

$$\sum_{i=1}^{n} \left[Z(x_i) - f(x_i) \right]^2 \quad \rightarrow \quad \min$$

where x_i is the vector notation of (X,Y).

linear $f(x) = 199.2 - 13.9X - 10.3Y$
quadratic $f(x) = 194.2 - 31.8X^2 - 6.2X + 26.7XY - 8.9Y + 37.5Y^2$

The difference between the linear and the quadratic fit is obvious: the range of Z values is much greater for the quadratic fit (quadratic 160-260, linear 180-220), but the surface passes through some data points. When applying second order or higher order polynomials, the surfaces may reach excessively large or small values because trend surfaces are very sensitive to outliers in the observed data. An undesirable feature of higher order trend surfaces is the tendency to wave the edges to fit points in the centre (Fig. III.5-5).

Trend surfaces are smoothing functions: they rarely pass through data points unless the order of the surface is large. In multiple regression it is implicit that the residuals from a regression line or trend surface are mostly to some degree spatial-dependent; in fact, one of the most useful applications of trend surface analysis is to reveal parts of a study area that show the largest deviations from a general trend. The main use of trend surface analysis does not deal with interpolation within a region, but as a way of removing broad features of data prior to using some other local interpolator.

Questions

1. What do you understand by "interpolation" and "extrapolation"?

2. Give the physical explanation of the theory behind interpolation methods

3. Which power would you give to the inverse distance method when interpolating precipitation data in case of:
    - cyclonic precipitation on daily basis,
    - convective precipitation on daily basis,
    - mean monthly data,
    - mean yearly data?

4. What are the differences between local and global interpolation methods?

References

Bras, R.L.: Hydrology - An Introduction to Hydrologic Science. Addison-Wesley Publishing Company, 1990

Bras, R.L., Rodriguez-Iturbe, I.: Random Functions and Hydrology. Addison-Wesley Publishing Company, 1985

Burrough, P.A.: Principles of Geographical Information Systems for Land Resources Assessment. Clarendon Press, Oxford, 1988

Chow, V.T., Maidment, D.R., Mays, L.W.: Applied Hydrology. McGraw-Hill Book Company, 1988

Chua, S.H., Bras, R.L.: Optimal estimators of mean areal precipitation in regions of orographic influence. Journal of Hydrology, Vol. 57, No. 1/2, May 1982

Creutin, J.D., Obled, C.: Objective analyses and mapping techniques for rainfall fields: an objective comparison. Water Resources Research, Vol. 18, No. 2, April 1982

David, M.: Geostatistical ore reserve estimation. Developments in geomathematics. 2. Elsevier Scientific Publishing Company, Amsterdam, 1977

Delhomme, J.P.: Kriging in the hydrosciences. Advances in Water Resources, Vol. 1, No. 5, 1978

Dijk, M.J. van, Kappel, R.R. van: Optimalization of the Cebu rainfall network. M.Sc. Thesis, Hydrology Section, Delft University of Technology, 1992

Dijk, M.J. van, Rientjes, T.H.M.: Geostatistics and hydrology: part 2, Estimation techniques, Publication of the Hydrology Section, Delft University of Technology, 1994

Lebel, T., Bastin, G., Obled, C., Creutin, J.: On the accuracy of areal rainfall estimation: A case study. Water Resources Research, Vol. 23, No. 11, 1987

III.6 Error Modelling in GIS Environment

Bela Márkus, Budapest

Goal:

This chapter focuses on the accuracy aspects of processing data using Geoinformation Systems. Greatest emphasis is given to the methods that are available as a pragmatic response to the wide spectrum of problems of error management.

Summary:

This chapter is founded on Chapter II.6 ('Data Quality in GIS Environment'), where error source identification, error detection and measurement errors of data collection are clearly explained. The structure of the chapter is based on a hierarchy of requirements to evaluate output product quality ([Veregin, 1989]): error propagation modelling, strategies for error management, and strategies for error reduction.

Keywords:

Errors, accuracy, metadata, error band, sliver polygons, error modelling, error propagation, size-probability function, cartographic errors, thematic errors

1. Introduction

This chapter discusses the accuracy aspect of data processing using Geoinformation Systems. Greatest emphasis is given to the methods that are available as a pragmatic response to the wide spectrum of problems of error management.

A GIS is a digital representation of the real word. Any abstraction of reality will lead to discrepancies from the original sources. With traditional cartographical methods many of the problems are visible and the skilled cartographer makes the necessary adjustments and knows how far the information can be relied upon. With a Geoinformation System the equivalent operations (adjustments) are not transparent (black box effect), usually the operators are not very experienced and the problems are more or less invisible. The digital modelling has the potential to dramatically increase both the magnitude and effect of errors in the models. The results may be used for decision making and planning despite the levels of uncertainty that are completely unknown and usually cannot even be estimated. That is why the accuracy analysis is one of the most important problems in the development and application of the system.

At several stages in the modelling process, errors are introduced and propagated. In order to simplify the problem from the point of view of system management, errors can be categorized into three groups (Fig.III.6-1). In the first group are the errors of data acquisition (positional and attribute errors: errors due to careless work, errors in measurement, inadequate sampling techniques, etc.) while building up the model of reality. The second group occurs when the information is derived from the data base (processing errors). At this level further abstractions are applied. At the end of the process user-errors are introduced. These are usually related to the inappropriate use of the final products and therefore, they could be considered the most difficult type of errors to trace (errors of information interpretation). One may conclude that the accuracy of a product is limited to some functions, dependent on two variables, to wit source errors and processing errors respectively. It is obvious, that user-errors can not be taken into account since, first of all, the user himself will be responsible for making sensible choices. If the value of these variables is estimated before the actual integration takes place, the product accuracy may be enhanced. Unfortunately, the ways in which the errors interact are not yet fully understood. Besides, information concerning the accuracy of source data is often lost in the data acquisition stage. As a consequence, it will be extremely difficult to estimate the final accuracy from the separate error sources ([van der Wel, 1991]).

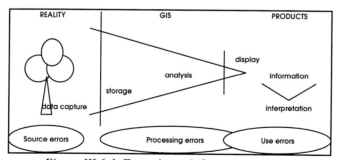

Figure III.6-1: Errors in geoinformation systems

These views are founded on Chapter II.6 ('Data Quality in GIS Environment'), where error source identification, error detection and measurement: errors of data collection are clearly explained. The structure of the chapter is based on a hierarchy of needs for evaluating output product quality ([Veregin, 1989]): error propagation modelling, strategies for error management, and strategies for error reduction.

2. Storing accuracy as metadata

Metadata have been defined as "data about data". Metadata relating to the geometry include the positional accuracy of the surveying method, the precision of the recorded coordinates, the date of the original survey, the coordinate system used, the geodetic datum, any transformations known to have been applied to the data, and especially for data acquired from existing maps, the map series, the scale of the map, the map projection, the positional error of digitizing, the date of the production of the map, etc. (see Chapter III.7).

Moving from the most general to the most specific, there is a logical hierarchy of objects in the spatial data base:

- The set of maps is a collection of data about a particular area. Usually all members of the set of maps are related to a common projection, and covering an identical area. It is the highest level of metadata which includes spatial and attributive data. One essential item of metadata at this level is the contents of the next level in the hierarchy, the data layers.

- The data layers are the individual themes about an area. They are variously referred to the image, the raster, the quad, the coverage, etc. in different systems.

- The object classes refer to the individual features on the thematic map. For instance, in a soil data layer the object classes would be the particular soil map units, while in a topographic layer they would include all the feature types present.

- The spatial objects are the elemental spatial features which GIS are designed to work with. These may be points, lines, areas or pixels, surface data being usually coded as line, points or pixels. Commonly one derivative layer can be the metadata for a source layer.

Table III.6-1: Raw Objects

Object types	Examples
Map set	The overall group of data (the study area)
Data layer	A class in the map set area (elevation, soil data, land use)
Object class	Individual map classes (soil type A, wetland, public buildings)
Spatial object	Objects actually displayed (points, lines, polygons, pixels)

2.1 Accuracy

Accuracy can be tracked from the highest level to the lowest. At the mapset level it refers to the accuracy of the global coordinates to which the mapset is registered, while at the data layer level it refers to the accuracy of any spatial transformations the layers may have undergone to become registered to the mapset area and projection; both being measures of *positional* accuracy. Occasionally all data layers will have the same error reports and so it may be recorded at the mapset level. At the data layer level, attribute data accuracy may also be recorded as the Percentage Correctly Classified, the Overall Accuracy, or the RMSE. These are all different broad measures of accuracy used for different data types including soils, remote sensing/land cover and elevation.

The object class may be associated with a number of different types of accuracy measures. In soil data this may include the types and frequency of map-unit inclusions, and in land cover data it may include the producer and user accuracy. Among the spatial objects in any polygon coverage this will be a unique value for the accuracy of the area of the polygon, while lines may have a unique value for the accuracy of the length of the line. In a classified remotely sensed image the fuzzy memberships of each pixel belonging to one of the land use classes may be recorded. Accuracy of viewed objects is very poorly researched, but may include the degree to which human observers may confuse the visualization/symbolism used for the different object classes, e.g. on a display. Metadata on the viewed spatial objects may give the degree to which the object has been deformed in the process of cartographic representation (the number of points removed, the ratio of the length shown and the actual length, etc.).

Table III.6-2: Accuracy metadata

Object types	Examples
Map set	Coordinates of control points on some wider area reference system
Data layer	Overall accuracy (percent correctly classified, RMSE)
Object class	Producer and user accuracy, map-unit inclusions
Spatial object	Accuracy of classification, positional accuracy, accuracy of area estimates

2.2 Statistics

Statistical reports should perhaps be regarded in the same general framework as the examples of metadata discussed above; going back to the definition of metadata, statistical summaries are information about the data. Indeed, many of the accuracy measures widely recognized as metadata are statistical summaries of the hierarchical level. For example, the correlations and regressions between different data layers are metadata at the level of the mapset, the spatial autocorrelation is at the level of the data layer, the object class or even the spatial object.

Table III.6-3: Statistical metadata

Object types	Example
Map set	Inter-layer correlation matrix
Data layer	Spatial autocorrelation
Object class	Join counts statistics
Spatial object	Local correlation

3. Error propagation modelling

This chapter examines two examples of geoinformation supplied by a GIS: annual soil loss per hectare and crop suitability rating. The first example uses a logic (or heuristic) model, which can be used to determine crop suitability ratings ([Drummond, 1987]). The second example uses a mathematical (arithmetic or functional) model, such as the USLE (Universal Soil Loss Equation). This approach provides a numeric answer which may aid in decision making ([Burrough, 1986]).

3.1 Logic models

Logic models are not based on mathematical formulae and do not lead to error propagation theory, but they do appear popular in GIS decision making environment, because in many application fields they represent the working of an *expert* more closely than mathematical models do. For example, a logic model can be used, to estimate the suitability rating of a land parcel for growing water melons. The suitability rating may be defined in terms of *very suitable, suitable* or *unsuitable*. The rating will have been derived from logical rules such as a land parcel will be very suitable only if the rainfall is greater than 600 mm/year, if the elevation is less than 200 m, and if the soil depth exceeds 100 cm.

Mathematical models comprise rules that can be used to estimate unknown variables from existing observations or other variables. The best estimate of a variable should derive from the weighted mean of several observations. Because observations vary, suitable estimations have a probability distribution which can be characterized by the standard deviation (*SD*). *SD* gives an indication about the quality of the estimate. It is thus possible to link *SD* to the concept of the probability that the value of a variable lies inside a confidence interval around the estimate.

To exploit the linkage indicated in the following section, some assumptions must be made: the estimate is unbiased, observations are uncorrelated, the probability distribution is known, the error related to an observation has a Gaussean distribution and the variance must be finite.

Example: Let us assume that the mean annual temperature for a land parcel to be very suitable for a crop must be between 14 and 22°C and that we have an estimated mean annual temperature (x) of 18°C. Associated with temperature estimates is an *SD* of 2°C. In this case the lower end of the confidence interval we are interested in is 14°C, which is $x - 2 \cdot SD$. The upper interval bound is 22°C, which is $x + 2 \cdot SD$. Briefly the multiplicator of *SD* is equal to 2. Normal probability tables can be used to determine the probability. In this case, there is a 95.4 percent probability that the land parcel is very suitable.

3.2 Mathematical models

This point is concerned with the propagation of errors in GIS operations that can be generally expressed as

$$R = F(Z_1, Z_2, \dots, Z_n, z_1, z_2, \dots, z_n)$$

where the resulting map R is obtained by applying expression F on input maps Z_i and model coefficients z_i. Here we restrict ourselves to situations where the Z_i contain quantitative data and where the expression F is a combination of arithmetical operators.

When error propagation is objective, it is assumed that not all parameters of F are exactly known. To take the presence of errors into account, a common approach is to represent the parameters of F by probability distributions rather than by deterministic quantities. The Z_i are thus modelled as random fields and the z_i as random variables. As a result, the output R will be a random field as well. The aim of an error propagation analysis is to determine the statistical properties of R. Recent publications in the GIS literature show that traditional methods of error propagation can be adopted for this purpose.

As an example, [Burrough, 1986] discusses the application of an empirical equation called USLE. In this mathematical model, annual soil loss (A) is determined

$$A = R \cdot K \cdot L \cdot S \cdot C \cdot P \quad [tonnes/hectare]$$

where R represents rainfall and runoff characteristics,
                      K represents soil erodibility,
                      L is the slope length,
                      S is the slope in percentage,
                      C represents cultivation practices, and
                      P reflects soil protection measures.

[Burrough, 1986] provides values for these six characteristics giving also the accuracy (standard deviation, *SD*) of data:

$$R = 297 \pm 72$$
$$K = 0.1 \pm 0.05$$
$$L = 2.13 \pm 0.045$$
$$S = 1.169 \pm 0.122$$
$$C = 0.5 \pm 0.15$$
$$P = 0.5 \pm 0.1$$

Thus, using the USLE the annual soil loss is 18.5 tons/hectare.

Because the parameters of this mathematical model are uncorrelated, they can be used with traditional error propagation techniques to derive an *SD* for the most probale value for the estimated soil loss. The application of error propagation techniques is outlined below:

$$SD_A^2 = \left(\frac{\partial A}{\partial R} \right)^2 SD_R^2 + \left(\frac{\partial A}{\partial K} \right)^2 SD_K^2 + \dots$$

In order to obtain a value for the *SD* of the most probable value of *A*, *SDs* for the parameters are used. The equation gives 12.4 tons/hectare as result for *SD* of *A*. The user of such a result will realize that the error in all contributing variables was considered to be normally distributed and therefore, there is a 68.2 percent probability that the soil loss will lie between 6.1 and 30.9 tonnes/hectare (i.e. $A \pm SD_A$), and there is a 95.4 percent probability that the soil loss will lie between -6.3 (!!!) and 43.3 tons/hectare (i.e. $A \pm 2 \cdot SD_A$). The parameters of the *F* operation are often correlated. So the following questions are important for applying correctly the error propagation techniques:

- What are the properties of the function?
- Which of the parameters are stochastic, which are deterministic?
- If a parameter is stochastic, what is the type of probability distribution?
- What are the mean values and *SD* of the stochastic parameters?
- What is the correlation between different parameters at the same location?
- What is the correlation between spatial parameters at different locations?
- Is every set of correlated parameters permissible?

4. Models of cartographic errors

4.1 Sliver polygons

Fig. III.6-2 illustrates the map overlay operation for two categorical data layers with a vector-based structure. Data layer 1 represents land use and data layer 2 represents agricultural capabilities. The objective of map overlay in this case might be to delineate areas for urban development that are non-urban and have low agricultural capability. Thus the Boolean AND operator would be applied.

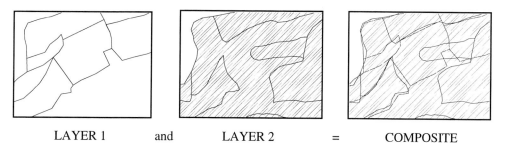

LAYER 1   and   LAYER 2   =   COMPOSITE

Figure III.6-2: Map overlay

Overlapping the two layers it can be realized that the areas of the same physical boundary in the real world does not coincide exactly. The cause of this discrepancy may be assumed, initially, to be cartographic. For example, the data layers may be of different scales, they may be based on different projections, or one of the layers may have been digitized with a higher density of points along the polygon boundaries. The main consequence of applying map overlay to data layers containing cartographic error is the creation of *sliver* (spurious) polygons on the composite map. Slivers are artifacts of the map overlay operation having no real-world interpretation.

The sliver polygon problem has received attention in the error modelling for a long time. [McAlpine, Cook, 1971] observed that the number of composite map polygons (n_C) is related to the number of polygons on the individual layers (n_i) by the equation

$$n_C = \left(\sum_{i=1}^{m} n_i^{1/2} \right)^2$$

where m is the number of layers. This equation shows that n_C tends to rise exponentially as m increases. Empirical tests revealed that small polygons tended to predominate on the composite map and that many of these polygons are indeed spurious.

4.2 Size and probability

Since this model is based on the number of polygons on each layer rather than the positional accuracy of these polygons, polygon size must be used as a measure of validity. [Cook, 1983] devised a model for identifying sliver polygons on the composite map under the assumption that polygon validity is inversely related to polygon size. This relationship may be expressed as the *size-probability function* (*SPF*) for each input layer. Assuming independence, one can calculate the joint *SPF* as the product of the *SPFs* of each layer. A simple hypothetical example is shown in Fig. III.6-3.

The joint *SPF* gives a probabilistic interpretation of the validity of a given composite map polygon-based solely on its size. A simple error reduction strategy might then be to delete those polygons having a probability below a certain threshold value.

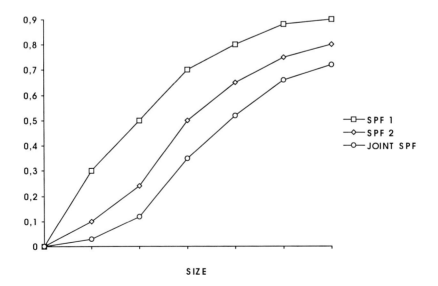

Figure III.6-3: The size probability function (*SPF*)

4.3 Error band

[MacDougall, 1975] approached the sliver problem in terms of positional or horizontal error. He argued that, assuming independence between layers, the total horizontal error on the composite map should be equal to the sum of the total horizontal error on each layer. The total horizontal error (H) of a given layer may be estimated as

$$H = h \cdot L / A$$

where h is an index of the horizontal error, L is the total length of all cartographic lines and A is the area of the layer. This equation gives an estimate of the proportion of the total area of the layer that is unreliable.

The model does not provide a means of identifying sliver polygons, although they conceivably could be identified as those polygons with a minimum width less than twice the horizontal error of the composite map.

A related strategy for reducing error in map overlay employs the *epsilon band* concept as a measure of positional accuracy. Cartographic lines on different layers are assumed to intersect if their epsilon band overlap. The band width or tolerance can be adjusted to accomodate different levels of positional accuracy. This approach can be applied to eliminate slivers having a minimum width less than twice the tolerance. However, detail may be lost if the distance between two different lines is less than the allowed tolerance.

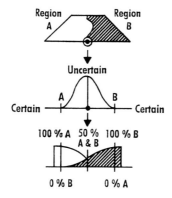

Figure III.6-4: What does position error mean

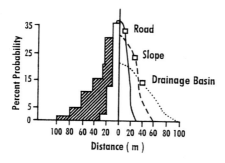

Figure III.6-5: Frequency diagrams to determine error bands

A common problem dealt with by GIS users is to select areas that possess certain properties. For example, let us identify all of the locations within a target area that are within 100 m of a road, have slopes greater than 15% and are contained within a drainage basin. The accuracy of the boundaries of geographic objects is different. For instance, the lines that define the road network likely represent the most accurate data layer. The location of the crest of the watershed is probably more uncertain than the location of the roads. The slope areas map is also probably not quite as accurate as the road, but it may be more accurate than the crest of watersheds. Fig. III.6-5 shows frequency diagrams to illustrate this concept.

5. Models for thematic errors

The preceding models are applicable if it may be assumed that the errors present in the input layers are cartographic in origin. As mentioned previously, errors may also be attributed to the thematic component of these data. The presence of thematic errors demands a much different approach to error modelling for map overlay. Moreover, such models may be differentiated according to their applicability to numerical and categorical data. For reasons of brevity the following discussion focuses on raster data, as the principles applied to error modelling are not necessarily applicable to vector data.

5.1 Numerical data

For numerical data, the accuracy of a given layer might be defined in terms of the deviations between the actual and estimated values associated with each pixel. One approach to error modelling for map overlay might then involve calculation of the error variance of each layer and the error covariance of each pair of layers. The error covariance (S_{ij}) for data layers i and j is defined as

$$S_{ij} = \frac{1}{N} \sum_{n=1}^{N} (z_{ni} - z_{ni}^{0})(z_{nj} - z_{nj}^{0})$$

where N is the number of cells in a data layer, and z_{ni} and z_{ni}^{0} are the actual and estimated values for cell n in layer i, respectively (and similarly for layer j). When $i=j$ the equation defines the error variance of a layer.

This approach facilitates the calculation of the error variance of the composite map as a function of the arithmetic operator applied in map overlay. For example, when m data layers are added, the error variance of the composite map is given by

$$S_{c}^{2} = \sum_{i=1}^{m} \sum_{j=1}^{m} S_{ij}$$

Although the error variance of each data layer is always positive, the error covariance may be either positive or negative. The presence of negative covariances implies that the error variance of composite map may be lower than the error variances of the individual data layers. This observation runs contrary to the notion that the accuracy of the composite map can never be higher than the accuracy of the least accurate data layer.

5.2 Categorical data

For categorical data, errors in the individual data layers might be measured using the *confusion matrix* (contingency table) approach often applied in remote sensing. In this approach the actual and estimated cover classes for a selected sample of cells on a map are cross-tabulated. It is possible to calculate the proportion of cells that are correctly classified and to estimate the accuracy of the entire map based on inferential statistics. A more sophisticated approach utilizes a random-stratified sampling method in which the same number of samples is chosen from each cover class. This has the advantage that minor classes are not under-represented in the sample which makes it possible to calculate the accuracy of individual classes.

As in the case of numerical data, error models for categorical data depend on the operator applied in the map overlay operation. [Newcomer, Szajgin, 1984] examined the effects of logical AND and OR operators and established very useful rules, as follows:

For the AND operator, the composite map portrays the presence of a particular cover class from data layer 1 *and* a particular cover class from data layer 2 (in the case of two layers, Fig. III.6-6). Thus, for a cell to be accurate on the composite map it must be accurate on all individual data layers. Composite map accuracy never can be higher than the accuracy of the least accurate data layer. In general, composite map accuracy tends to decline as the number of data layers increases.

```
00100              00000              00100
00100              00010              00110
10000   AND        00001    >         10001
00000              00000              00000
00010              10010              10010


LAYER 1            LAYER 2            COMPOSITE
```

Figure III.6-6. Errors of the composite map using the AND operator

For the OR operator, in contrast, the composite map portrays the presence of a particular cover class from data layer 1 *or* a particular cover class from data layer 2 (Fig. III.6-7). In this case a cell needs to be accurate on only one layer in order to be accurate on the composite map. The accuracy of the composite map can never fall below that of the most accurate data layer. As the number of data layers increases, composite map accuracy will tend to increase.

```
00100              00000              00000
00100              00010              00000
10000   OR         00001    >         00000
00000              00000              00000
00010              10010              00010

LAYER 1            LAYER 2            COMPOSITE
```

Figure III.6- 7: Errors of the composite map using the OR operator

Geographisches Institut
der Universität Kiel

Questions

1.   "In manual map analysis, precision and accuracy are similar, but in GIS processing, precision frequently exceeds the accuracy of the data". Discuss this statement!

2.   What is meant by metadata, and why is it important in understanding the accuracy of spatial data bases?

3.   Design an experiment to measure the accuracy achieved by an agency in its digitizing operations. How would you measure the accuracy?

4.   Compare the methods available in any GIS to which you have access, to those discussed in this chapter. Does your system offer any significant advantages?

5.   Some GIS processes can be very sensitive to small errors in data. Give examples of such processes, and discuss ways in which the effects of errors can be managed.

References

Burrough, P.A.: Principles of Geographical Information Systems for Land Resources Assessment. Clarendon, Oxford, 1986 (Chapter III.6: 'Error Modelling in GIS Environment')

Drummond, J.A.: Framework for handling error in geographic data manipulation. ITC Journal, No. 1, 1987, p.73-82

Fisher, P.F.: Conveying object-based meta-information. Proceedings, Autocarto 11, Minneapolis, 1993

Goodchild, M.F., Gopal, S. (eds.): The Accuracy of Spatial Data Bases. Taylor & Francis, Basingstoke, 1989 (edited papers from a conference on error in spatial data bases)

MacAlpine, J., Cook, B.: Data reliability from map overlay. Proceedings, Australian and New Zeeland Association for the Advancement of Science, 43rd Congress, 1971

MacDougall, E.: The accuracy of map overlays. Landscape and Planning, No.2, 1975, p.23-30

Newcomer, J., Szjgin, J.: Accumulation of thematic map errors in digital overlay analysis. The American Cartographer, 11, 1984, p.58-62

Veregin, H.: Error modelling for the map overlay operation. In: Goodchild, Gopal (eds.): Accuracy of spatial data bases. Taylor & Francis, London, New York, Philadelphia, 1989, p.3-18

van der Wel, F.J.M.: The integration of remotely sensed data into a GIS: estimation of the accuracy, EGIS Conference, Brussels, 1991, p.1219-1227

III.7 Digitizing Maps

Thomas Vögtle, Klaus-Jürgen Schilling, Karlsruhe

Goal:

This chapter will give an introduction to different methods of digitizing maps. Most data for a GIS are still in analogous form and have to be digitized for further processing in such a computer based system. The most important aspects which have to be taken into account during the digitizing process shall be explained by means of simple examples.

Summary:

Three principle methods of digitizing maps will be explained: manual, semi-automatic and automatic digitizing. Until today the first method is still the mostly used and traditional one, where interactive work of a human operator is necessary. The digitizing process will be explained step by step, beginning at a short hardware description, the preparation phase, special aspects of manual digitizing, up to post-processing and editing of data layers. Semi-automatic digitizing requires fewer interactions of an operator, who will only start and control this process. A totally automatic approach - based on a preceding scanning of the map - uses raster/vector-conversion procedures and pattern recognition methods for a software-driven interpretation of the map contents. At last a short estimation of accuracy will be given, taking the main error components into account.

Keywords:

Data acquisition, digitizing hardware, manual digitizing, automatic digitizing, raster/vector-conversion, data editing, digitizing accuracy.

1. Introduction

Digitizing maps is one of the most common methods for data acquisition and data input to a Geoinformation System (GIS), because a large amount of relevant spatial information is still in analogous form, i.e. printed maps. However, this information is needed in digital form for automated data processing in a computer-based system. A digital database allows complex analyses, modelling and simulation procedures, including the realization of numerous alternatives and variations in an easy way.

The contents of a line map can be defined as equivalent to "analogous" vector information, i.e. the geometry information is related to the well-known graphical primitives *points/point-symbols, lines* and *polygons* (areas), while the semantic meaning is assigned to these graphical primitives by *attributes*. A few examples for the graphical realization of such attributes should be mentioned here: specific colours and width of lines, different patterns inside polygons or well-defined characteristics of cartographic elements (symbols, dash-length etc.) in analogous maps (see Chapter III.1). This corresponds to specific code numbers (feature codes, labels) in digital databases.

All more or less complex objects drawn in a map, e.g. streets, buildings etc., are built up by these graphical primitives and their attributes (see also Chapters II.3 and III.1). To register map information, all these cartographic elements (points, lines, symbols, numbers, colours, patterns etc.) have to be digitized, i.e. a digital database with all the above-mentioned information (geometry, semantics) must be created ([Johannsen, 1978]). Therefore, it is not a direct (primary) data acquisition procedure like e.g. geodetic field survey (Chapter II.2) or geo-related in-situ measurements, but a secondary acquisition method, by which all information is derived from already existing maps.

The main advantages of this method are that it is (mostly) faster and cheaper than primary data acquisition and no field work is necessary, but we have to take into account that only a lower degree of accuracy can be attained (see Section 5).

In general, three different approaches can be distinguished ([Bill, Fritsch, 1991]):
- manual digitizing
- semiautomatic digitizing
- automatic digitizing

2. Manual digitizing

Compared with the other two methods, manual digitizing is a widely used and conventional one, because it has been proved in practice for many years, the systems (hard- and software) have reached a high technical level and are supported by comfortable user interfaces ([Schrader, 1990]). The contents of a map, in this case the semantic meaning of objects and their interrelations, has to be interpreted and the object positions have to be recorded by a human operator. Therefore, the process depends on his/her individual ability and experience in recognizing the map elements. The operator has to move the digitizing device along the map objects by hand and he/she has to control - e.g. on a monitor - the correctness and completeness of digitization. The great advantage of manual digitizing may be the high efficiency and extreme intelligence of the human visual recognition system, but on the other hand a human operator may become tired after some hours of work, this may introduce errors into digitizing results and reduce the effective working time.

2.1 Hardware

The digitizing process can be defined as a conversion of any analogous source (e.g. point, line and polygon elements in maps, texts, numbers etc.) into digital values (e.g., coordinates, alpha-numerical data, etc.) ([Bill, Fritsch, 1991]). For this Analog/Digital-Conversion (A/D-Conv.)

special hardware devices have been developed; concerning maps these are *digitizing tables* (or *digitizers*) and *scanners* resp., which are also used for digitizing images (see Chapter III.8).

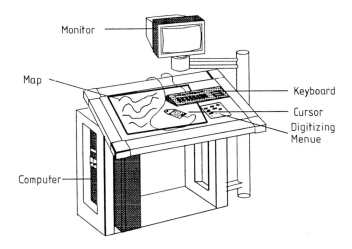

Figure III.7-1: Digitizing table connected to a computer system (from [Bill, Fritsch, 1991])

Fig. III.7-1 shows an example of a digitizer, which consists of mainly 3 basic components:
- table
- movable registration unit (cursor)
- interface to the computer

The table has a plane surface called working area, where the maps can be fixed. A magnifying glass with a cross-hair ensembled in a movable registration unit with different buttons for command or data input is called *cursor*. It has to be centred manually on the points of the maps which have to be digitized. When the operator presses the registration button, the coordinates of the actual cursor position are measured and transferred through the interface to the computer system ([Johannsen, 1978]).

It can be distinguished among several types of such digitizers based on different technical principles to measure the cursor position:
- Orthogonal digitizers based on
  - electrical loops
  - mechanically driven sensors
  - cross-slide systems
- Polar digitizers

Orthogonal digitizers use a rectangular table coordinate system in x- and y-direction. Those with electric loops beneath the table surface are nowadays the most common ones and became a standard solution for manual digitizing. Therefore, they will be explained in more detail. All other systems are of less importance from the practical point of view; detailed explanations can be found particularly in [Johannsen, 1978].

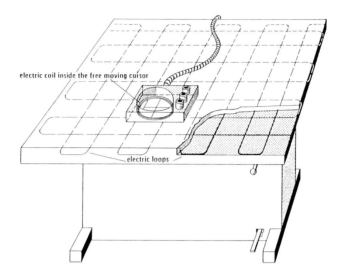

Figure III.7-2: Measurement principle of orthogonal digitizers with electrical loops (from [Johannsen, 1978])

Digitizers with electrical loops contain conductors inserted beneath the surface of the table (Fig. III.7-2). They are regularly distributed like meanders and build up the measuring grid. The cursor, which is the only movable part in this technical construction, consists of a coil with a cross-hair in its centre and a keypad. An alternating current applied to this coil creates, by induction within the conductors, an alternating current, too, which varies when the cursor is moved. These electric variations are continuously registered in electronic counters, so that the actual cursor position can be calculated. The intervals of the conductor lines are placed to be exactly half an inch. These rather large intervals are electronically subdivided, thereby a better resolution can be attained.

The resolution of a digitizer can be defined as the smallest physical measurement unit (e.g. in dots/inch) inside the digitizing area ([Bill, Fritsch, 1991]). Resolution commonly ranges from 0.2 mm for non-cartographic digitizers to approximately 0.025 mm for high precision digitizers. The dimension of the digitizing area mostly lies between DIN A2 (about 0.6 m × 0.4 m) and DIN A0 (about 1.2 m × 0.8 m). The cursor normally contains several buttons for command and data input, e.g. a registration button, a button to finish digitization or another to cancel the last registration. Most digitizer systems offer two alternative registration modes:
- single point registration: Every point to digitize is registered by the operator manually, i.e. he has to push the registration button to record the actual cursor position. The point-distribution and density along a graphical element can be chosen very individually.
- stream mode registration: The registration impulses are given by the system electronics within prefixed intervals. It can be switched between time-mode (constant time intervals e.g. every 0.1 second) and distance-mode (constant length intervals e.g. every 0.3 mm)

Another common method of manual digitizing is based on *scanners*. In this approach analogous maps are treated like other images, e.g. photographs (see Chapter III.8). After scanning the analogous map information[1] a raster matrix is obtained which can be displayed on the monitor. Presuming a suitable digitizing software (e.g. AutoCAD, MicroStation, ARC/INFO, MapInfo, IDRISI etc.) objects can be determined by means of visual interpretation and manual registration by an operator. This method is called *on-screen digitizing*.

2.2 Digitizing principles

2.2.1 Preparation

After scanning the map or fixing it onto the digitizing table a transformation must be calculated because the output data of the measuring system are internal image or table coordinates resp. (see Section 2.1.), which must be transformed to world coordinates (see Chapter II.1). The transformation describes the relation between image/table and map coordinates. This relation is influenced mainly by two factors: the position (translation, rotation) of the map inside the working area and, on the other hand, the slight distortions of the map itself, e.g. small deviations in scale caused by paper shrinking.

If a similarity transformation is performed, two translations (in x- and y-direction), one rotation and one (uniform) scale are calculated (see also Chapters II.1 and III.2):

$$X = a_0 + a_1 x' - b_1 y'$$
$$Y = b_0 + b_1 x' + a_1 y'$$
$$(III.7-1)$$

These equations contain 4 unknown parameters (a_i, b_i); therefore, at least 2 control points (tic-marks) are necessary to solve the equations, i.e. calculating the 4 parameters. Tic-marks are well-defined points on the map, the world coordinates (X, Y) of which are known, e.g. the corner points of a map. These tic-marks have to be digitized (table coordinates x', y'), while the corresponding world coordinates (X, Y) have to be typed (e.g. with keyboard) by the operator.

In most cases more than only the 2 necessary tic-marks are measured; hence the system can perform a least squares adjustment procedure, which gives not only the mean parameter values, but also the accuracy of the transformation. It is important to have such an error control, because a rough error in only one coordinate destroys the validity of all the following digitizations. Because of the characteristics of the paper on which maps are printed (shrinking process), often a non-uniform distortion may appear, which cannot be handled by a similarity transformation. Therefore, most systems offer additionally transformations using polynoms of first, second, or even higher degree. An affine transformation (first degree) can correct - in extension to a similarity transformation - different scales in different directions, which fits much better to the physical shrinking process of paper:

$$X = a_0 + a_1 x' + a_2 y'$$
$$Y = b_0 + b_1 x' + b_2 y'$$
$$(III.7-2)$$

To determine this affine transformation, 6 parameters have to be calculated now; therefore, at least 3 tic-marks are necessary. But as explained above, it is better to digitize more than the minimal number of tic-marks (reliability, accuracy).

---

[1] The scanning process is described in more detail in Section 3 and Chapter III.8

2.2.2 Data registration

There are different approaches to digitize maps ([Bill, Fritsch, 1991]). A relatively simple method is the line approach, where the operator records line by line as they may appear on the map without a special order or without taking object structures and their relationship (topology) into account. Creation of topology - i.e. to build up complex objects like buildings, streets etc. - has to be done by a post-processing module, which has not yet reached a high level of development. Therefore, the most common approach is to digitize polygon structures, i.e. object topology and semantic meaning. Here, a smaller number of points has to be digitized, the system closes polygons (see Section 2.2.3.) and calculates intersections etc. automatically.

The easiest map elements to digitize are those with point character (e.g. top of a hill, point-symbols, measurement locations like meteorology stations etc.) or objects which consist of a sequence of straight lines (e.g. squares, buildings, agricultural fields etc.). Here the centre of these point elements resp. breakpoints are digitized and stored.

Figure III.7-3: Approximation of irregularly shaped curves by straight lines (from [Gottschalk, 1978])

The digitizing of irregularly shaped curves may be more complicated, e.g. topographic contour lines, rivers or traffic lines. Such curves can be approximated by connecting representative points with straight lines (Fig. III.7-3). Thus, the density of digitized points along a curve will be a critical value. It must be high enough that the maximum error Δ (difference between original curve and the straight connection line of two adjacent points) does not exceed a predefined limit or threshold (Fig. III.7-3)

$$\Delta < t \qquad\qquad t = \text{threshold} \qquad\qquad\qquad (III.7-3)$$

where t can be defined - for instance - by the cartographical accuracy of a map (e.g. $t = 0.1$ mm). Fig. III.7-4 shows the same curve digitized with different density of registration points.

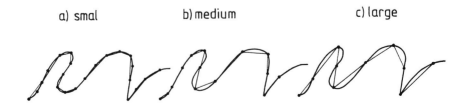

Figure III.7-4: Irregular curve digitized with different density of registration points (from [Johannsen, 1978])

If digitizing is not carried out in *single point mode*, where the operator can choose the position of registration points individually, but in *stream mode*, quite different results may occur (Fig. III.7-5). A disadvantage of distance mode may be that the constant length between digitizing points does not depend on the curvature of the line. Therefore, at locations with higher curvature greater errors than desired will occur, so this method seems to be more suitable for relatively straight curves. An alternative would be to adjust the (constant) distance interval to the maximum curvature, but then too many points would be stored in straight parts of the curve.

a) distance – mode   b) time – mode

Figure III.7-5: Digitizing the same curve with *distance mode* and *time mode* (from [Johannsen, 1978])

Time mode (Fig. III.7-5) is more adequate for the (above defined) condition, that point density should depend on the curvature of the line. The operator normally moves the cursor more slowly at those parts of a line with higher curvature, so automatically a higher point density will be created at such segments. At approximately straight parts of the curve he moves the cursor faster, so a lower point density can be obtained. The maximum error will be more or less constant along the line.

To connect semantic meaning to a digitized point, line or polygon (attribution), a *code-number* (or *feature code, label*) can be transmitted to the system during the digitizing process by typing it on the cursor buttons (if available) or on the keyboard. It has to be taken into account, that a (in most cases hierarchical) structure of code numbering has to be designed *before* starting the digitizing process, containing specific code numbers for all elements to be registered. Subsequent changes or forgotten object classes may cause great problems and a lot of editing work afterwards. Tab. III.7-1 shows a simple example of such structured code numbering. The dimension (length) of these code numbers depends on the number of different classes which have to be distinguished.

Table III.7-1: Structured code numbering for attribution of digitized elements

code	semantics
2 000	**Forest**
2 100	Deciduous Forest
2 110	< 10 years
2 120	> 10 years
2 200	Coniferous Forest
:	:
:	:
2 300	Mixed Forest
3 000	**Cities**
3 100	> 1.000.000 inh.
3 200	> 100.000 inh.
:	:
:	:

The storage of digitized data is organized mostly in relational database systems (see Chapters II.7 and II.8) which are structured by logically connected tables containing the attributive and geometrical information. Nowadays digitizing is normally an online process, i.e. the digitized data are transferred online to the computer system and are drawn simultaneously on the monitor, so that the operator has an immediate control of his actions. He can cancel false registrations and continue the digitizing process.

2.3 Post-processing and editing

Post-processing of digitized data aims at a reduction of redundance and the correction of slight errors or deviations, e.g. small gaps in digitization. In contrast, editing may describe the procedure of correcting gross errors (e.g. confusion of object points in polygon digitizing, beginning a new object without closing the old one or the procedure of changing an existing data set, because the real-world objects have changed (e.g. new buildings, changed course of a street, etc.). While post-processing can be executed almost automatically, editing has to be done mostly manually.

2.3.1 Post-processing

There are four main tasks in post-processing of digitized data: 1) Reduction of redundance (data reduction), 2) Closing open polygons/lines, 3) Corrections by specific constraints, 4) Smoothing of digitized lines.

Reduction of redundance in digitized lines or polygons is the removal of all registered points which contain redundant information. If, for instance, the digitization has been carried out in the stream mode with constant time intervals of e.g. $t = 0.1\ sec$, a lot of points are stored when the cursor is moved slowly, more than are necessary to keep the predefined accuracy. Therefore, such a data reduction is useful to avoid problems in storage space and - quite more important - problems in performance. Considering Eq. (III.7-3) digitized points can be removed iteratively by the system (without interaction), as long as the defined error condition is fulfilled.

The closing function corrects automatically a slight deviation of an end point of a line from another element, on which it should lie; for instance the end point of a closed polygon does not hit exactly the starting point. Fig. III.7-6 shows some examples. Closing algorithms use a small search radius (*snap circle*) to find an adjacent digitized element in the surrounding area of such an end-point (Fig. III.7-6). Its position will be corrected, so both elements coincide. If there are more than two elements inside this snap circle, the system will ask the operator for manual interaction to solve these ambiguities.

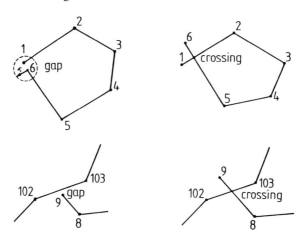

Figure III.7-6: Examples of slight deviations in digitizing corrected by the closing function

Corrections by means of specific constraints can be helpful if geometrical (manmade, artificial) objects have been digitized. For instance, there may be a constraint of rectangularity of a building or a constraint of parallelism of street boundary lines. This knowledge can be used to correct those digitized elements (Fig. III.7-7).

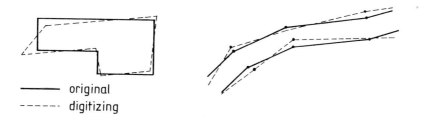

Figure III.7-7: Use of constraints (rectangularity, parallelism) to correct digitized elements

Smoothing of digitized lines should reduce random digitizing errors. It can be performed by calculating a local trend function using two or more neighbouring points on the line (e.g. linear, spline etc.). But it has to be taken into account that this procedure will generalize also the (real) contours of cartographic objects. Smaller details which may be correct and important are eliminated. Additionally, significant displacements may occur at sharp corner points. Therefore, this procedure seems to be more suitable for lower accuracy requirements.

2.3.2 Editing

The editing process, i.e. correction of gross errors or subsequent changes (up-dating) of existing databases, requires some specific interactive software functions. At first there must be an identification of the concerning object by the operator. In most systems this is realized by "clicking" the object with the mouse (on the monitor) or with the cursor (on the map). The system will locate it in the storage. Afterwards that part of the line which should be changed (not always the whole line) will be marked by the operator. The old status can be cancelled and a new digitizing process can be started as described above (Fig. III.7-8).

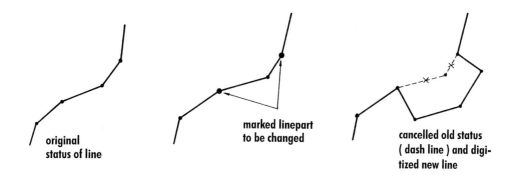

Figure III.7-8: Editing process of existing digitized elements

3. Semiautomatic digitizing

Manual digitizing of maps is a relatively time-consuming procedure and a great charge for a human operator with increasing fatigue and decreasing concentration. Therefore, semiautomatic or automatic methods have been developed to unburden the operator. In semiautomatic solutions the process of digitizing lines or polygons is performed by the system itself. The operator will only start this process by setting the start position of an automatic line-tracking procedure, typing the related feature code (attribute) and controlling the system during operation.

There are two different approaches in semiautomatic digitizing, a hardware-driven method and a software-driven one. The first method mentioned uses special hardware components to perform the tracking process. The analogous original has to be transferred to micro-film, then a CCD sensor is used to track the lines, e.g. system "Lasertrack", LASERSCAN (GB). The costs of such systems are extremely high ([Bill, Fritsch, 1991]). In the software-driven method the analogous original has to be digitized first by a *scanner* (see Section 4.1 and Chapter III.8). Now this image will be displayed on a monitor and the operator has to define the starting point of a line (by mouse) and can proceed as described above. The line-tracking algorithm is here a pure software solution without interactions of specific hardware components. Out of numerous approaches, two currently used methods should be shortly presented here.

3.1 Active contour models (Snakes)

Active contour models, also known as "snakes", are "energy minimizing" splines, guided by shape and radiometry forces to fit to, and thus identify edges in digital images ([Kass et al., 1988]). Snakes combine radiometric comparisons and geometric smoothness constraints to extract object outlines. The operator manually provides the necessary initial edge approximations in the form of a few speed points. Assuming that local energy minima correspond to object boundaries, edges are identified by minimizing an energy function. This function comprises two terms, one radiometric and another geometric. The radiometric energy part uses the first derivative of image intensity (grey values) which is maximal in direction of the gradient at step edge points. The geometric part defines the form of the edge contour by means of a cubic spline function, continuous in first and second derivative (enforcing smoothness). The total energy is minimized through an optimization procedure which forces the snake to approach the actual edge contour ([Fua, Leclerc, 1990]).

3.2 Least square matching (LSM)

This well-known method can be adapted for edge detection in local areas. A typical (representative) edge pattern is introduced as a reference template which has to be subsequently matched with image patches containing actual edge segments. As a result of this process, edge locations are identified in the image as the conjugate positions of the a-priori known template edge positions, i.e. the operator places the cursor approximately to a line and the LSM-algorithm will search other edge elements within the surrounding area ([Grün, Agouris, 1994]).

Both methods need permanent guidance by a human operator, but the charging digitization process can be transmitted to the computer-based system. The semiautomatic approach has no advantage in digitizing maps with a large scale, but if we have separate originals with only one object class respectively - e.g. only contour lines or only waters - this method can be up to ten times faster than manual digitizing ([Lichtner, 1986]).

4. Automatic digitizing

4.1 Scanning maps

The first step in automatic digitizing will be the scanning process of the analogous map. In this case the map is treated like other analogous images (e.g. photographs) to be digitized. The map contents will be divided by a scanner in a matrix of small raster elements (*picture elements, pixel*) with different digital values (grey values, e.g. 8 bit/pixel) - proportional to the original brightness of the map at the related pixel area. In case of coloured maps each pixel contains (normally three) different digital values representing the colour components (e.g. red/green/blue or I/H/S). One of the most significant disadvantages is that the scanning process do not conserve topological relations.

To obtain a separation of different map objects, the scanning process can be restricted to specific map objects by optical filters, i.e. all objects printed in the same colour - e.g. vegetation symbols and forest areas (green), isolines (brown) or water areas (blue) - can be digitized and stored in separate layers during such an repetitive scanning process. This approach will simplify the further very complex procedure of vectorization, creation of topology and pattern recognition (see Section 4.3).

4.2 Raster data pre-processing

The most common characteristic of a line map are black/coloured cartographical objects on a bright background. Therefore, without a great loss of accuracy, the digital raster image can be transferred to a binary image by means of a local threshold operation. The advantage will be a simplification of the present data (Fig. III.7-9) and a reduction of the necessary storage. Basic image processing algorithms should be performed to enhance image quality, e.g. elimination of single pixels introduced by scanner noise, filling of slight gaps or line smoothing procedures ([Lichtner, Illert, 1989]). These digital map information in raster format may be suitable in a GIS for a visual interpretation by combining it with other data sources (e.g. aerial or satellite images) on the monitor. But for a numerical data modelling with several other data layers inside a GIS, structured vector format is often required and a raster-vector conversion will be necessary.

4.3 Raster/vector conversion

If maps are scanned, the raster size should be at least about two times smaller than the smallest line width, e.g. 0.05 mm in cartographic maps ([Lichtner, Illert, 1989]). Therefore, different line widths are covered by different numbers of pixel, a fact which introduces problems in the vectorization process. The first step will be to thin the raster lines to only one pixel line width, i.e. to build up *skeletons* (Fig. III.7-9). In this procedure the topology of the line elements - nodes and starting points - will be created. A tracking algorithm will follow the line elements beginning at the starting points and nodes; the extracted coordinates are stored in vector format ([Lichtner, Illert, 1989]). Now, the geometry and the connections between the map elements are determined, but no semantics were related. Therefore, these vector data are still unstructured.

To build up structured data, i.e. objects with valid descriptions, a *classification* process is necessary by means of *pattern recognition* procedures: objects should be assigned to predefined classes (of the same semantic meaning) using specific characteristics (features, descriptors), which are able to describe the objects sufficiently and allow to distinguish them unambiguously from all other objects.

The map elements to be recognized can be divided into 2 groups ([Lichtner, Illert, 1989]): Isolated objects which can be classified without context information, e.g. letters, numbers and symbols.

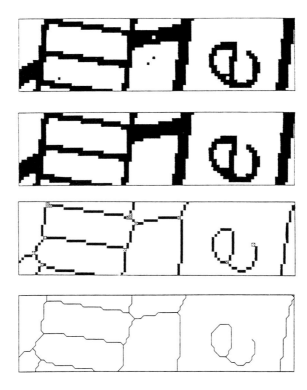

Figure III.7-9: Procedure of automatic digitizing: a) Binary operation after scanning the map, b) Post-processing by noise reduction, c) Creation of skeletons with marked nodes and starting points, d) Vectorization by line tracking algorithm including topology (from [Illert, 1990])

Other elements (complex objects like buildings, streets etc.) cannot be recognized by the line itself, because the whole context information about form, patterns, structures and relations is needed. In the first step isolated elements can be separated easily by means of their dimension compared to the remaining line network. Because of a great variability of those map elements rotation- and dimension-invariant features are used by special pattern recognition algorithms to classify the different signs/patterns. A lot of such recognition methods exist, e.g. grid-/sector-method, Fourier descriptors, structure-based algorithm calculating Levenshtein distance or knowledge-based approaches. A detailed explanation can be found in the related literature (e.g. [Zahn, Roskies, 1972], [Niemann, 1983], [De Simone, 1986], [Holbaek-Hansen et al., 1986], [Illert, 1988], [Kilpeläinen, 1989]).

To recognize more complex objects, line elements and isolated symbols have to be combined to structures of higher degree. A knowledge-based system uses facts and rules about these objects and their interrelations to neighbouring objects for classification (expert system using artificial intelligence techniques), e.g. single letters are combined to a text and will be interpreted - for instance as name of a street - or a pattern of parallel lines inside a rectangular polygon will be interpreted as a building.

At the moment, automatic digitizing is still under investigation because not the whole map contents can be transferred totally automatical into structured digital vector data. The remaining amount of unclassified elements has to be determined by manual post-processing. Therefore, the suitability of this method depends on the graphical complexity of the map. If analogous originals are processed which consist of only one or two object classes (e.g. forest coverage, street map), this method delivers a good recognition quality. On the other hand, automatic digitizing of a common topographic map can not yet satisfy the requirements of the users in practice.

5. Accuracy

The resulting accuracy of a digitizing process is influenced by several factors. At first, the accuracy of the map itself has to be taken into account. Additionally, there will be a contribution of the hardware which is used - either a digitizing table or a scanner. At last, the registration procedure of map elements - by human operators as well as software algorithms - introduces random errors into the final results.

5.1 Map accuracy

The geometrical accuracy of a map is limited by the cartographical and printing procedure. As experience shows, pure drawing accuracy can be assumed to be about 0.1 mm to 0.2 mm. In addition, shrinking processes may have to be taken into account. This effect can exceed the drawing accuracy significantly, but commonly its displacements are corrected by the coordinate transformation mentioned in Section 2.2.1. Therefore, only the accuracy of this transformation (i.e. the control points) is relevant for the digitizing process.

If a map scale of 1:25.000 or smaller is used, cartographic generalization will occur in the graphical elements, e.g. smoothing of object contours or extending of adjacent elements. This effects depend on the scale: the smaller the scale, the greater the amount of generalization, which cannot be reversed. Fig. III.7-10 shows an example of generalization in different scales.

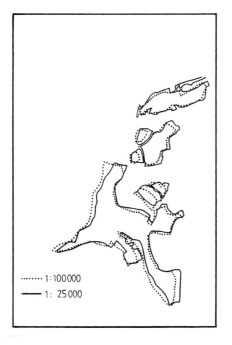
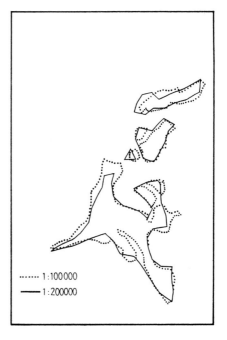

Figure III.7-10: Generalization of the same objects in different scales

It is difficult to estimate the dimension of those displacements, because they depend on numerous factors, e.g. local density of (important) map elements, straightness of the original object contours or the absolute area of objects. Some investigations at our institute ([Scheef, 1994]) with the official topographic map series of Germany may give a rough impression of these displacements: about 25 m (1:50.000), 35 m (1:100.000) up to 80 m (1:200.000) in generalized map elements. For important objects which need a special signature (e.g. highways) the displacement of adjacent objects can be up to 3 times higher.

5.2 Accuracy of manual digitizing

5.2.1 Hardware accuracy

In a digitizing table a specific (finite) geometric resolution is realized, e.g. 0.025 mm (see Section 2.1.). Because of slight manufacturing deviations it can be assumed that the real table accuracy could be 3 to 6 times lower, i.e. 0.07 mm to 0.15 mm ([Johannsen, 1978]). Additionally, in digitizers based on electric loops some special effects have to be taken into account ([Vögtle, 1983]). Test series at our institute have shown that there may be a dependence on the rotation angle of the cursor (Fig. III.7-11), i.e. measuring the same point with different cursor rotations will result in different coordinates.

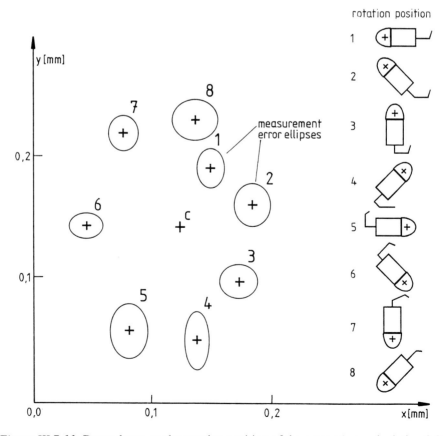

Figure III.7-11: Dependence on the rotation position of the cursor (max. deviation 0.2 mm)

A mismatch (decentering) of the cross-hair and the electric coin (no correct coincidence) may be assumed in this case. Also a degradation of the accuracy induced by metallic objects near the cursor (e.g. wrist watch) up to 0.1 mm could be proved.

For accuracy of scanners see Section 5.3.1.

5.2.2 Measurement accuracy

In manual digitizing measurement accuracy can be derived from the correctness of centring the cursor on a map element by hand and from the approximation quality of a irregular curve. The manual centring accuracy depends on the characteristics of the map elements itself. If single points or point symbols - which are graphically well-defined in a map - have to be digitized, an accuracy of about 0.03 mm can be obtained. A slightly lower accuracy can be assumed if line elements are digitized: On one hand the cursor normally will be moved while digitizing, on the other hand it is more difficult for a human operator to centre the cursor on a line element than on a point.

The quality of approximation of irregular curves by straight line elements was already discussed in Section 2.2.2., where the operator has to estimate the local digitizing distance for each curvature to fulfill the predefined condition Eq. (III.7-3).

5.3 Accuracy of automatic digitizing

5.3.1 Hardware accuracy

The accuracy of scanners depends on the technical construction (see Chapter III.8). The geometrical resolution of a scanner is not identical with its accuracy, which is often significantly lower. In literature different values for the accuracy of the scanning process can be found, e.g. 0.25 mm ... 0.005 mm ([Bill, Fritsch, 1991]) or about 0.5 mm / 1 m ([Hofer-Alfeis, 1989]).

5.3.2. Software accuracy

Line tracking algorithms can yield a theoretical accuracy of about 0.5 to 1 pixel. If the influence of noise (see Section 4.2.) is taken into account, the values increase to 1 to 3 pixel, i.e. approx. 0.03 mm to 0.08 mm (800 dpi). Depending on the pixel size of the scanned map the accuracy may be better than manual digitizing.

Questions

1. What are the advantages (and disadvantages) of manual digitizing ?

2. Which hardware components are necessary for a digitizing table ?

3. How can existing data layers be updated ?

4. What are the main steps of automatic digitizing ?

5. What error sources may occur in digitizing process ?

References

Bill, R., Fritsch, D.: Grundlagen der Geo-Informationssysteme. Band 1 Hardware, Software und Daten, Wichmann Verlag Karlsruhe, 1991, 416 p.

De Simone, M.: Automatic Structuring and Feature Recognition for Large Scale Digital Mapping. Proceedings Auto Carto London, 1986, pp. 86-95

Fua, Leclerc: Model Driven Edge Detection. Machine Vision & Applications, Vol. 3, 1990, pp. 45-56

Gottschalk, H.-J.: Automatic processing of digital cartographic data. International Cartographic Association, Computer-assisted Cartography, papers presented at the seminar 6-11 Nov. 1978, Nairobi, Kenya, pp. 86-112

Grün, Agouris: Linear Feature Extraction by Least Squares Template Matching constrained by internal shape forces. Proceedings ISPRS Commission III, Working Group 2, 1994, pp. 316-323

Hofer-Alfeis, J.: Scannen - und dann ? Stand der Technik der computergestützten Erfassung und Interpretation grafischer Vorlagen. In: Schilcher, Fritsch (ed.): Geo-Informationssysteme. Wichmann Verlag Karlsruhe, 1989, pp. 271-282

Holbaek-Hansen, Braten, Taxt: A General Software System for Supervised Statistical Classification of Symbols. Techn. Report, Norwegian Computing Center, Oslo, 1986

Illert, A.: Automatic Recognition of Texts and Symbols in Scanned Maps. Proceedings EUROCARTO SEVEN, Enschede, The Netherlands, Sept. 1988, ITC Publications, 1988, pp. 32-41

Illert, A.: Automatische Erfassung von Kartenschrift, Symbolen und Grundrißobjekten aus der Deutschen Grundkarte 1 : 5000. Wissenschaftliche Arbeiten der Fachrichtung Vermessungswesen der Universität Hannover (ISSN 0174-1454), Heft Nr. 166, Hannover 1990, 130 p.

Johannsen, T.M.: How to get cartographic data in computer systems. International Cartographic Association, Computer-assisted Cartography, papers presented at the seminar 6-11 Nov. 1978, Nairobi, Kenya, pp. 37-85

Kass, Witkin, Terzopoulos: Snakes: Active Contour Models. International Journal of Computer Vision, Vol. 1, No. 4, 1988, pp. 321-331

Kilpeläinen, T.: Automatic Object Recognition from Vectorized Line Maps. The Royal Institute of Technology, Department of Photogrammetry, Stockholm, 1989

Lichtner, W.: Investigations and Experiences of Automatic Digitization of Maps. Internationales Jahrbuch für Kartographie, Bonn 1986, pp. 101-107

Lichtner, W., Illert, A.: Entwicklungen zur kartographischen Mustererkennung. in: Schilcher, Fritsch (ed.): Geo-Informationssysteme. Wichmann Verlag Karlsruhe, 1989, pp. 283-291

Niemann, H.: Klassifikation von Mustern. Springer Verlag, Berlin-Heidelberg-New-York-Tokyo, 1983

Scheef, P.: Untersuchungen zum Generalisierungs- und Verdrängungsgrad in amtlichen Topographischen Karten unterschiedlicher Maßstäbe (1:25.000 - 1:200.000). Unveröff. Studienarbeit am Institut für Photogrammetrie und Fernerkundung, Universität Karlsruhe, 1994, 30p.

Schrader, B. (ed.): Digitale Leitungsdokumentation. Beiträge und konzeptionelle Vorstellungen des Vermessungswesen. Zeitschrift für Vermessungswesen, Sonderheft 24, 115. Jahrgang, DVW Arbeitskreis 6 Ingenieurvermessung, 1990

Vögtle, T.: Untersuchungen des Digitalisiertisches 'COMPLOT' (Bausch & Lomb). Unveröff. Studie am Institut für Photogrammetrie und Fernerkundung, Universität Karlsruhe, 1983

Zahn, Roskies: Fourier Descriptors for Plane Closed Curves. IEEE Transactions on Computers, Vol. C-21, No. 3, March 1972, pp. 269-281

III.8 Digitizing Imagery

Hans-Peter Bähr, Karsruhe

Goal:

Demonstration of theory and practice of A/D transformation

The great amount of alternatives of hardware solutions are systematically analysed.

Summary:

Shannon's Sampling Theorem is the theoretical base for A/D transformation, starting from band limitation of analogous signals. The question of 'correct' (suitable) pixel size is critically discussed, and the volume of digital data for different image types are shown.

Hardware options are presented in a table and then individually discussed, separating 'high precision-' and 'low cost-' systems.

Keywords:

Sampling theorem, resampling, data volume, drum based systems, CCD based systems, réseau scanner, desk top scanner, macro scanning, Photo-CD, still video cameras

1. Introduction

The *real world* is, there is no doubt about it, analogous. A photographic image, a transformation of this *real world,* is analogous, too. But, as soon as we want to process imagery by digital computers, we must transform the analogous into digital data. Digital imagery obviously implies a hybrid process: the physical process of data acquisition is always done in the analogous domain which is then followed by the digital conversion.

It should be pointed out that in the next sections we shall only deal with the second step. There are, by the way, no technical means which transform the analogous *real world* directly into the digital domain. In a way, all sensors - including those which are called "digital" - use analogous components at the interface to the *real world.* The output of a sensor signal is digital only if an analogous to digital conversion is performed.

2. Theoretical considerations

2.1 Quantization and pixel size

The analogous domain is continuous whereas the digital domain is discrete. Fig. III.8-1 shows a simple digitizing process for a one-dimensional continuous (analogous) curve. Here we have to distinguish between the sampling process on the abscissa and the quantization process on the ordinate. For imagery, the curve would represent an image line. The ordinate corresponds to the grey values and the abscissa to the distance between the individual picture elements (= "pixels").

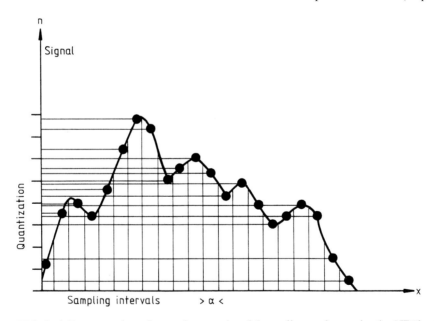

Figure III.8-1: A/D conversion of a continuous signal (sampling and quantization)[Bähr, 1985]

Digitizing in spatial domain is done by sampling the curve in equal steps of distance α. A grey value is attributed to all these "points" and these values are available as ordinal numbers from 1 to infinity (∞). Every pixel has an 8 bit correspondency to values between 1 and 256 or 0 and 255, respectively. There is no problem as far as the quantization is concerned, because expanding the range would not present problem for data storage, though it means a considerable increase of data volume.

This is, however, different for the sampling procedure. The amount of data to be stored is very much sensitive to the sampling interval: dividing the sampling intervals by a factor of 2 would increase the data volume of an image by a factor of 4.

The question of the adequate sampling interval, i.e. for the adequate pixel size, is fundamental for digitizing imagery.

2.2 Band limitation, Nyquist frequency and Shannon's sampling theorem

This section provides a *theoretical* answer to the question about the adequate sampling interval. For this purpose we have to introduce the "band limitation" concept.

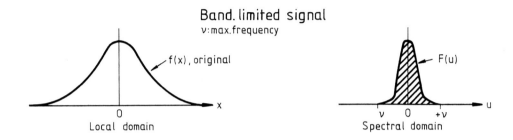

Figure III.8-2: Band limitation ([Bähr, 1985])

Fig. III.8-2 shows a continuous signal on the left in the so-called *local domain*. The signal type corresponds to the signal in Fig. III.8- 1. Any continuous signal may be written as a function of its frequency instead of a function of distance or time (in Fig. III.8-2 we took x for this parameter). The conversion is done by Fourier analysis which will not be discussed here (FFT = Fast Fourier Transform) ([Gonzalez, 1987]).

The curve on the right has the same meaning but it is given as a function of its frequencies. We call this the representation of a function in its *spectral domain*.

Any signal produced by a "physical process" will yield band limited signals. This means, for the spectral domain, that there is a limitation for the existent frequencies. The example shows that the curve is only defined within the range -v to +v, no higher frequencies are involved. It is easily understood that in principle any "physical" signal has such a limitation. For *imagery*, without any doubt, the limitation is its geometric resolution. We have to accept that resolution of a photographic image is not infinitely high but limited by the lens, motion of the sensor, film granularity, development process, spectral range used, atmospheric conditions etc.

The highest frequency that is present in a signal, i.e., the value which defines the band limitation is called *Nyquist Frequency* (v) after the physicist Nyquist. In case we know the Nyquist Frequency of a signal - for instance an image - there is a rule which gives the adequate sampling size, the *Sampling Theorem* ([Lüke, 1990]) (formulated by Shannon in the late forties). The relationship between the sampling interval α and the Nyquist Frequency is:

$$\alpha \leq 1/2 \ v \qquad\qquad\qquad\qquad\qquad\qquad\qquad \text{(III.8-1)}$$

The rule behind this formula may be visualized very simply. Taking an image with a resolution of n lp/mm (line pairs / mm), the equation demands a sampling interval *smaller* than or - for a particular case - *equal* to half a line pair. This means that, in general, per line pair *more* than two samples should be available.

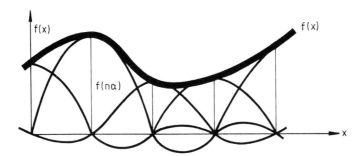

Figure III.8-3: Resampling by *sin(x) / x*

If the pixel size for an image satisfies Eq. (III.8-1), the original curve can be represented without losing any information. The "analogous" image can then be reconstructed by interpolating its digital representation between the discrete sampled values. We call this procedure "resampling". In order to get the correct original function which has been sampled, resampling has to be done by a *sin(x) / x* function (Fig. III.8-3). As this interpolation function will give values even for infinite, every discrete value (every pixel) has theoretically an impact on any other value. In practise, resampling is done using simpler interpolation functions (up to cubic splines) and interpolation is performed in a "window" of about 3×3 pixels only.

3. The "correct" pixel size

The sampling theory is a theoretical rule to determine appropriate pixel sizes. However, in practise, this has to be discussed again a bit more critically and from another viewpoint. In the preceding section, we saw that all "physical" signals are band limited. Starting from this value (the Nyquist Frequency ν), we may simply apply the sampling theorem and get the correct pixel size. For practical considerations, however, the computed theoretical pixel size may be too small. "Too small" means that we would store much more data than necessary for the respective application. This is not economic.

On the other hand we must point out that an "ideal" straight line is not band limited: two adjacent neighbouring points on that line will always give room for another point. This example shows that resolution or band limitation for an image is not equal for the whole area, but depends on the signals ("features shown"), too. The hardware performance of the imaging system, e.g. lens, film etc. may allow a homogenous resolution, but the signals that compose the image do not necessarily correspond to this resolution ([Bähr, 1992]).

Consequently, there are more *pragmatic* views to define a "correct" pixel size than applying sampling theory. One approach takes into account the mechanical limitations of a scanner. For drum scanners, it has been shown ([Ehlers, 1984]) that a 20 μm pixel size is the limit of useful resolution in case of colour films; smaller pixels, if applied, add noise instead of signal only. In these cases, it makes no sense to use pixel sizes smaller than 20 μm (colour films) or ca. 5 μm (b/w films), no matter what the resolution of the scanned image would be.

Another viewpoint starts from the individual application, for example digital orthophoto production. If we want to produce a result comparable to conventional analogous processing, we may use a pixel size of 50 μm. This pixel size corresponds to a raster image done by photoprint. "Photoprint" means rastering of the original half-tone image. This raster corresponds in general to a pixel size of 50 μm in digital imagery.

Figure III.8-4: a) Original   b) Digitalization with varying pixel sizes from 0.05mm to 2mm

As the frequencies are not identical in the whole image (dependent on the image content), it may not be useful to digitize the whole image by the same pixel size. The effect of this approach may be not really "visible" because image quality should not suffer from this procedure. The procedure of varying pixel size is shown in Fig. III.8-4. Fig. III.8-4a presents the original, Fig. III.8-4b the procedure of digitalization by extremely varying pixel sizes from 0.05 mm to 2 mm, and finally Fig. III.8-4c provides a visualization of these pixel sizes in vector form.

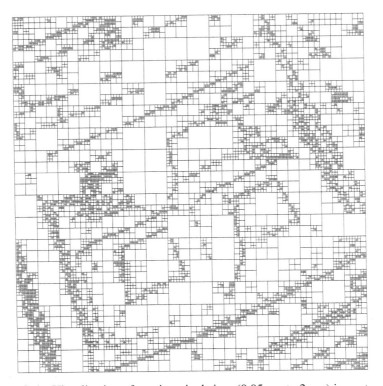

Figure III.8-4c: Visualization of varying pixel sizes (0.05mm to 2mm) in vector format

Tab. III.8-1 gives an overview for the data volume if the sampling interval is put to *one third* of v. This is, of course, a bit finer than the Shannon Theorem would require. The LANDSAT-MSS and TM scenes have been put into that table in order to compare the data volume with the conventional photogrammetric imagery. Two types of photogrammetric cameras with different resolution characteristics are listed. 80 lp/mm may be obtained when measuring in the centre of an image if forward motion compensation is provided. In this case we see that the enormous amount of data of more than 3 Gigabyte will be necessary to store a single frame. We have to say that the resulting pixel size of 4 µm is extremely small and that 80 lp/mm (the corresponding value for the resolution) is a very high one.

Finally Fig. III.8-5 shows a diagram where the data volume of a photogrammetric image in *23 cm × 23 cm* is presented as a function of pixel size.

Table III.8-1: Number of data (pixels) for selected sensor systems ($\alpha = 1/3\ v$)

Sensor	RMK 30/23 (convent.)	RMK 30/23 (FMC)	MFK - 6	KFK-1000	Landsat MSS-Scene (50 μm)	Landsat TM-Scene (50 μm)	Hitachi VK-M98E Digital Camera
	1	2	3	4	5	6	7
Image format (cm)	23 x 23	23 x 23	5,5 x 8,1	30 x 30	16 x 16	35 x 35	0.88 x 0.66
Resolution (Lp/mm)	30	80	80	80	6	6	27 (h) 32 (v)
Pixels	$428{,}5*10^6$	$3047*10^6$	$256{,}6*10^6$	$4800*10^6$	$7{,}4*10^6$	$39{,}3*10^6$	$0.22*10^6$

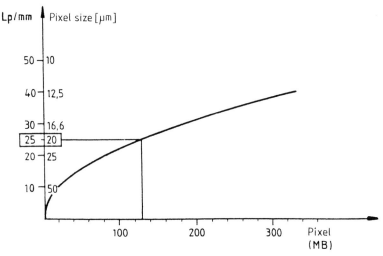

Figure III.8-5: Data volume as a function of pixel size for digitizing a standard photogrammetric image (23 cm × 23 cm)

4. Hardware considerations

Tab. III.8-2 lists different scanner types available for image digitization. In the 1970's there was only one option, the high precision but very expensive *Drum-based Systems*. Besides the "big machines" designed for cartographic and reproduction purposes like the HELL instruments, the market was dominated by Optronics. Optronics instruments are still in use in many institutions. By the expanding use of digital image processing in the 1980's, new types of A/D conversion options appeared on the market. They are based on CCD (=charge coupled device) techniques. CCD arrays proved to be very useful for digitizing due to their very good geometric stability, low price and simple integration in hard- and software systems. Products were developed both for the high precision line, (e.g. ZEISS PS) and for the low cost market (*Desk Top Scanner*).

Table III.8-2: Scanner types for image digitization

	High precision	Low Cost
Drum-based Systems	Optronics	
CCD-based Systems	Zeiss PS 1 Rollei Réseau Scanner (RS)	Desk Top Scanners
Camera-based Systems	Macroscanning (UMK-Highscan)	Photo-CD; Still Video Cameras

Finally we should include *Camera-based Systems* in our listing. In this group, sensor and product (for instance photographic camera and image) are no longer separated. These systems are the latest development which started in the 1990's (*Still Video Cameras, Photo-CD* etc.).

The next sections will present the listed scanner types (Tab. III.8-2) in more detail.

4.1 High precision drum-based systems

We take the Optronics as an example (see Fig. III.8-6). The image to be digitized is fixed on a rotating drum; the advancing carriage allows 2-D digitization.

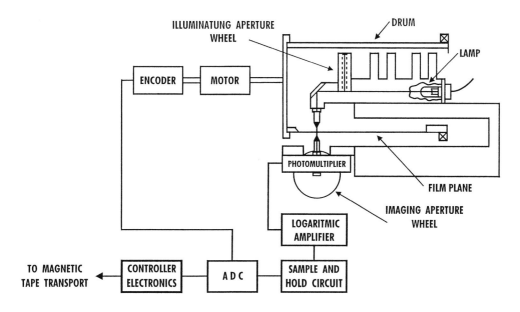

Figure III.8-6: OPTRONICS unit for image scanning (from: [Ehlers, 1984])

Fig. III.8-6 shows very clearly that the drum-based systems require a lot of mechanics. The A/D conversion of the analogous signal of the photomultiplier is done on the very end of the processing sequence. The system allows scanning at resolutions between 12.5 µm to 100 µm. However, as experiments have shown, 12.5 µm is not acceptable due to mechanical restrictions. The Optronics Company offered a large variety of instrumental alternatives including colour scanners. The company finally was absorbed by Intergraph. Beside Optronics there are more companies offering similar instruments (the Japanese/British Joyce-Loebl Scandic, the French Vizicolor and the Canadian MDA Fire 240, for example).

4.2 High precision CCD-based systems

In this group we will show two different solutions: the Zeiss *PS1 Scanner* and the Rolleimetric *Réseau Scanning System*.

Both systems are specially designed for photogrammetric imagery having a 260×260 mm^2 standard format. The Zeiss instrument - in coproduction with Intergraph - offers a basic component for digital orthophoto production, whereas the Rollei design is more suitable for industrial application and architectural photogrammetry.

Fig. III.8-7 shows the principle of the Rolleimetric Réseau Scanner RS 1. It uses a CCD array different from the CCD *line sensor* applied by the Zeiss instrument. The metric fidelity is guaranteed by the *reseau principle*. Consequently the mechanical components are less important than for the Zeiss system. On the other hand, the velocity for scanning a full frame photogrammetric image is low for the Rolleimetric Instrument compared to Zeiss PS 1.

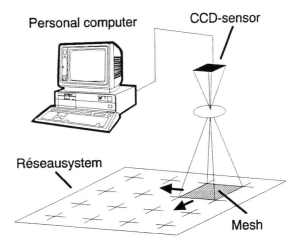

Figure III.8-7: Reseau scanning principle ([Knobloch, Rosenthal, 1992])

Tab. III.8-3 lists the technical specifications of the Zeiss PS1 Photoscan. The system uses a CCD line sensor with 2048 sensor elements. Pixel size ranges from 7.5 to 120 µm. The CCD-based technical realisation guarantees a geometric resolution of 1 µm and high geometric fidelity. There is still another advantage if compared with the drum-based principle, the very fast performance: 7.5 µm scanning of a whole photogrammetric image which yields about 1 Gigabyte takes less than 20 minutes. In the Optronics case (Fig. III.8-6) the same procedure takes hours.

Table III.8-3: Technical specifications of ZEISS PS 1 Photo Scan

Maximum scan area	260 mm x 260 mm
Geometrical resolution	1 µm
Pixel sizes	7.5, 15, 30, 60, 120 µm
Maximum scan rate	7.5 µm – up to 2 megapixels/s 15 µm – up to 1 megapixel/s
Throughput for a 230 x 230 mm monochrome photo	7.5 µm – less than 20 min 15 µm – less than 10 min 120 µm – less than 3 min
Scanning mode	variable
Data quantity from a 230 x 230 mm monochrome photo	approx. 1 GB with 7.5 µm pixels
Scan rotation range	$\pm 10^{grads}$
Swath width	15.36 mm
CCD line sensor	2048 pixels
Radiometric resolution	256 grey levels
Data representation	linear with transmittance or linear with film density
Spectral characteristics	black-and-white or colour with 3 channels
Coloured filters	red, green, blue spectral characteristics see figure on previous page
Output data scanner to disc disc to device	raster: tiled in real time conversion from tiled format to another raster format

4.3 Desk top scanner

Desk top scanners are typical low cost A/D conversion instruments. These instruments were neither designed for scientific nor for technical applications. A characteristic environment is the office, where halftone documents, pictures etc. have to be digitized, stored on tape or disc and used for further access. In this respect Desk Top Scanners can be considered substitutes for microfilm recorders.

The working principle here is again a CCD chip or a CCD line array. Compared to the solutions given in the preceding chapter, Desk Top Scanners do not yield full geometric stability. On the other hand cost are very low: it ranges from 1000 to 10.000 US $ per unit.

[Leberl, 1991] lists 22 scanners available on the market in 1991. Compared to the high precision option there is an enormous variety. Their resolution is very modest because they were not designed for applications in Photogrammetry and Remote Sensing.

However, technical development goes on. A very useful feature for the image processing environment is the ability to scan transparencies and digitizing colour. Desk Top Scanners now allow formats up to DIN A3 (approx. 40 cm × 30 cm) to be scanned.

"Geometric Resolution" is specified in *dots per inch* [dpi]. This notation is originally from printing environment where resolution in general is given in this form. Consequently 100 dots per inch are equal to a pixel size of 250 µm. The resolution offered by Desk Top Scanners, in a range of 85...300 dots per inch, is (in most cases) not adequate for our purposes. This bottleneck may be overcome by magnifying the original imagery before scanning. For instance, when scanning an image which has been magnified before by the factor of 5, 100 dots per inch would be equivalent to a pixel size of 50 µm in the original image.

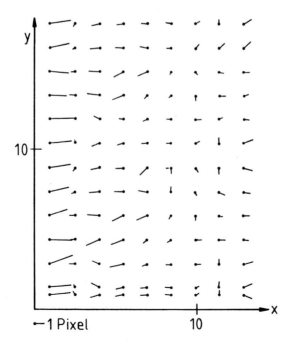

*Figure III.8-8:*Geometric quality of the EPSON GT 6000 Desk Top Scanner (residuals after affine transformation [Ehmig, 1993])

Additional facts restricting geometric stability exist. Fig. III.8-8 shows residuals (i.e. small geometric deviations between a measured point and its adjusted position) at a scanned réseau plate. These residuals (vectors) visualize distortions introduced by the scanning process i.e. by the equipment itself. This figure of residuals demonstrates very clearly the artefacts introduced by the scanning process: the discontinuity in vertical columns may be produced by the vertical movement of the CCD line. The result shown in Fig. III.8-8 is from a study applying digital images to architectural photogrammetry. Here it was necessary to skip the marginal part of the image where distortions are extremely high.

4.4 Photo-CD

At the end of 1992 at the International Photokina Fair in Cologne, the KODAK Photo-CD System was presented. It was soon accepted by the commercial photo market. The Photo-CD System represents a hybrid approach; hybrid because films and conventional photographic cameras are still used for image aquisition. In a second step the analogous images of the (negative) film are transformed digitally and stored on a conventional CD-ROM disc. The capacity of such a Photo-CD disc is 100 images of 18 Megabytes each.

In the meantime, the market offers a wide range of necessary hardware components to produce Photo-CDs. They consist of a scanner, the data manager (which is a high powered image workstation), the Photo-CD writer, a colour printer for hard copies and finally a separate ROM. Up to now, this hardware configuration is typical for the *commercial* environment. It can be expected that in the near future, individual photoenthusiasts will do this sort of digital image processing on their own. This development will have, no doubt, a strong impact on scientific digital image processing, too.

Tab. III.8-4 lists different types of photo-CD presently on the market. In all cases the features are listed for colour images which is the standard material. The four types differ in their resolution. It starts from the 128×192 "Indexprint" to 4096×6144 which yields 72 MB per colour print. The selected resolution during the scanning process depends of course on the application. If the user wants to display the images on a TV screen, 512×768 pixels would be enough according to the TV standard. For the future digital TV, the number of pixels per line and column may be doubled (1024×1536). "Photoquality" is maintained when digitizing by 2048×3072, this would mean 18 MB per 35 mm colour image. This is the standard for a 35 mm slide and produces 100 images per photo-CD.

Table III.8-4: Types and characteristics of Photo-CD's (Photo-CD Newsletter, No. 1, 1993)

	Photo-CD Master	Pro Photo-CD Master	Photo-CD Portfolio	Photo-CD Catalog
Digitizing Film	35 mm	up to 100×125 mm		
Digitizing Prints	x	x	x	x
Resolutions (Pixels)				
128×192 Indexprint	x	x	x	x
256×384 Contact	x	x	x	x
512×768 CRT/TV	x	x	x	x
1024×1536 HDTV	x	x		
2048×3072 18 MB	x	x		
4096×6144 72 MB		x		
Images / CD	100	25 - 100	800	> 3 000

The standard for photoquality CD and 35 mm film is based on a pixel size of 12 μm × 12 μm. This can be accepted as an average value for the highest frequency in an image. If this is not sufficient, there is still higher resolution available which produces 72 MB per image. This, however, can only be obtained by the Pro Photo-CD Master (second column in the table). Here we may scan images up to a format of 100 mm × 125 mm.

At the other extreme of the resolution range, we get 128 × 192 elements per image. This is well suited for archiving and cataloguing applications. More than 3000 images may be stored on one CD-ROM.

What about the geometric fidelity of the Photo-CD system? Up to now, the process seems to be a bit like a "black box" because, in general, there is no access to the processing unit and no specification for the hardware available. The systems are designed for "amateur" camera images only: geometric stability of the digitizing process is not so important because the geometry of the original images is relatively poor. This, however, is not the case if the original image presents a better geometric stability, produced for instance by photogrammetric cameras.

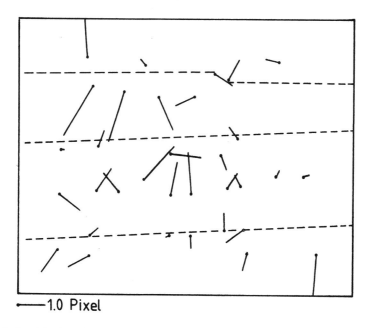

Figure III.8-9: Residuals [pixel] after transformation of the original analogous image to the respective digital Photo-CD representation (analysis performed at the IPF, Karlsruhe Univ.)

Fig. III.8-9 shows the residuals [pixel] after a transformation of 33 control point coordinates measured in an analogous test photo and their corresponding coordinates measured in the same image after digitization on a CD photodisc. The geometric distortions are within the order of magnitude of 1 pixel, i.e. 12 μm. The absolute values obtained here are, however, not so important as the fact that there is obviously a *systematic* behaviour in the residuals, marked by the horizontal lines in the figure. It could be assumed that these 4 areas of the digitized image were shifted vertically against each other during digitization, but a more detailed analysis of this systematic behaviour can not be performed as there is no information about the specific scanning process available.

4.5. Still video cameras

This again is, like the Photo-CD System, a very new development. It uses conventional phototechniques but instead of film, a CCD chip is taken for image acquisition. As an example we take the NIKON 8 008 S. Tab. III.8-5 shows the technical data of this camera combined with a KODAK DCS 200 chip device. The number of 1524 × 1012 sensor elements is relatively small and the sensitive area covers only 14 mm × 9.3 mm which is a considerable reduction in relation to the conventional 36 mm × 24 mm film. In order to get about the same viewing angle of the camera, one has to use an extremely small focal length in the range of superwide angle.

Table III.8-5: DCS200 technical data

camera body:	Nikon 8008s
sensor:	1524 × 1012 full frame CCD, 14 mm × 9.3 mm, black-and-white
frame grabber:	in camera body
storage:	50 images, uncompressed, on harddisk in camera body
interface:	SCSI port
softwarre:	Adobe Photoshop (Mac) / Photostyler (PC)
weight:	1.7 kg
power supply:	AC adaptor/charger
lens mount:	Nikon bajonet

The big advantage and the big difference to conventional "CCD Cameras" is the possibility to digitize and store imagery directly on a 80 MB hard disc. This allows storage of 50 images without data reduction. The system is fully autonomous as no host computer is necessary and A/D transformation is done by the system itself. The price is below 10.000 US $. A *DCS 200* interface allows the transmission of data to other systems.

Still Video systems are purely digital for image acquisition and do not follow the hybrid line like the Photo-CD treated in the previous chapter. The image quality from the photographer's point of view is, of course, not comparable to what a Photo-CD offers, as the resolution is really poor. While it remains in the order of 1.5 MB, it will probably not be accepted by the commercial photomarket. Applications, however, are very interesting for close range photogrammetry where the first publications show high geometric fidelity and very comfortable handling ([Peipe et al., 1994], [Maas et al., 1994]). Fig. III.8-10 shows the Still Video Camera System as it is linked to a computer and open to software.

4.6 Macroscanning

Camera-based systems like Photo-CD or Still Video Cameras take images by means of conventional photographic cameras. This is also true, in a way, for the UMK-HighScan Macrocanning camera. The UMK (Universal-Meßkamera) is a worldwide well known high precision photogrammetric camera for close range applications. It uses a relatively large photo format of 13 cm × 18 cm. Highscan is a modification of this camera offering a camera back which contains four CCD arrays of 739 × 512 sensor elements each. Similar to Still Video Cameras, the system again replaces the conventional film by CCD arrays. The differences compared to Still Video is not only the use of a special camera but also the overall resolution obtained: Macroscanning of the whole format (166 mm × 120 mm) yields 170 MB for each frame. This is *two* orders of magnitude more than for the NIKON system in Tab. III.8-5.

Fig. III.8-11 shows the UMK-Highscan System, both camera body and camera back. A single sensor element is about 11 μm × 11 μm. The geometric fidelity is in the order of ± 1 μm, this means excessive geometric stability. The price for this very special development is very high and justifies applications only for industrial measurements of the highest possible precision.

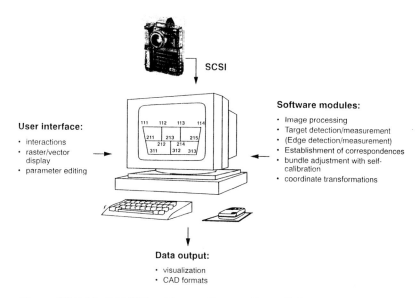

Figure III.8-10: Still Video Camera System (from: [Maas, Kersten, 1994])

Figure III.8-11: UMK HighScan (from: [Goldschmidt, Richter, 1993])

A/D conversion and digital image processing is done outside the UMK-Highscan System, different to Still Video Camera Systems.

5. Conclusion

The "paradigmatic step" in Photogrammetry from analogous to digital ([Ackermann, 1995]) is also reflected in hardware design for A/D conversion. The recent developments like Photo-CD, Still Video and UMK-Highscan all present very different technologies: the "philosophy" behind the respective developments implies different applications. First results are available; it will be very interesting to watch future tendencies.

References

Ackermann, F.: Digitale Photogrammetrie - Ein Paradigma-Sprung. Zeitschrift für Photogrammetrie und Fernerkundung, Heft 3/1995, p. 106

Bähr, H.-P.: Appropriate Pixel Size for Orthophotography. ISPRS Comm.III, Washington D.C., 1992

Bähr, H.-P.: Grundlagen der digitalen Bildverarbeitung. In: Bähr, H.-P.(Hrsg.), Digitale Bildverarbeitung, Anwendung in Photogrammetrie und Fernerkundung (1. Auflage), Herbert Wichmann Verlag, Karlsruhe, 1985

Ehlers, M.: Digitale Bildverarbeitung. Institut für Photogrammetrie und Ingenieurvermessungen, Universität Hannover, Heft 9, 1984, 146 p.

Emig, B.: Untersuchung zur Verwendbarkeit eines Büroscanners und von Standardbildverarbeitungssoftware zur geometrischen Umbildung photogrammetrischer Nahaufnahmen. Unveröff. Diplomarbeit am Institut für Photogrammetrie und Fernerkundung, Universität Karlsruhe, 1993

Goldschmidt, R., Richter, U.: Neue Hardwarekomponenten für die digitale Nahbereichs- photogrammetrie. Zeitschrift für Photogrammetrie und Fernerkundung, Heft 2/1993, Wichmann Verlag, Heidelberg, 1993, S. 71-75

Gonzalez, R.C.: Digital Image Processing. Addison-Wesley Publishing Company, 1987

Kazmierczak, H.: Erfassung und maschinelle Verarbeitung von Bilddaten - Grundlagen und Anwendungen. Springer Verlag, Wien, New York, 1980

Knobloch, M., Rosenthal, T.: MIROS - A new software for Rollei RS1 digital monocomparator. ISPRS Comm.V, Washington D.C., 1992

KODAK: KODAK - Newsletter 12/1992 and 1/1993

Leberl, F.: The Promise of Softcopy Photogrammetry. In: Ebner, Fritsch, Heipke (eds.): Digital Photogrammetric Systems. Wichmann Verlag, 1991

Lüke, H.D.: Signalübertragung - Grundlagen der digitalen und analogen Nachrichtenübertragungssysteme. Springer-Verlag Berlin, New York, 1990

Maas, H.-G., Kersten, T.: Digitale Nahbereichsphotogrammetrie bei der Endmontage im Schiffsbau. Zeitschrift für Photogrammetrie und Fernerkundung 3/1994, p. 96ff.

Bösemann, W., Peipe, J., Scheider, G.-T.: Zur Anwendung von Still Video Kameras in der digitalen Nahbereichsphotogrammetrie. Zeitschrift für Photogrammetrie und Fernerkundung, Heft 3/1994, Wichmann Verlag, Heidelberg, 1994, p. 90-96

III.9 Image Classification (Examples)

Christiane Weber, Strasbourg

Goal:

The whole classification procedure - pre-processing, segmentation, validation - should be explained, based on the theoretical background of Chapter III.3.

A concrete example should give a better understanding of the methodology of the classification process.

Summary:

Different typical treatment schemes (like image improvement, channel combination selection etc.) and the classification procedure (selection of training areas, statistical information processing, validation etc.) are presented by means of an example (periurban zone of Strasbourg). Different alternatives and the results are discussed.

Keywords:

Image classification, classification examples, normalized vegetation index, principal component analysis, training area selection, statistical information, validation.

1. Introduction

As explained in the Chapter III.3 ('Image Classification'), this segmentation principle enables the researcher to associate the knowledge of the field under study and the results of statistical treatments of the reflectance data obtained from the sensor.

Before applying any treatment, it is important to put the methodological approach in its conceptual context. Allocating pixel aggregates to a specific class results from a methodological progression which, on the basis of the knowledge of the characteristics of the landscape under study, combines it with a certain type of coherence with a pre-defined or at least prospective nomenclature. This nomenclature may correspond to natural phenomena:
  - typology of the superficial layers of a desert zone
    or functional ones:
  - typology of land uses (sports facilities, public bodies, garbage dumps etc.)

It is obvious that the association carried out during classification process cannot occur without a good knowledge of both the characteristics of the landscape, in order to be able to refine the results, and the characteristics of the typology used in the framework of the study.

This preliminary knowledge is absolutely necessary; it is composed in particular of the contributions made by other sources of information (maps, diagrams, or even literary descriptions, aerial photographs etc.). This is a precious support for it makes it possible to approach the reality of the field on the basis of both interpretation and communication rules and its physical representation. As stressed by [Clos-Arceduc, 1987]: "photographical (and satellite) recording is constantly biased", and underlines "appearances of what is visible at the surface of the Earth".

The first step consists in creating "a link between appearance and real nature". It is the passage from a representation III.of reality (aerial photography or satellite image) according to reflectance characteristics, to an interpretation of this reality, "recognized", i.e. various characteristics are linked through a recognition process (a catalogue) leading to a hermeneutics of the space under study.

In order to reach that point, it is necessary to be able to isolate and identify spatial objects within an image, according to attributes such as size, shape, colour and aspect of the surface, and the links between them. "The interpretation starts at the point where deductive reasoning is needed to explain the presence and the shapes of the objects and structures within a set" ([Clos-Arceduc, 1987]).

The reading and deciphering necessary for interpretation depend on two basic elements: the scale of the appearance of phenomena and the scale of the image; the spatial resolution of the images used interferes in this case with the researcher's visual sensing abilities ([Bonn, Rochon, 1992]) and the types of treatments applied to improve the legibility of the product (contrast, filtering etc.), but it also interferes with the characteristics of those spatial objects which are more or less detectable. Therefore, in order to reach this space hermeneutics it is necessary to use the spatial and spectral characteristics of an image, and - in addition - a catalogue of the associations needed to identify the available objects.

Several difficulties may arise while treating images; therefore, it is quite tricky - notwith-standing the increasing geometric quality of sensors (SPOT Panchromatic) - to restore the shapes of even important objects, due to the discernability conditions as well as the imaging conditions, the visual angle, hour and time conditions (date of the image). Just as one cannot expect to find homogeneous spatial objects in nature, one cannot expect either to find them in an image. Spectral heterogeneity forms one of the limitations of the interpretation of satellite images. Under similar reflectance characteristics one can find dissimilar objects, just as similar objects may have different reflectances. Moreover the measure on which the analysis is based relies on a certain ground element on which an energy content is registered. The heterogeneity of the surface leads in most cases to the creation of mixels (=mixed picture elements), which are difficult to deal with during classification process.

Therefore, it is necessary to use different treatments judiciously in aiming at an optimum interpretation according to several possible schemes (Fig. III.9-1). Depending on which approach is chosen, preference will be given to :

- either extracting elements according to the spectral characteristics of the image in the first place (1), i.e. a synthesis of the information included in the image followed by an analysis of the spatial organization
- or obtaining the elements of the spatial organization of the landscape by using spatial characteristics of the image, and then identifying them according to the spectral characteristics of the elements which have been extracted.

Raw Data

Channel combination

1   2

Information synthesis   **Spatial Organization of the landscape**
Spectral **Filters**   **Spatial Filtering (texture)**
Neo-Channels (NDVI)   (1st and 2nd rank operators)
Classification   Elements Extraction

Spatial Organization of the landscape   **Information synthesis**
Spatial filters   **Spectral filters**
Extraction of one or several dimensions   **Classification**
of the landscape (green spaces, building density, etc.)   Arithmetic Operations
to be used alone or in combination   Indexes

Figure III.9-1: Schemes for image treatment (according to [de Keersmaecher, 1989])

2. Schemes of treatment and applications

In order to define the methodological progression of a classification process in a better way, an example will be given that will bear on a periurban zone of Strasbourg, which includes forest areas. This example will serve as a basis to understand the different stages of the process, identification of objects according to spectral and spatial characteristics, interpretation according to different treatment schemes (pre- and post-classification treatment) and finally assessment of the results.

Certain types of preliminary treatments make it easier for one to get a first grip of the space under study according, of course, to the topic of the study which ought to be defined as precisely as possible. In the case presented below, the aim is to identify the periurban zones of the Strasbourg agglomeration.

It is important to define the potentially interesting data for the study (among all remote sensing data available), taking into account the data of image acquisition, type of channels and the desirable resolution (according to what was said above and to the scale of the work). Periurban zones are characterized by a strong interweaving of vegetable and mineral surfaces; they are transition zones between an anthropological object (the town) and a more or less natural environment (its surroundings). Therefore, it is necessary to select among the potential satellite data those which make it possible to individualize natural spaces (forests, meadows, cultures, etc.) in the best way. The contrast between these two different types of land use allows the interpretation of the interweaving zones in periurban areas, possibly according to information not obtainable from the image as far as the functionality of these spaces under study is concerned.

2.1 A constrained optimization model

The choice of data corresponds to a succession of conditions to be fulfilled:
- need for chlorophyllous activity, so as to be able to distinguish different types of vegetables, and hence the date of satellite image should be at the end of spring - beginning of summer;
- since the chlorophyllous activity can be identified best in the near-infrared domain, the sensor to be chosen must possess a channel in this spectral band;
- since the aim is to work on periurban zones, a fine spatial resolution will be necessary in order to obtain a better identification of different objects, in particular the infrastructures;
- finally since the object of study is an urban agglomeration, a scale of analysis of 1:25.000 should be sufficient.

Considering this constrained optimization model, a choice has to be made between two systems, namely SPOT and LANDSAT and two sensors, namely HRV and Thematic Mapper (see Chapter II.4). Among the available images, we chose the one from June 28th, 1986, SPOT HRV System.

2.1.1 Characteristics of the image

Spot HRV1 Multispectral data (Fig. III.9-2)
KJ: 50-252; date: 28/6/86;
Size of the extract (560 × 450); location: Strasbourg

Table III.9-1: Statistical data of the SPOT channels

	Minimum	Maximum	Mean	St. deviation
XS1	27	254	54.55	9.66
XS2	12	254	44.65	12.41
XS3	18	246	77.19	23.77

2.1.2 Geometric Correction

It is important to note that before operating any spatial or spectral treatment on images, the geographical dimension of the space under study has to be taken into account. Actually the geometric rectification of images with regard to a geographical reference (i.e. a specific local, national or international projection system, Chapter II.1) can be carried out either *before* treatment (e.g. if satellite images have to be integrated into a GIS) or *after* treatment (e.g. if several scenes are being used, in order to constitute a mosaic).

This rectification can be done in two ways, either the images are transformed according to (mathematical) models introducing sensor and shooting parameters ([Bonn, Rochon, 1992]) and, if available, a *Digital Terrain Model,* or the images are rectified according to a general (coordinate) transformation based on control points without a-priori knowledge of the acquisition system and its parameters (see Chapter III.2).

It is obvious that - according to the resampling algorithm chosen (nearest neighbour, bi-linear or bicubic interpolation, truncated sinusoidal) - this procedure causes more or less a spectral degradation of the raw data used later on in the classification process.

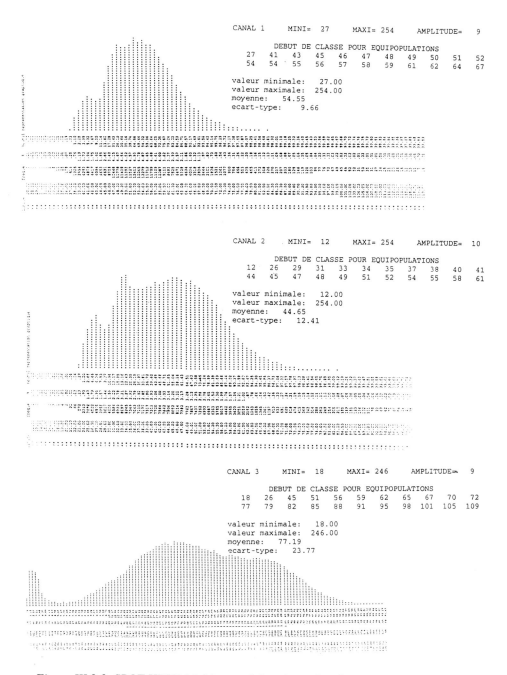

CANAL 1   MINI= 27   MAXI= 254   AMPLITUDE=   9

DEBUT DE CLASSE POUR EQUIPOPULATIONS
27   41   43   45   46   47   48   49   50   51   52
54   54   55   56   57   58   59   61   62   64   67

valeur minimale: 27.00
valeur maximale: 254.00
moyenne: 54.55
ecart-type: 9.66

CANAL 2   MINI= 12   MAXI= 254   AMPLITUDE= 10

DEBUT DE CLASSE POUR EQUIPOPULATIONS
12   26   29   31   33   34   35   37   38   40   41
44   45   47   48   49   51   52   54   55   58   61

valeur minimale: 12.00
valeur maximale: 254.00
moyenne: 44.65
ecart-type: 12.41

CANAL 3   MINI= 18   MAXI= 246   AMPLITUDE=   9

DEBUT DE CLASSE POUR EQUIPOPULATIONS
18   26   45   51   56   59   62   65   67   70   72
77   79   82   85   88   91   95   98   101   105   109

valeur minimale: 18.00
valeur maximale: 246.00
moyenne: 77.19
ecart-type: 23.77

Figure III.9-2: SPOT HRV1 Multispectral data (zone Strasbourg): XS1, XS2, XS3

2.2 Improvement of the image

In order to facilitate the interpretation of the image, it is necessary that it first undergoes a contrast enhencement by means of a linear or non-linear transformation, stretching the histogram of greyvalues from 0 to 255. These transformation can be carried out according to mathematical or statistical functions (see Chapter III.3).

In order to further improve the image analysis process, it is interesting to manipulate the data, e.g. superpositions (diachronic compositions) or mathematical manipulations such as means (albedo), ratios (biomass index) or differences (vegetation index NDVI, Fig.III.9-3).

Table III.9-2: Values of NDVI channel

NDVI SPOT 86	Minimum	Maximum	Mean	St. deviation
	98	220	158.3	26.06

One of the visualization and hence interpretation constraints are the physical ability of the human eye which limits the amount of information the brain is able to decipher. Therefore, using a different primary colour for each image in a series of three channels within a given space provides a good approach to synthesizing information. Therefore, it is interesting to enter into combinations of spectral bands which lead to enriched information by adding the specific characteristics of all channels. Colour composite imagery makes it possible:
-     on one hand, to identify main thematical sets and to obtain a global information on the space under study,
-     on the other hand, to identify internal variations within identified sets and hence to have access to local information on the space under study.

Colour composite imagery on the screen may look natural (true colour composite imagery) or may have the characteristics of an infra red colour photography (false colour composite imagery), depending on the requirements of the application.

2.3 Search for potential channels

In order to improve the detection of individual spatial objects within an image, it is necessary to choose the channels to be used from the available ones. The above mentioned constraint model has already presented some necessary priorities of our example: a red channel or near infrared is necessary. The optimal combination of three channels has to be selected to obtain maximal discrimination of the ground locations.

It must be noted that - in the case of Thematic Mapper - 35 combinations are possible, which means 210 potential colour composites. Consequently, it might be necessary to resort to selection procedures in order to choose the optimal combination of channels according to the topic of the study.

[Sheffield, 1985] proposed an algorithm based on the variance-covariance matrix, which checks the most interesting of the channel triplets. Applying this to a study on Brussels (Belgium) with Landsat TM channels, [de Kersmaecher, 1989] considered that the best results were provided by combining channels in the visible range (1 to 3, maximal reflectance of buildings) and channels in the infrared range (4 and 5, maximal reflectance of vegetation and barren grounds). Other methods can be used in order to define optimal separability between various types of occupancy of the ground in the zone under study.

In our case the choice was limited to the number of channels, XS1, XS2 and XS3 of SPOT HRV.

VEGETATION INDEX SPOT - STRASBOURG 28/6/86

LOGICIEL CARTEL   URA 902 - CNRS - GSTS

```
            DEBUT DE CLASSE POUR EQUIPOPULATIONS
   98  120  123   126  130  133  136  139  143  146  150  153
  157  161  165   168  173  177  181  184  188  192  196  201
```

NDVI Channel
Minimum : 98.   Maximum : 220.
Mean : 158.30   St. dev. : 26.06

Figure III.9-3:   Normalized vegetation index NDVI (satellite image as Fig. III.9-2)

2.4 Combination of channels for a better understanding

Special spectral filters like the *Principal Components Analysis* in particular allow channels to be de-correlated, and the volume of data to be reduced while the informational content is not (Fig.III.9-4 and III.9-5).

Table III.9-3: Correlation matrix between factors

	XS 1	XS 2	XS 3
XS 1	1		
XS 2	0.955	1	
XS 3	-0.217	-0.323	1

The factor loadings are given in Tab. III.9-4, with the variance explained. The first factor is the most important showing a strong opposition between the first channel and the last one. The third factor could be considered as noise in this case.

Table III.9-4: Factor loadings

	Factor 1	Factor 2	Factor 3
1	0.976	0.163	0.142
2	0.952	0.275	-0.137
3	-0.480	0.877	0.017
VP	20.89 %	8.72 %	3.9 %

These transformations allow the extraction of elongation axes from the cloud of points, which represent the reflectance values, corresponding to the components of the image according to the maximum variance of input data. According to the properties defined by the sensor and the observed environment, each component represents the characteristics of the image at stake: brilliance, tree coverage etc. The factorial colour composition makes it possible to isolate and identify interesting spatial elements which would not necessarily be visible on the raw image.

Other image improvements can be envisaged by combining sensors of different spatial resolution, for instance by combining the SPOT panchromatic channel (10 m) and the SPOT HRV multispectral channels (20 m), or the Landsat Thematic Mapper channels (30 m); several combinations can be envisaged ([Cliche et al., 1985], [Weber et al., 1993]), in order to improve the geometric representation without damaging the spectral content.

This improved legibility of the image is important when constituting recognizable zones and interpreting objects of a landscape. Whatever choice is made, for the classification method, three steps are necessary:
- defining signature classes
- classifying each pixel according to its spectral signature
- checking the results.

Whether or not one has a-priori knowledge of the objects, a good understanding of the spatial organization of elements is necessary to obtain coherent results within the framework of either a supervised or a non-supervised classification.

3. Segmentation of the spectral space and classification

In order to understand one of the difficulties of this procedure it is necessary to recall two fundamental concepts which are to be taken into consideration at this stage, namely:
- that all objects belonging to the same class have identical signatures (almost);
- that the signatures of all object classes differ.

PRINCIPAL COMPONENT ANALYSIS
STRASBOURG 28/6/86

0	91	103	115	126	137	148	163

LOGICIEL CARTEL   URA 902 - CNRS - GSTS

```
            DEBUT DE CLASSE POUR EQUIPOPULATIONS
    39    78    86    91    95    99   103   107   111   115   118   122
   126   130   133   137   141   144   148   152   157   163   171   186
```

Factor 1 Channel
Minimum : 98.   Maximum : 220.
Mean : 158.30   St. dev. : 26.06

Figure III.9-4: Principle component analysis, factor 1   (Satellite image as Fig. III.9-2)

PRINCIPAL COMPONENT ANALYSIS
STRASBOURG 28/6/86

DEBUT DE CLASSE POUR EQUIPOPULATIONS

6	33	71	86	94	100	105	108	112	115	118	121
124	127	130	132	135	138	142	145	149	154	160	171

Factor 2 Channel
Minimum : 98.   Maximum : 220.
Mean : 158.30   St. dev. : 26.06

Figure III.9-5: Principle component analysis, factor 2   (Satellite image as Fig. III.9-2)

Of course it is deluding to think that the segmentation of the spectral space can occur without mixing up classes. It becomes necessary to overcome possible overlaps between signatures and hence potentially ambiguous class limits. The distances used to measure divergences between signatures can belong to three types ([Bonn, Rochon, 1992]): circular (euclidean), elliptic (quadratic correlation) or conical (chisquare). Each category characterizes the algorithm which is being used and the methodological performances which are obtained. The a-priori knowledge of the statistical parameters of classes (distribution law) is the basis of discriminating parametric functions; in contrast, non-parametric functions do not imply this type of knowledge (see Chapter III.3).

Information on object classes leading to a segmentation of spectral space results from a learning process based on image zones delineated depending on their homogeneity, their coherence without any assessment of the nature of these zones, or a preliminary knowledge of their localization and precise nature.

3.1 Definition of training sets

Several conditions have to be met in choosing training sets:
- link with the topic of the study, in this case the choice, bears on mineral zones and vegetable zones which allow a distinction between transitional periurban spaces
- stratification of the distribution of training areas according to the heterogeneity of the zone under study
- similarity of the statistical distributions of sets compared to the total population
- definition of the type of sampling: random, systematic or stratified
- amount of training sets which is not neutral; it has to correspond to a statistical validity of the results ([Kershaw, 1974])
- size and shape of the sampling unit: if the square or rectangular shape is empirically satisfying. It is nevertheless necessary to meet two requirements: a minimal number of pixels per set and the maximal homogeneity, both spectral and spatial, (first and second condition of [Justice, 1978]); this last point has of course to be analysed according to the field under study.

In this case, a stratified sampling project (Colour-Page 3) was defined according to three components of the landscape: water, woodland or cultivated zones and built zones. Since the identification of periurban zones is our aim, it is necessary, at least in a first stage, to dissociate vegetation and built territory. Tab. III.9-5 presents the sampling sets and their landscape characteristics. The outlines of their spectral signatures will ensure subsequent thematic groupings.

Table III.9-5: Training set (25 samples)

Number	Landscape Description	Landscape Location	Mean (XS1,XS2,XS3)	St. Deviation (XS1,XS2,XS3)
1	Canal	Rhine	33,0 33,5 22,0	0,70 0,52 0,00
2	Water surface fit out	Ballastière pond	48,0 25,5 24,4	0,50 1,01 0,52
3	Water surface fit out	Gerig pond	39,4 24,3 21,7	0,52 0,50 0,66
4	Water surface fit out	Baggersee pond	40,8 23,3 22,4	0,33 0,70 0,52
5	Densified built unit	Down town north	59,3 51,4 47,4	3,50 2,83 3,43

6	Densified built unit	Down town	49,8 41,0 40,3	4,31 3,46 2,17
7	Densified built unit	Downtown south	52,4 45,7 42,7	3,77 4,89 3,66
8	Suburb densified built unit	Neudorf center	57,1 52,3 55,4	2,93 3,24 3,04
9	High rise buildings with some vegetation surface	Collective dwelling units	60,3 51,2 63,7	2,17 3,23 4,86
10	One house built unit	One dwelling unit - Neudorf	57,2 52,6 62,7	2,16 4,03 3,73
11	Densified built unit with large parcels and vegetation	German district	53,8 48,0 51,0	1,26 2,44 1,58
12	One house built unit with garden	One dwelling unit - Meinau	56,6 50,5 76,1	4,00 5,19 4,42
13	One house built unit	One dwelling unit - Cronenbourg	54,6 44,4 85,2	1,73 1,81 2,63
14	Old industrial buildings	Industrial Meinau	72,3 67,3 61,3	8,83 9,82 6,24
15	Station sorting out surface	Industrial area Station	62,5 58,0 53,7	2,12 2,50 1,71
16	Harbour surface	Industrial port	62,5 56,5 57,7	7,19 7,77 5,91
17	Exhibition surface	Industrial Exhibition park	61,8 56,5 57,7	1,05 1,16 1,50
18	Specific site	Shaving site	100,3 122,2 139,4	9,50 11,65 11,54
19	Mixed Rhine valley forest	Robertsau forest	46,1 33,2 107,0	0,92 1,20 5,43
20	Mixed Rhine valley forest	Neuhof Forest E	41,3 26,7 103,3	0,70 0,97 15,13
21	Wood surface	Vegetation SE	40,5 26,6 95,2	0,72 0,50 3,03
22	Mixed Rhine valley forest	Neuhof Forest W	46,2 32,0 106,6	1,20 1,80 1,50
23	Cultivation	Culture NW	45,3 32,5 110,5	0,70 1,23 2,50
24	Cultivation	Culture W	48,1 34,5 108,4	1,76 1,65 1,33
25	Cultivation	Culture SW	52,5 36,0 130,5	1,50 1,11 7,66

According to their calculated statistics (last 2 columns in Tab. III.9-5), these 25 samples seem to be unique and homogeneous enough to give good results in a classification process. They were chosen according to the 3 main classes mentioned above (4 samples for class *water*, 13 for *inhabited and industrial areas* and 8 for *vegetation surfaces*) which include all different aspects of a city landscape. All samples are of the same size (3 × 3 pixels).

3.2 Information synthesis and analysis of spatial organization

These training sets were introduced into a stepwise discriminant analysis (BMDP software) in order to test them. The checking means provided by this analysis are of statistical and graphical types.

3.2.1 Available statistical information

The importance of the first two channels, which provided (in added proportions) 98 of the explained variance of the canonical axes, is confirmed; consequently the spectral space of values can be analysed according to only two dimensions, namely the first two canonical axes, as they reduce the information most. The classification matrix is provided and it is easy to extract omission and commission errors (Tab. III.9-6: Jacknifed Classification Matrix):

● Commission error is given by the equation:

$$P_1(i) = \frac{X_{ii}}{L_i}$$

where $i = i^{th}$ class
X_{ii} = well-classifed pixels

$$L_i = \sum_{j=1}^{s} X_{ij} = \textit{Sum of pixels classified in the } i° \textit{ category on the image (sum on line)}$$

● Ommission error is given by the equation:

$$P_2(i) = \frac{X_{ii}}{C_i}$$

where $i = i^{th}$ class
X_{ii} = well-classifed pixels

$$C_i = \sum_{k=1}^{r} X_{ki} = \textit{Sum of pixels belonging to the } i° \textit{ category in reality (sum on column)}$$

Two classification matrices are given: the classification matrix according to the highest posteriori probability and the *jacknifed classification method*. This method classifies each pixel into the class with the highest posteriori probability according to the classification functions computed for all data except the case being classified. This is a special case of the general cross validation method in which the classification functions are computed on a subset of the case, and the probability of misclassification is assessed on the basis of the remaining cases.

Table III.9-6: Jacknifed classification matrix

	Jacknifed %	1	2	3	4	5	6	7	8	9	10	11	12	13	14	15	16	17	18	19	20	21	22	23	24	25	Σ	% Commission
1	100	9	0	0	0	0	0	0	0	0	0	0	0	0	0	0	0	0	0	0	0	0	0	0	0	0	9	100,0
2	100	0	9	0	0	0	0	0	0	0	0	0	0	0	0	0	0	0	0	0	0	0	0	0	0	0	9	100,0
3	100	0	0	9	0	0	0	0	0	0	0	0	0	0	0	0	0	0	0	0	0	0	0	0	0	0	9	100,0
4	100	0	0	0	9	0	0	0	0	0	0	0	0	0	0	0	0	0	0	0	0	0	0	0	0	0	9	100,0
5	33,3	0	0	0	0	3	0	1	0	0	0	0	0	0	0	2	0	3	0	0	0	0	0	0	0	0	9	75,0
6	55,6	0	0	0	0	1	5	2	0	0	0	0	0	0	0	0	0	1	0	0	0	0	0	0	0	0	9	100,0
7	22,2	0	0	0	0	0	4	2	3	0	0	0	0	0	0	0	0	0	0	0	0	0	0	0	0	0	9	66,7
8	44,4	0	0	0	0	1	0	0	4	0	2	2	0	0	0	0	0	0	0	0	0	0	0	0	0	0	9	100,0
9	66,7	0	0	0	0	0	0	0	0	6	2	0	0	0	0	0	1	0	0	0	0	0	0	0	0	0	9	75,0
10	44,4	0	0	0	0	0	0	0	2	0	4	0	2	0	1	0	0	0	0	0	0	0	0	0	0	0	9	100,0
11	66,7	0	0	0	0	0	1	0	0	0	0	6	2	0	0	0	0	0	0	0	0	0	0	0	0	0	9	100,0
12	55,6	0	0	0	0	0	0	0	0	0	0	0	5	2	0	0	2	0	0	0	0	0	0	0	0	0	9	71,4
13	100	0	0	0	0	0	0	0	0	0	0	0	0	9	0	0	0	0	0	0	0	0	0	0	0	0	9	100,0
14	55,6	0	0	0	0	1	0	0	1	0	0	0	0	0	5	1	1	0	0	0	0	0	0	0	0	0	9	100,0
15	77,8	0	0	0	0	0	0	0	0	1	0	0	0	0	0	7	1	0	0	0	0	0	0	0	0	0	9	87,5
16	55,6	0	0	0	0	0	0	0	0	1	0	0	0	0	0	0	5	3	0	0	0	0	0	0	0	0	9	62,5
17	100	0	0	0	0	0	0	0	0	0	0	0	0	0	0	0	0	9	0	0	0	0	0	0	0	0	9	100,0
18	100	0	0	0	0	0	0	0	0	0	0	0	0	0	0	0	0	0	9	0	0	0	0	0	0	0	9	100,0
19	66,7	0	0	0	0	0	0	0	0	0	0	0	0	0	0	0	0	0	0	6	0	0	1	2	0	0	9	85,7
20	22,2	0	0	0	0	0	0	0	0	0	0	0	0	0	0	0	0	0	0	6	2	1	0	0	0	0	9	33,3
21	88,9	0	0	0	0	0	0	0	0	0	0	0	0	0	0	0	0	0	0	0	0	8	0	0	0	1	9	88,9
22	66,7	0	0	0	0	0	0	0	0	0	0	0	0	0	0	0	0	0	0	2	0	1	6	0	0	0	9	85,7
23	66,7	0	0	0	0	0	0	0	0	0	0	0	0	0	0	0	0	0	0	0	0	0	2	6	1	0	9	75,0
24	22,2	0	0	0	0	0	0	0	0	0	0	0	0	0	0	0	0	0	0	0	1	2	1	1	2	0	9	100,0
25	88,9	0	0	0	0	0	0	0	0	0	0	0	0	0	0	0	0	0	0	0	0	0	0	0	1	8	9	100,0
Σ		9	9	9	9	7	10	5	11	7	8	9	8	11	6	10	11	14	9	14	3	12	10	9	6	9	225	
% Omission		100	100	100	100	42,9	50	40	36,4	85,7	50	66,7	62,5	81,8	83,3	70	45,5	64,3	100	42,9	66,7	66,7	60	66,7	33,3	88,9		
% Total																												68,0

Table III.9-7: Gathered matrix

Categories	N°	1	2	3	4	5	6	7	8	9	Σ	% Commission
Water	1	36	0	0	0	0	0	0	0	0	36	100
Desified buildings	2	0	23	14	5	0	0	7	0	0	36	64
High rise buildings	3	0	0	6	0	1	0	2	0	0	9	67
Suburbs	4	0	6	0	10	2	0	0	0	0	18	56
One dwelling unit	5	0	0	0	2	5	2	0	0	0	9	56
Mixed vegetation & buildings	6	0	0	0	0	0	9	0	0	0	9	100
Industrial zones	7	0	4	0	0	0	0	41	0	0	45	91
Forest	8	0	0	0	0	0	0	0	21	6	27	78
Cultivation	9	0	0	0	0	0	0	0	8	28	36	78
Σ		36	33	7	17	8	11	50	29	34	225	
% Omission		100	70	86	59	63	82	82	72	82		80

One can note that the general percentage of correctly classified pixels is 68%. Although omission errors are not taken into account, information on the general validity of the classification is provided without including possible mistakes. However, it has to be noted that some classes do not seem to be representative enough. This is no major inconvenience if the samples are chosen in such a way as to cover the largest spectral distribution of values; then sometimes it is easier to take pixels into account which would not have been considered otherwise while they lie tangential to the set of points of the chosen categories, for instance. Therefore, in a second step, it is interesting to gather these "weak" classes according to common topics (Tab. III.9-7).

3.2.2 Graphic information

Two graphs, following the first canonical axes, make it possible to analyse (1) the relative localization of the sample centres and (2) the distribution of different sets in the spectral space (Fig. III.9-6 and III.9-7).

Fig. III.9-6 positions the sample centres according to two canonical axes, which split into three groups and one element. The centres of the vegetation samples, with high values on axis 2, form a compact aggregate, while those corresponding to water samples with high values on both axes have a clear positioning, as well as the element corresponding to the shaving residue of a sawmill, with an indeniable reflectance quality. The centres of the settlement samples can be split into four subsets, corresponding to (1) private house dwelling with gardens, (2) denser and lower house dwelling and more recent collective buildings, with green surroundings; (3) specific parcelling of the beginning of this century with rather large plots, broad streets at right angle, namely the German district and the Neudorf suburbs. In contrast however, the centre of the industrial harbour areas can be found for those samples characterized by relatively old buildings and are located in a wood surrounding. This proximity is not surprising if one looks at the outlines of the spectral signatures of these groups. This sample builds actually a link with the sample group of the industrial and train built area. The last subset (4) corresponds to Strasbourg city centre, characterized by the ellipse of the old fabric with narrow and winding parcels, with no opening onto green spaces. The centre of the sample of the fair zone (north of the city) joins this subset, characterized by buildings with an average ground basis, surrounded by reflecting surfaces such as parking lots.

The second graph helps to localize all sampling points so as to localize the coalescence zones between the groups. The sets can be discerned and it is obvious that the confusion zones arise from the above mentioned proximities. These are well-known difficulties in the urban milieu as soon as one tries to refine the typology of the built areas. However, the local characteristics linked with the Ill and Aar canals in the city centre, the oldness of industrial zones close to the city centre, as well as the harbour zones, should also to be taken into account and related to the discriminating possibilities of the sensor used.

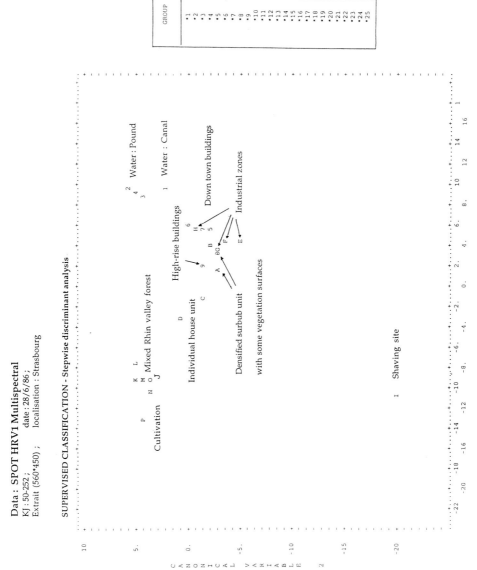

Figure III.9-6: Stepwise discriminant analysis

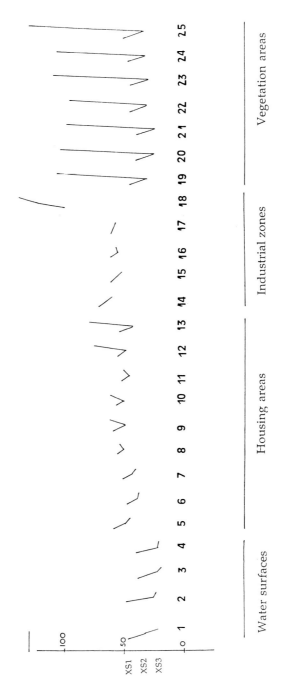

Figure III.9-7: Spectral signatures

The graphic result of the classification (Colour-Page 4) makes it possible to identify the built-up zones according to different categories, the forest and vegetation zones as well as the free waters of the Rhine canal and those of the ponds and gravel extraction zones around Strasbourg. The classification within the group of densely built areas, of gravelled Rhine and canal banks is one of the striking points. However, the green spaces are clearly identified and the connecting zones between built areas and peripheral vegetation allow a clear identification of developing zones to the north, near the Robertsau forest and even more to the west, in particular along the traffic axes. A second classified image results from the topical gathering of the 25 original classes (Colour-Page 5). It improves the legibility of the results, as it is shown by the resulting grouping matrix (Tab. III.9-7).

In order to underline this classified information for a topical cartography or its use within a GIS, it is often necessary to generalize the information to facilitate its interpretation by erasing local heterogeneities or to try and erase imperfections resulting from classification errors. To this end, a neighbouring pixel characterized by classification modalities should be taken into account following operational transition rules defined by the operator. Postclassification filters provide a better homogenization of landscapes, eliminating isolated pixels according to the prevailing immediate surrounding (Colour-Page 6), if it is coherent with the topic.

4. Validation

"The quality of the results can be judged on the basis of the context within which the study takes place".

A precise analysis of classification is carried out step by step, according to the various possible sources of error. In some cases an error results from a confusion between the radiometric values of two allocations. It can be punctual, i.e. corresponding to a given moment in a cycle (vegetation cycle). A multitemporal analysis may raise some of these ambiguities, but it may also be that the materials are different but have the same spectral response; in such a case, the difficulty cannot be overcome on the radiometric level.

Unsatisfactory results may also be caused by the heterogeneity of the milieu. The urban milieu, with mainly mixels, does not always allow classification results of very high precision, especially if the requested level of detail is incoherent with the spatial resolution.

More practical errors are the source of a low precision level, e.g. so called localization errors may arise when there is a deviation between references and field data, hence the need to examine the whole data very closely. Other errors are connected with an inappropriate adequation between the model of the choice of images and the topic.

5. Conclusion

On the basis of this example of satellite image analysis, it is possible to follow the progress of an application in an urban environment. It has to be noted that the methodological progress has to be based on a decision tree with numerous retroactive loops. It is always necessary to be in a position to check the choices which were made in order to have a good mastery of the results and to be able to validate them later on. From identification to interpretation of spatial objects and their relationships, a deductive progression has to be followed in the succession of statistical steps but an inductive opening has to be maintained, to provide for intuition in order to be able - if necessary - to allow new shapes and spatial schemes to appear leading to new questioning.

Questions

1. What are the different approaches in a classification process?

2. Why do we need to have miscellenous data in order to make a classification?

3. What is the difference between commission and omission errors? What kind of information can you extract from these data?

4. What are the different sources of confusion in a classification process?

References

Bonn, F., Rochon, G.: Précis de Télédétection. Vol. 1, Principes et méthodes, Presses universitaires du Québec, AUPELF, 1992, p. 485

Clos-Arceduc, A.: L'interprétation des photographies aériennes. IGN, Paris, 1987, p. 36

de Kersmaecher, M.: Potentialités de la Télédétection satellitaire pour l'étude de la structure interne des villes. Application au cas de Bruxelles, Thèse. Louvain la Neuve, 1987, p. 479

Justice, O.: The effect of ground conditions on Landsat multispectral scanner data for an area of complex terrain in Southern Italy. In: Justice, Townshend (eds.): Terrain analysis and remote sensing. G. Allen & Unwin, London, 1978

Kershaw, K.A.: Quantitative and dynamic plant ecology. Edwerd Arnold, London, 1974

Sheffield, C.: Selecting band combinations from multispectral data. Photogrammetric Engineering and Remote Sensing, 51, 1985, pp. 681-688

Serradj, A.: Traitement d'Images satéllitaires d'Alger et de Strasbourg. Vol. 12, Thèse de 3° cycle, Université Louis Pasteur, Strasbourg, 1985

Cliche et al.: Integration of the SPOT Panchromatic Channel into its Multispectral Mode for Image sharpness Enhancement. Photogrammetric Engineering & Remote Sensing, Vol. LI, N°3, 1985, pp. 311-316

Weber, C., Serradj, A., Hirsch, J.: Cartographie d'une forêt spécifique: la forêt alluviale de la Robertsau. In: Télédétection et Cartographie. Ed. AUPELF-UREF, Les presses de l'Université de Québec, 1993, pp. 275-282

III.10 Digital Terrain Models

Gábor Mélykúti, Budapest

Goal:

The digital description of terrain surfaces.

This chapter shows how geomorphologic, geometrical, logical and mathematical knowledge can be used for digital modelling of terrain surfaces.

Summary:

The digital representation of a terrain surface is called a Digital Terrain Model (DTM) or a Digital Elevation Model (DEM). Terrain surfaces are continuous surfaces with continuously varying relief. Normally, digital terrain models comprise various arrangements of individual points in x,y,z coordinates called original points. The basic purpose is to compute new spot heights from the original ones.

Keywords:

Digital terrain model, topographical structure, TIN structure, interpolation, finite element method

1. Introduction

In accordance with the variability of the relief and the great variety of elevation data acquisition techniques which result in different data structures, there have been evolved several methods to create Digital Terrain Models (DTM). Data describing the relief may be basically divided into two main types: the first type includes what is called mass elevation data, which define the altitude of the relief at any place of the terrain; the other type of data records the elevation values which determine the topographic features. These two types of data need separate treatment both during data acquisition and data storage and also during data processing. Thus, the topographic require-ments can be satisfied by products, mainly contour maps, derived from digital data.

2. Geometrical structures

A DTM cannot describe a continuously varying terrain surface using three discrete variables, so all descriptions are necessarily approximations to reality. The geometrical structure of a DTM is based on geometrical elements: points, lines and areas.

2.1 Point model

A systematic grid of spot heights is often used to describe a terrain. Elevation is assumed to be constant within each corner of the grid. In general, the raster size of the grid is constant in this model and the lines of the grid are parallel within the coordinate system.

2.2 Line model

There are three kinds of line models, each of them with a different data structure:
- **Isoline models** (isolines are continuous lines connecting points of the same elevation).
- **Parallel profile lines model** (profile lines are straight lines in the xy-plane, connecting points of varying distances and elevation)
- **Model of both isolines and individual points** (individual points can be used additionally to characterize the surface at its extreme points)

2.3 Area model

There are two kinds of area models:
- The first is an **array of triangular areas**, each corner being a selected point of great impor- tance, the elevation of which is known. The resulting model is called the **Triangulated Irregular Network (TIN)**. For each triangle area a $z = f(x,y)$ function is defined for determination of elevation z.
- The other is a **raster model**, where the elevation is assumed to be constant within each cell of the raster. The size of the raster is relatively small and always constant in the model. Raster models are based on the fundamentals of discrete geometry (raster-based triangulation, transformations between vector and raster data, etc.).

3. Logical structure

3.1 Geomorphologic structure

Skeleton lines, i.e. ridge-, drainage- and breaklines as well as spot heights are of essential importance for the description of terrain surfaces. Data defining the topographic characteristics of a relief are also called *structure data*. Digital terrain models which are able to consider such characteristics are called structural models.

3.2 Mathematical aspects

The main mathematical task concerning DTM is the interpolation which is the process of transferring information (elevations) from measuring points into space, i.e. to points between the surrounding measurement points. To build a structural DTM, the mathematical method should also be able to handle the geomorphologic structures of a relief.

3.3 Database aspects

DTM data are connected to the kind of information of the terrain. So data sets of 2+1, 2.5 and 3 dimensions can be built. The grade of dimension depends on the data structures of the geometrical database being used. If in a 2D system DTM functions can be used and the elevation of 2D points can be interpolated, it is a 2+1D system; if the elevation is an attribute of each point in the database, it is a 2.5D system. In case of a 3D system, every point has three coordinates (x,y,z), and it allows the use of real 3D functions.

4. Interpolation methods

By means of the elevation data stored as a point field, it is possible to compute the elevation of a new point as the weighted median of the surrounding points, the closest points having the greatest weight. Neighbouring points are searched within a local specified area around the interpolation point. The elevation of a new point can be computed with different methods (see also Chapter III.5).

4.1 Linear interpolation

Linear interpolation is the simplest method of height determination. The elevation of a new point is computed by linear functions of the selected points:

$$z \ = \ \sum_{k=0}^{n} \sum_{i+j=k} a_{ij} \, x^i \, y^j \qquad\qquad \text{(III.10-1)}$$

where if:
$n = 0 \quad\Rightarrow\quad$ horizontal plane
$n = 1 \quad\Rightarrow\quad$ tilted plane

4.2 Polynomial interpolation

In this case, the elevation of a new point is computed by a polynomial function of higher degree determined by means of the selected (surrounding) points (see also Chapter III.5, Section 3):

$$z \ = \ \sum_{k=0}^{n} \sum_{i+j=k} a_{ij} \, x^i \, y^j \qquad\qquad \text{(III.10-2)}$$

where if:
$n = 2 \quad\Rightarrow\quad$ quadratic parabola
$n = 3 \quad\Rightarrow\quad$ cubical parabola
$n > 3 \quad\Rightarrow\quad$ polynomial of higher degree

4.3 Collocation

The basis of this technique is the rate at which the variance of point measurements changes over space. This is expressed in the socalled *variogram* which shows how the average difference between values at different points changes with the distance between the points. The interpolated values are the sum of the weighted values of some number of known points where weights depend on the variance between the interpolated and the known points.

$$\check{z} = c_{ni}^{*} \; C_{vv}^{-1} \; (Aa\text{-}z) \qquad\qquad \text{(III.10-3)}$$

where

$\check{z}$	- estimated elevation
c_{ni}^{*}	- covariance vector between the new and the original points
C_{vv}^{-1}	- covariance matrix of original points
A	- form matrix, depending on the x,y coordinates of the original points
a	- coefficient vector
z	- elevation of the original points

4.4 Finite element method

A bicubic spline function generates a surface in each triangle which is tangential to the previously defined planes. If one of the triangle sides is a breakline (a line along which there is an abrupt change of slope, e.g. a discontinuity of the ground) the tangential condition is not applied, i.e. there will be a discontinuity in the mathematical surface at the breakline.

From the view of a users, the problem is the solution of a linear equation system. The unknowns in the equation system are quantities assigned to arbitrary points. In surface modelling, these quantities and their topographic interpretation - depending on the order of the approximate polynomial used in the mathematical solution - may be the following:

function value	$\Rightarrow$	height of the terrain
first derivatives	$\Rightarrow$	slope of the terrain
second derivatives	$\Rightarrow$	curvature of the terrain

The relief of a terrain, i.e. its characteristics, can be unambiguously described and clearly expressed by such quantities and the user can interpret as well as modify them. This duality is attainable, because these quantities may be considered known or unknown depending on the available information. Therefore, if the height of a terrain is measured at certain points, the function values for these points are known quantities while they are unknown for the others. The same can also be stated about other quantities (slopes, curvatures).

4.5 Matching

The automatic determination of conjugate points in digital images, called "digital image matching", is one of the key factors towards a complete automation in photogrammetry. A general model for digital photogrammetry integrates area-based multi image matching, point determination, object surface reconstruction and orthphoto generation in a model.

5. Applications

5.1 Profiles

One of the most important tasks in DTM analysis is to calculate a profile of the relief between two points. Profiles are of interest especially for telecommunication purposes and for volume computation in road design and open pit mining. Discrete points on a profile can be obtained by intersecting the vertical plane of the profile with DTM. Straight line connections or cubic spline curves between the discrete points are considered to give a sufficiently accurate terrain representation.

5.2 Slope, slope directions

A terrain's slope and slope direction may be computed relative to a plane through the elevation model points. The triangle terrain model, which comprises triangular surfaces, is ideal for simulating terrain surface conditions. Slope and slope direction are computed directly from the coordinates of the corners of a triangle. Slope is usually expressed in degrees or percentage terms with respect to the horizon. Slope direction is almost expressed in degrees from North.

5.3 Differential DTM

If the terrain surface is changed, it is possible to compute the differences between two DTMs in two different dates but at the same place. These differences can be used like elevations for the new differential DTM to analyse the changes.

5.4 Volumes

A DTM offers the possibility to compute the volume of a vertical prismoid defined by a closed polygon on the surface and a horizontal reference height. Accurate surveys of excavation volumes are crucial in planning road works, major industrial buildings, and so on. Soil borings are used to calculate cover quantities, such as down to bedrock, and replacement quantities, as down to usable soil. From these quantities, haulage can be calculated down to the project level. Quantitative surveys may be conducted in GIS by entering data in a terrain model program, which includes the extent of the excavations in elevation and ground plan, soil conditions, slopes for planned levelling, volume expansions due to blasting, etc.

5.5 Shading and draping

A terrain model may be used to produce automatically relief maps which use shading to effect the appearance of the third dimension of height. These processes demand the computation of shaded areas from an assumed solar position. The simplest approach is to colour the cells of a raster model. The colours can be varied with elevation, as on ordinary topographic maps. The colour intensity can be varied with cell slope, to give the impression of sun and shadow.

Draping a digital aerial photo, satellite image or thematic map over the surface of a terrain model produces a more realistic terrain image.

6. Accuracy

The accuracy of terrain descriptions are determined primarily
- by the accuracy, amounts and distributions of basic data,
- by the method of generating the model grid or triangular surfaces, and
- by the method used to interpolate between points in the model.

The accuracy is defined as the closeness of measurements to the true values. Normally, the accuracy is characterized by the standard deviations or root-mean-squares. If the result is characterized by the random variable x, then the standard deviation is

$$s = \sqrt{E(x-E[x])^2}$$

where E is the symbol of the statistical expectation value.

In the case of DEM the accuracy theoretically can be given as:

$$s^2 = \frac{1}{(b-a)(d-c)} \int_a^b \int_c^d [f(x,y) - s(x,y)]^2 dx\,dy$$

where $f(x,y)$ is the terrain surface function,
               $s(x,y)$ is the DEM function
               a, b, c, d are the boundaries of a surface.

The elevations of the points can be determined by sampling. In this case the Eq. (III.10-2) has the form:

$$\check{s}^2 = \frac{1}{mn} \sum_{i=1}^n \sum_{j=1}^m [f(i,j) - h(i,j)]^2$$

where $f(i,j)$ are points of the terrain surface
               $h(i,j)$ are measured points of the DEM function

The term

$$e(i,j) = f(i,j) - h(i,j)$$

is the unknown error of the DEM function.

7. User interface

7.1 Functionality

A good DTM system offers different possibilities to the user to choose the optimal solution in the field:
          - data input
          - data management
          - database structure
          - mathematical solution
          - applications
          - visualization
          - data output.

7.2 Interactive possibilities

It is necessary to follow the data flow step by step. It makes an interactive management of DTM system possible. The user can
- change and manipulate the data,
- show immediately the results,
- modify the parameters of calculation and visualization.

7.3 Connection to GIS

A DTM can be integrated into 2D GIS with the result that queries and analyses with regard to x,y and z are possible. As visualization is desirable before to construction in many engineering applications, some GIS terrain models are integrated into three-dimensional Computer Aided Design (CAD) systems. Such combination can present realistic images of planned constructions.

Questions

1. Which kind of geometrical structure can have a DTM?

2. What means *geomorphologic structure*?

3. Which kind of interpolation methods can be used for DTM?

4. Which are the most important applications of DTM?

References

Bernhardsen, T.: Geographic Information System. VIAK IT, Arendal, 1992

Ebner, H., Hossler, R., Wurlander, R.: Integration von digitalen Geländemodellen in Geo-Informationssysteme - Konzept und Realisierung. NaKaVerm 105, 1990

Ecker, R.: Rastergraphische Visualisierung mittels digitaler Geländemodelle. TU Wien Geowissenschaftliche Mitteilungen, Heft 38, 1991

Fritsch, D.: Raumbezogene Informationssysteme und Digitale Geländemodelle. DGK Reihe C, Nr. 369, München, 1991

Kennie, T.J.M., McLaren, R.A.: Modelling for digital terrain and landscape visualisation. Photogrammetric Record, Vol. XII, Nr. 72, 1988

Lee, J.: Analyses of visibility sites on topographic surfaces. International Jurnal of GIS Vol. 5, Nr. 4, 1991

Maguire, D.J., Goodchild, M.F., Rhind, D.V.: Geographical Information System: principles and application. Longman, London, Vol. 2, 1991

Melykuti, G.: Flächenbestimmung mit der Methode der Finiten Elemente. Bildmessung und Luftbildwesen, Vol. 50, 1982, p. 48-49

Melykuti, G.: The Structural Digital Terrain Model. Hungarian Cartographical Studies, ICA 14th World Conference, Budapest, 1989, p. 259-264

Reiss, P.: Aufbau digitaler Höhenmodelle auf der Grundlage einfacher finiter Elemente. DGK Reihe C, Nr. 315, München, 1985

IV.1 Erosion Modelling

M.J. van Dijk, R.H. Boekelman, T.H.M. Rientjes, Delft

Goal:

The goal of this chapter is to show how to use a GIS as a preprocessor and postprocessor for erosion modelling.

Summary:

The empirically based and physically based model approaches are mainly used to model soil erosion. A case study is made about the estimation of potential erosion in the Mananga basin (The Philippines). Calculations are made with the empirically based model USLE.

Keywords:

Erosion, Universal Soil Loss Equation, distributed modelling, erosion control, agricultural non-point-source pollution model

1. Introduction

Erosion is as old as the Earth. A serious study of erosion, however, has only developed during the last century and research shows that a gap still exists between theoretical understanding of various erosion processes and practical implementation of soil conservation measures to prevent or control erosion. The tremendous impact of erosion processes on nature needs to be analysed before it is too late.

Poor farming practices, for example those which expose soil to erosion, result in loss of top soil and agrochemicals. Sediment and (often adsorbed) chemicals are released into the river system and cause sedimentation problems. This combined process of erosion and sedimentation must be seen as an integrated phenomenon: the total erosion (and sedimentation) does not only have an impact on agricultural productivity of the fields, but also on urban and rural infrastructure and loss of land (geological).

Understanding of the mechanisms of erosion and sedimentation is required if one aims at improving the present situation. With this knowledge, predictions of future erosion-sedimentation processes arising from preventive measures can be made.

The main goal of erosion control is to keep the balance between soil formation and soil loss. Erosion control itself can be defined as the measures which aim at stabilizing existing erosion processes and restoring the results of the erosion that has already occurred.

The general approach towards erosion control is a structural one, in which preventive measures are preferred to curative measures.

2. Erosion and sedimentation

Erosion and sedimentation by water involve the detachment of soil particles by raindrop impact and by water flow, their transport by raindrops and by flow, and their deposition. Detachment is caused by the kinetic energy of raindrops impinging on the soil surface and by the mechanical force of surface runoff. The greatest detachment of soil particles is generally caused by raindrop impact. This detachment process removes soil particles from the soil mass and produces a sediment which in turn is added to the sediment load being transported. The transport processes move the sediment from its place of origin, mainly by overland flow. Deposition occurs at those places where the transport capacity of the fluid decreases to so that it loses part of the sediment load which is added to the soil mass.

The two most commonly modelled forms of water-caused erosion are sheet erosion and rill erosion; modules for splash-, gully-, and channel erosion are also sometimes added to this model. Many variables and interactions influence sheet and rill erosion.

Erosion process can generally be related to two effects: soil erosivity and soil erodibility. Erosivity is defined as the potential ability of rainfall or flow to cause erosion. Erodibility is the vulnerability of soil to erosion or soil's resistance to erosion.

Erosion control measures aim at reducing the effect of erosivity and erodibility by respectively limiting the quantity and velocity of surface runoff, and optimizing the protective soil cover. The most important factors that influence the mechanism of water-caused erosion are:
- rainfall impact,
- surface runoff,
- soil characteristics,
- land use (vegetation),
- topography.

3. Measurement of erosion

The measurement of erosion is very time consuming and expensive. This is due to the complexity of the erosion process, its extent, and its spatial and temporal distribution.

There are different ways to measure erosion, some of which are:
- surveying; the surface level is measured at different time intervals,
- measuring of the sediment concentration and water flow rate or volume past some point,
- analysing remote sensing images.

To acquire an insight of the potential erosion and the actual erosion without long-lasting measuring campaigns (which are still indispensable), models are developed which give reasonable estimates of the erosion process.

4. Modelling of erosion

The final purpose of erosion models is to obtain a comprehensive model for simulating all the effects of proposed or existing control measures. A catchment erosion model must be able to describe accurately the effects of changing topography, land use, management, soil responses and meteorologic inputs ([Beasley, 1986]).
    To obtain an estimation of erosion and sediment yield, a variety of models are available. There are models of statistical origin like the *Universal Soil Loss Equation* (USLE) and the *Soil Loss Estimator for Southern Africa* (SLEMA), which try to quantify the influence of the most important factors of erosion. In the past, these empirical models were widely used. At the moment, physically-based models are emerging and are likely to become *the* erosion prediction technology of the future. These kinds of models which deliver better descriptions of the erosion process are generally more complex (e.g. SHE, ANSWERS, KINEROS) since they include more parameters and, therefore, need also an extensive data base of values.
    It must be kept in mind that every model is a simplified representation of a complex mechanism. This simplification, caused by the equations used to describe the processes, and by the lumping and averaging of spatial and temporal processes, introduces errors in the erosion estimates (see Chapter IV.4). This is why every model, even physically-based ones, should be verified by comparing predicted and measured values so as to evaluate modelling errors. Studies of several erosion models have shown that the models underestimate the sediment yield of larger events ([Wu et al., 1993]).

4.1 Universal Soil Loss Equation (USLE)

The USLE is an empirically-based model developed in 1977 at the Purdue University to predict soil loss caused by sheet and rill erosion. This equation is designed to predict the long-term average soil losses of field areas under specified cropping and management systems. The USLE is an equation which does not account for deposition nor sedimentation yield but groups the variables under six major erosion factors: *R, K, L, S, C, P* (Fig. IV.1-1).
    The essence of the USLE is to isolate each factor causing soil erosion and to express its effect in a number. This factor number is multiplied by numbers related to other factors, the result being the amount of soil lost. The difficulty is the quantification of each variable. The USLE is presented in the following form:

$$A = 0.224 \cdot R \cdot K \cdot L \cdot S \cdot C \cdot P \qquad\qquad \text{(IV.1-1)}$$

where
A = soil loss, in [tonnes per acre]
R = rainfall erosivity factor
K = soil erodibility factor
L = slope length factor
S = slope angle factor
C = crop management factor
P = conservation practice factor.

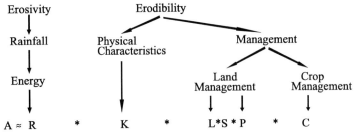

Figure IV.1-1: The Universal Soil Loss Equation

The soil loss is an estimated annual average. Factors L, S, C and P are ratios of soil loss corresponding to a unit area of a standard plot having the following characteristics:

length:	22.1 m (72.6 ft),
slope:	9%
cover factor:	bare soil
practice factor:	tilled condition.

The USLE is based on a statistical analysis of field measurements carried out over many years on hundreds of test fields in the U.S. The USLE only considers fields with moderate slopes and medium soil textures, located in regions where the erosive forces are primarily from raindrop impact. At the moment it continues to be one of the most applied erosion models due to its simplicity and low data requirements.

4.2 Agricultural Non-Point-Source Pollution Model (AGNPS)

The AGNPS is a distributed, single event model that calculates runoff, erosion, and agricultural chemical loss, generated within each grid cell. The water, sediments, and chemicals are routed down slope from one cell to the next until the watershed outlet is reached.

AGNPS uses the Soil Conservation Service (SCS) curve to estimate runoff, and a modified version of the USLE to compute detachment. SCS curve numbers are selected based on antecedent moisture, as well as on soil and land use factors. The erosion equations used in AGNPS are wholly empirical.

5. Case study

The objective of this case study is to use a GIS software package in order to calculate soil loss in the Mananga basin (the Philippines) with aid of the USLE model. Some of the collected data are in form of punctual data, some in form of values over a polygon. In order to compute soil loss it is necessary to combine the individual factors of the USLE for each area of the map. This was achieved by presenting the factors on a unit grid system whereby the maps of the individual factors could be easily overlain for computational purposes. For the Mananga basin a grid size of 50 m × 50 m was chosen.

The R-factor:
The erosivity factor R, includes erosivity of both rainfall and runoff. The calculation of the annual R-factors is done as follows in a few steps ([Renard, 1994]):

-       The R-factors for individual rainstorms is calculated at a station with a recording rain gauge, by taking the product of two rainstorm characteristics, kinetic energy (E) and the maximum 30-minute intensity (I_{30}).

-       A transfer function is established between the calculated R-factors and more readily available precipitation data; here annual totals are used.

-       The transfer function is applied to estimate R-factors for rainfall stations with annual precipitation data.

-       The R-factors are interpolated over the whole watershed with the Kriging interpolation technique (Fig. IV.1-2).

The K-factor:
The soil erodibility (K-factor) gives a quantitative description of the inherent erodibility of a particular soil. It is determined by the structural and hydrological properties of the soil (i.e. permeability, particle size, etc.). The K-factor map is constructed using a geological map of the Mananga basin (Colour-Page 7) together with the results of a soil survey.

The L-factor and the S-factor:
The slope length factor L and the slope gradient factor S are commonly represented as a single topographic factor LS. The L and S factors are calculated from the topographic map of the Mananga basin; slope gradients are calculated with special filters (Colour-Page 8).

The C-factor:
The cropping management factor C represents the role of crop or other ground covering types facing erosive forces. Crops act as rainfall interceptors that dissipate the kinetic energy of raindrops, increase infiltration, reduce runoff, increment soil moisture loss, bind of soil particles with their roots, increase organic matter in the soil, and slow down or obstruct the overland flow.
       For application in the USLE, the C-factor is defined as a dimensionless ratio between the soil loss of a land cropped under specified conditions and the corresponding loss of a clean-tilled, continuous fallow ([Wischmeier, Smith, 1978]). The C-factors are calculated from land use maps produced by the Water Resources Center in 1990 (Colour-Page 9).

The P-factor:
The conservation practice factor P is the ratio of soil loss resulting from a specific land management such as contouring and terracing, compared with soil loss resulting from up- and down-hill culture. In the Mananga basin, contouring is practised almost all over the area. The P-factor is calculated based on contouring and depending only on the land slope percentage.

Calculation of the total erosion factor with the USLE:
From the maps of the factors of the USLE (Fig. IV.1-2, Colour-Pages 7-10) soil loss A, is estimated. The resulting map (Colour-Page 10) shows, as expected, areas with high soil loss values and areas with low soil loss values. Comparing Colour-Pages 8 and 10 a high correlation can be observed: the soil loss is highly determined by the topography of the area.

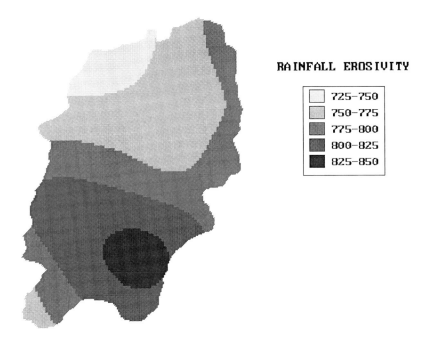

Figure IV.1-2: Rainfall erosivity factor

References

Beasley, D.B.: Distributed parameter hydrologic and water quality modelling. In: Giorgini, Zingales (eds.): Agricultural non-point source pollution, 1986

El-Swaify, S.A., Dangler, E.W., Armstrong, C.L.: Soil erosion by water in the tropics. University of Hawaii, HITAHR-CTAHR Research and Erosion series, No. 24, 1982

Mitchel, J.K., Bubenzer, G.D.: Soil loss estimation. In: Kirkby, Morgan (eds.): Soil erosion, 1980

Renard, K.G., Freimund, J.R.: Using monthly precipitation data to estimate the R-factor in the revised USLE. Journal of Hydrology 157, 1994, pp. 287-306

Wischmeier, W.H., Smith, D.D.: Predicting rainfall erosion losses. Agricultural handbook 537, USDA, Washington DC, 1978

Wu, T.H., Hall, J.A., Bonta, J.V.: Evaluation of runoff and erosion models. Journal of Irrigation and Drainage Engineering, Vol. 119 (4), 1993, pp. 364-381

IV.2 Decision Making in GIS

B. Márkus, Budapest

Goal:

Introduction to multiple criteria decision making

Description of some of the simpler strategies developed to solve multiple criteria problems

Demonstration of the potential applicability of Geoinformation Systems and Remote Sensing in Spatial Decision Support Systems

Presentation of methods to translate GIS concepts into GIS commands

Summary:

Multiple criteria methods allow the presence of more than one objective or goal in a complex spatial problem. They assume, however, that the problem is sufficiently precise for the definition of goals and objectives. Many problems are ill-structured in the sense that goals and objectives are not fully defined. Such problems require a flexible approach. The system should assist the user by providing a problem solving environment.

This chapter is an introduction to the topic of multiple criteria analysis and deals with the potential integration of quantitative multiple criteria analysis and GIS. The decision process is explained by a case study 'waste disposal site selection'.

Keywords:

Decision, criterion, factors, constraints, decision rule, problem solving, multiple criteria methods, Spatial Decision Support Systems (SSDS)

1. Introduction

Multiple criteria methods allow the presence of more than one objective or goal in a complex spatial problem. They assume, however, that the problem is sufficiently precise for the definition of goals and objectives. Many problems are ill-structured in the sense that goals and objectives are not fully defined. Such problems require a flexible approach. The system should assist the user by providing a problem solving environment.

Spatial Decision Support Systems(SDSS):
SDSS are designed to help decision makers in solving spatial problems. GIS fall short of the goals of SDSS for a number of reasons:
  - often analytical modelling capabilities are not part of a GIS
  - many GIS databases have been designed solely for the cartographic display of results - SDSS goals require flexibility in the way information is communicated to the user
  - the set of variables or layers in the database may be insufficient for complex modelling
  - the scale or resolution of data may be insufficient
  - GIS designs are not flexible enough to accommodate to variations in either the context or the process of spatial decision making

SDSS provide a framework for integrating:
  1. analytical modelling capabilities
  2. database management systems
  3. graphical display capabilities
  4. tabular reporting capabilities
  5. the decision maker's expert knowledge

GIS normally provide 2, 3 and 4; the addition of 1 and 5 create a SDSS

2. Multiple criteria decision making - definitions

2.1 Decision

A decision is a choice between alternatives. The alternatives may represent different courses of action, different hypotheses about the character of a feature, different sets of features, and so on.

2.2 Criterion

A criterion is a basis for a decision that can be measured and evaluated. It is the evidence upon which a decision is based. Criteria can be of two kinds: factors and constraints.

Factors:
 A factor is a criterion that enhances or detracts from the suitability of a specific alternative for the activity under consideration. Therefore, it is measured on a continuous scale. For example, a forestry company may determine that the steeper the slope the more costly it is to transport timber. As a result, more suitable areas for logging would be those on shallow slopes - the shallower the better. Factors are also known as decision variables in the mathematical programming literature (see [Fiering, 1986]) and structural variables in the linear goal programming literature (see [Ignazio, 1985]).

Constraints:
 A constraint serves to limit the alternatives under consideration. A good example of a constraint would be the exclusion of areas designated as wildlife reserves. In cases where the alternatives consist of features, a constraint will form a boolean (logical true/false)

map. However, where the choice is to be made between alternative sets of features, the constraint will be expressed as a characteristic the acceptable set should or should not possess. Constraints such as these are often called goals ([Ignazio, 1985]) or targets ([Rosenthal, 1985]).

2.3 Decision rule

The procedure by which criteria are combined to arrive to a particular evaluation, and by which evaluations are compared and acted upon, is known as a "decision rule". Decision rules typically contain procedures for combining criteria into a single composite index and a statement of how alternatives are to be compared using this index. For example, we might define a composite suitability map for waste disposal sites based on a weighted linear combination of information about land use, soils, slope, vegetation and distance from roads and rivers. The rule might further state that the best 5 hectares are to be selected. This could be achieved by choosing that sets of polygons (ARC/INFO) or raster cells (IDRISI) in which the sum of suitabilities is maximized. It could equally be achieved by rank ordering the cells and taking enough of the highest ranked cells to produce a total of 5 hectares. The former might be called a "choice function" (known as an objective function or performance index in the mathematical programming literature ([Diamond, Wright, 1989]) while the latter might be called a "choice heuristic".

Choice functions provide a mathematical means of comparing alternatives. Since they involve some form of optimization (such as maximizing or mini-mizing some measureable characteristic), they theoretically require that each alternative be evaluated in turn. However, in some instances, techniques do exist to limit the evaluation only to likely alternatives. For example, the Simplex Method in linear programming ([Fiering, 1986]) has been specifically designed to avoid unnecessary evaluations.

Choice heuristics, on the other hand, specify a procedure to be followed rather than a function to be evaluated. In some cases they will produce an identical result to a choice function (such as the ranking example above), while in other cases they may simply provide a close approximation.

2.4 Objective

Decision rules are structured in the context of a specific objective. The nature of that objective, and how it is viewed by the decision makers (i.e., their motives) will serve as a strong guiding force in the development of a specific decision rule. An objective is thus a perspective that serves to guide the structuring of decison rules.

2.5 Multiple Criteria Evaluations

It is frequently the case that several criteria will need to be evaluated to meet a specific objective. Such procedures are called multiple criteria evaluations. However, the weights chosen are not free from the underlying motives of the decision making group (e.g., profit maximization, optimal resource utilization, etc.). Thus the manner in which the criteria are combined is very much dictated by the objective in question.

Two of the most common procedures for multiple criteria evaluation are weighted linear combination and concordance analysis ([Eastman, 1993]). In the former, each factor is multiplied by a weight and then added to arrive at a final suitability index. In the latter, each pair of alternatives is analysed for the degree to which one outranks the other on the specified criteria. Unfortunately, concordance-discordance analysis is computationally impractical when a large number of alternatives is present (such as with raster data).

2.6 Multiple Objective Decisions

While many decisions we make are prompted by a single objective, it may also occur that we need to make decisions that satisfy several objectives. These objectives may be mutually complementary or conflicting ([Eastman, 1993]) in nature:

Complementary Objectives

With complementary or non-conflicting objectives land areas may satisfy more than one objective. Desirable areas will thus be those which jointly serve these objectives in some specified manner. For example, we might wish to allocate a certain amount of land for combined recreation and wildlife preservation uses. Optimal areas would thus finally be those that satisfy both of these objectives to the maximum possible degree.

Conflicting Objectives

With conflicting objectives, they compete for the available land since it could be used for either objective but not for both. For example, we may need to resolve the problem of allocating land for waste disposal and wildlife preservation. Clearly the two cannot coexist. Exactly how they compete, and on what basis one will prevail over the other will depend upon the nature of the decision rule that is developed.

In cases of complementary objectives, multiple objective decisions can often be made through a hierarchical extension of the multiple criteria evaluation process. For example, we might assign a weight to each of the objectives and use these along with the suitability maps developed for each to combine them into a single suitability map indicating the degree to which areas meet all of the objectives considered. However, with conflicting objectives the procedure is more involved.

With conflicting objectives it is sometimes possible to rank order the objectives and reach a priorized solution ([Rosenthal, 1985]). In such cases, the needs of higher-ranked objectives are to be satisfied before lower-ranked objectives are dealt with. However, this is often not possible, and the most common solution to conflicting objectives is the development of a compromise solution. Undoubtedly the most commonly employed techniques for resolving conflicting objectives are those involving optimization of a choice function such as mathematical programming ([Fiering, 1986]) or goal programming ([Ignazio, 1985]). In both the concern is to develop an allocation of the land that maximizes or minimizes an objective function subject to a series of constraints.

2.7 Uncertainty and Risk

It is obvious that information is vital to the process of decision making. However, we rarely have perfect information. This leads to uncertainty, of which two sources can be identified: database and decision rule uncertainty. Database uncertainty is that which resides in our assessments of the criteria which are enumerated in the decision rule. Measurement error is the primary (but not exclusive) source of such database uncertainty. Decision rule uncertainty is that which arises from the manner in which criteria are combined and evaluated to reach a decision. Both sources contribute to the risk that the decision reached will be incorrect.

When uncertainty is present, the decision rule will need to incorporate modifications to the choice function or heuristic to accommodate the propagation of uncertainty through the rule and replace the hard decision procedures of certain data with the soft ones of uncertainty. A variety of theoretical constructs have been developed to accommodate this uncertainty, including Bayesian Probability Theory, Fuzzy Set Theory, and Dempster-Shafer Theory ([Lee et al., 1987], [Stoms, 1987]). Here the issues of uncertainty and risk are not addressed since they would detract from the main issue of multiple criteria decision making for waste disposal site selection.

3. Waste disposal site selection - a case study

Criteria: Simplified but realistic criteria to find a site for waste disposal are:

1. Road orientation
   - site should be near a paved road
   - a limited distance from such a road should be set

2. Climate
   - the site should not be upwind of major residential areas much of the time
   - the least common wind direction should be determined so as to locate the facility downwind

3. Ownership
   - building on public land will probably be less expensive and produce less litigation

4. Land use
   - avoiding residential areas and minimising distance from major producers of waste material

5. Soils
   - avoiding soils where it would be expensive to build on

6. Water orientation
   - many statewide zoning laws regulate contstruction activities near water bodies

7. Geology
   - material stored on the site may leach into the groundwater
   - groundwater depth and permeability of surficial material should be taken into consideration

8. Topography
   - avoidance of excessively steep areas
   - this information may be used to select a site that is not readily visible to the community, using a viewshed algorithm

9. Vegetation
   - use of vegetation for screening
   - use of vegetation as a filtering buffer for the dust and noise caused by the activities
   - vegetation may be considered a low priority piece of information that could be used if the other data leave an abundance of sites on the regional map - to break ties

4. Problem definition and existing data

General steps involved in spatial decision making are:
1. identification of the issue
2. collection of the necessary data
3. rigorous definition of the problem by stating:
   - objectives
   - assumptions
   - constraints

if there is more than one objective:

  • the relationship between objectives is to be defined by quantifying them in commensurate terms, i.e. all objectives have to be expressed in the same units, usually in dollars, e.g. if we wish to minimize both the cost of construction and the environmental impact we must express the environmental impact in dollars, i.e. how much it would cost to avoid the impact

  • then collapse the objectives into one objective, e.g. minimize the sum of construction and environmental costs

4. find the appropriate solution procedure
5. solve the problem by finding an optimal solution

Many spatial problems are semi-structured or ill-defined because not all their aspects can be measured or modelled. The waste disposal site selection problem itself is also ill-structured. Some of the factors are difficult to evaluate or predict. Relative impacts of each of these factors on return may be unknown (except the last - direct cost). It is impossible to structure the problem completely, i.e, define and precisely measure the objective for every possible solution.

   The case study simulates the procedure to integrate topographic map (scale 1:10.000) information (contours, land use, roads, surface water) with soil map (scale 1:25.000) and aerial photo interpretation (height of vegetation), combined with climatic and statistical data. Environmental aspects are included. The case is situated in the northern part of the *Lake Balaton*.

Problem
  •   Suitability evaluation of a number of sites. Selection of suitable locations for communal waste disposal.
Objectives
  •   maximization of economic efficiency
  •   minimization of the environmental impact

What is suitable (expert-knowledge):

Environmental aspects
  •   distance from water bodies - distance buffer
  •   not visible from main roads - use vegetation in the visibility study
Economical aspects
  •   distance to paved roads < 300 m
  •   nearly flat areas - slope < 17 %
  •   low land use value
  •   suitable soils
  •   low permeability

The procedure to support site selection must be flexible and
  •   allow new factors to be introduced
  •   allow the relative importance of factors to be changed in order to evaluate sensitivity or reflect differences of opinion
  •   diplay results of analysis in informative ways
Solutions to this class of problems are often obtained by generating a set of alternatives and selecting from among them those which appear to be viable. Thus, the decision making process is iterative, integrative and participative
  •   iterative because a set of alternative solutions is generated which the decision maker evaluates; insights gained are fed back and used to define in further analyses
  •   participative because the decision maker plays an active role in defining the problem, carrying out analyses and evaluating the outcomes

- integrative because value judgement that materially affects the final outcome is made by decision makers whose expert knowledge must be integrated with the quantitative data in the models

Conflicts
- decision makers have to express environmental quality in terms of economic efficiency (monetary values)
- different interest groups will value the environment differently
- no consensus, therefore, environmental quality cannot be assessed in monetary terms
- objectives are noncommensurate

Solution
- Identify and map the different land uses, soils, vegetation, etc. and environmental impacts on separate layers
- construct several combinations of overlays based on various priorities
- determine suitability surfaces for the different combinations of priorities
- let politicians make the ultimate choice

5. Existing data and data base design

Existing data
- topographic maps
- soil maps
- aerial photos
- statistical data from local government
- hydrogeological data
- climate data (wind, temperature, rainfall)
- regional plans

Data base design
- Provides a comprehensive framework for the data base. Permits identification of potential problems and design alternatives. Without a good data base design, there may be
- irrelevant data that will not be used
- omitted data
- no update potential
- inappropriate representation of entities
- lack of integration between various parts of the data base
- unsupported applications
- major additional costs to revise (update) the database

Stages in database design
1. Conceptual
    - software and hardware independent
    - describes and defines included entities and spatial objects
2. Logical
    - software specific but hardware independent
    - determined by the data base management system
3. Physical
    - both hardware and software specific
    - related to issues of file structure, memory size and access requirements

Tiles and layers
- many spatial data bases are internally partitioned
- partitions may be spatially defined (like map sheets) or thematically defined, or both
- the term tile is often used to refer to a geographical partition of a data base and layer to a thematic partition

Reasons for partitioning
- the storage capacity of devices may limit the amount of data that the system can handle as one physical unit
- it is easier to update one partition (e.g. map sheet) at a time

"Seamless" data bases
- despite the presence of partitioning, system designers may choose to hide partitions from the user and present a homogeneous, seamless view of the data base
- in seamless data bases, data must be fully edgematched
- features which extend across tile boundaries must have identical geographic coordinates and attributes at the adjacent edge.

Organizing data into layers
- the source documents (maps) generally determine the initial thematic division of data into layers. These initial layers need not coincide with the way the data are structured internally, e.g. the application may consider lakes and streams as one layer while the data structure may see them as two different objects - polygons and lines several distinct layers may be available from the same map sheet, e.g. topographic maps may provide contours, lakes and streams (hydrography), roads. The data base manager may choose to store these as different thematic partitions in the data base.

When deciding how to partition the data by theme, it is necessary to consider:
- data relationships:
    which types of data have relationships that need to be stored in the database. These will need to be on the same layer or stored in such a way that relationships between them can be quickly determined
- functional requirements:
    what sets of data tend to be used together during the creation of products. It may be more efficient to store these on one layer

Data conversion
- the process of data input to create the database is often called data conversion
- it involves the conversion of data from various sources to the format required by the selected system
- often there are several alternative sources and input methods available for a single set of data.

Database requirements
We have to consider data base requirements in terms of:
- scale:
    will determine the largest scale that is required for sets of data; may not need to go to added expenses and time to input data at larger scales
- accuracy:
    the required accuracy will determine the quality of input and the amount of data that may be created, e.g. coarse scanning or digitizing vs. very careful and detailed digitizing, or field data collection vs. satellite image interpretation
- cost

Alternatives for creating the database include:
- obtaining and converting existing digital data
- manual or automated input from maps and field sources
- contracting data conversion to consultants

In-house conversion
- data entry is labor intensive and time consuming
- some GIS vendors assist in the conversion effort, and there are a number of companies specialized in conversion
- some agencies do their conversion in-house, but there is a reluctance to do so, in many cases, as the added personnel may not be needed once the initial conversion has been completed
- advantages of in-house conversion
  - agency personnel, who are familiar with the "ground truth" and unique situations of the areas of interest, are able to supervise the conversion effort
  - this can be important for unanticipated situations in which general rules cannot be uniformly applied
  - auxiliary maps and data are available if needed for interpretation
  - if the maps are sent out for digitizing, what you send is all you get
  - in-house validity checks can be made more easily

- disadvantages of in-house conversion
  - additional equipment and personnel need to be added to the project plan
  - long-term commitment to full-time employees can be expensive

6. GIS as a decision making / problem solving tool

GIS is an ideal tool to use for the analysis and solution of multiple criteria problems
- GIS data bases combine spatial and non-spatial information
- a GIS generally has ideal data viewing capabilities - it allows the efficient and effective visual examination of solutions
- a GIS generally allows users to interactively modify solutions to perform sensitivity analysis
- a GIS, by definition, should also contain spatial query and analytical capabilities such as measurement of area, distance measurement, overlay capability, corridor analysis
- GIS has the potential to become a very powerful tool to assist in multiple criteria spatial decision making and conflict resolution
- some GIS have already integrated multiple criteria methods with reasonable success, for example TYDAC's SPANS system or IDRISI ([Eastman, 1993])

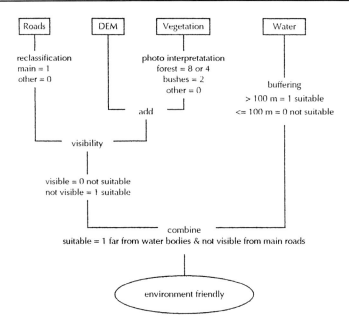

Figure IV.2-1: Flow diagram of decision making process in waste disposal site selection (constraints)

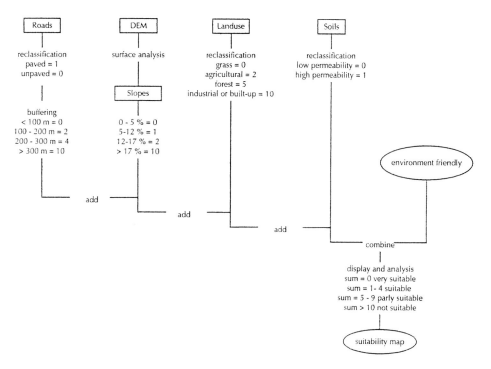

Figure IV.2-2: Flow diagram of decision making process in waste disposal site selection (factors)

Questions

1. Compare the traditional and GIS decision making in terms of technique, practicality and effectiveness at reaching to solutions of difficult problems.

2. One of the advantages of decision making by using GIS is that the effects of changes in criteria can be seen almost immediately, e.g., in the search for the best site for an activity. Discuss the impact that this capability might have on the decision making process. Do you regard this impact as positive or negative?

3. Select a current local planning issue and discuss the decision making criteria being promoted by various interest groups and individuals.

4. What kinds of models might be included in an SDSS generator for each of the following applications: monitoring of ground-water quality; emergency evacuation from areas surrounding nuclear power stations; monitoring and fighting of forest fires? Are there any similarities in these models?

References

Barber, G.: Land Use Plan Design via Interactive Multi-Objective Programming. Environment and Planning, Vol. 8, 1976, pp. 239-245

Cohon, Jared L.: Multiobjective Programming and Planning. Academic Press, Mathematics in Science and Engineering, Vol. 140, 1978

Courtney, J. F., Jr., Klastorin, T.D., Ruefli, T.W.: A Goal Programming Approach to Urban-Suburban Location Preference, Management Science, Vol. 18, 1972, pp. 258-268

Dane, C.W., Meador, N.C., White, J.B.: Goal Programming in Land Use Planning. Journal of Forestry, Vol. 75, 1977, pp. 325-329

Diamond, J.T., Wright, J.R.: Efficient Land allocation. Journal of Urban Planning and development, Vol. 115, No. 2, 1989, pp. 81-96

Eastman, J.R., Kyem, P.A.K., Toledamo, J.: A procedure for multi-objective decision making in GIS under conditions of conflicting objectives. Proceedings of EGIS'93, 1993, pp. 438-447

Fiering, B.R.: Linear programming: An introduction. Quantitative applications in the social sciences, Vol. 60, Sage Publications, London, 1986

Ignazio, J.P.: Introduction to linear goal programming. Quantitative applications in the social sciences, Vol. 60, Sage Publications, London, 1985

Lee, N.S., Grize, Y.L., Dehnad, K.: Quantitative models for reasoning under uncertainty in knowledge-based expert systems. International Journal of Intelligent Systems, Vol. II, 1987, pp. 15-38

Lee, S. M.: Goal Programming for Decision Analysis. Auerbach, Philadelphia. A general introduction to Goal Programming, 1972

Massam, B.H.: Spatial Search. Pergamon, London, 1980
(It gives many examples of applications of multi-criteria methods, in addition to the North Bay study used in this chapter)

NCGIA: Core Curriculum. University of Santa Barbara, 1989

Rietveld, P.: Multiple Objective Decision Methods and Regional Planning. Studies in Regional Science and Urban Economics, Volume 7, North Holland Publishing Company, 1980

Rosenthal, R.E.: Concepts, theory and techniques: Principles of multiobjective optimization. Decision Sciences, Vol. 16, No. 2, 1985, pp. 133-152

IV.3 Regional Monitoring of Environment

Gusztáv Winkler, Budapest

Goal:

Understanding the methods of monitoring by means of remote sensing

Introduction of remote sensing into regional analysis

Summary:

The role of environmental information systems (EIS) is growing at present and will continue to do so in the near future.

Remote sensing seems to be the most important procedure for loading data into a small-scale system suitable for regional data supply.

The chapter presents the possibilities of remote sensing for gathering information in space and time through the EIS developed at the Department of Photogrammetry, Technical University of Budapest.

Keywords:

Environmental monitoring, remote sensing, Environmental Information System (EIS)

1. Introduction

The boom of information flow and information acquisition during the last decade implied the necessity of systematic data processing. This development has given rise to a special field of Geoinformation Systems (**GIS**), the so-called *environmental information system* (**EIS**). The basic reasons for creating an **EIS** are:

- **to deal with objects on earth surface using spatial information technology** (2.5D or 3D modelling);

- **the possibility to safe spatial and temporal data**;

- **the processing of various data;**

- **the availability of proper functions for environmental descriptions;**

- **the possibility to simulate environmental processes.**

One of the most important steps in creating an **EIS** is the definition of the purpose. It will be decisive for the type of data to be collected and the basic materials to be used. Furthermore, this defines the "view" scale of the database. On this basis, the following possibilities are available:

- **National wide vector systems**:
  Mainly for central administration. A specific spatial database. Many local and map-related data (view scale: 1:100.000);

- **National wide raster systems:**
  Mainly for central administration. A specific spatial database. Many surface and map-related data (view scale: 1:100.000, pixel size approx. 100 m × 100 m);

- **Regional/local vector systems:**
  For municipalities and regional organizations. A general database. Few map and surface data (view scale: 1:10.000);

- **Regional/local raster systems:**
  For municipalities and local organizations. A special database, generally for ecology and flora inspection use (view scale: 1:100.000, pixel size approx. 10 m × 10 m);

As can be seen from the above mentioned aspects, data acquisition requires different exploitations of remote sensing and other data. In some cases it is indispensable to use measured and calculated point-related data in mass, while in other cases the general environmental description is required. As data from remote sensing manifests controversy in its content, purpose, and ability of application with the data measured on site, it may be useful to overview the possibilities of integrated processing, from the point of view of remote sensing:

- **Environmental monitoring:**
  Assignment of environmental control points.
  Area extension of data from point-related information sources.
  Investigation of spaces between point-related information sources.

- **Environmental description:**
  Local displaying of changes in the environment.

Identification of anthropogene areas (landscape injuries, litter).
Discovery of habitats (localisation) of natural eco-systems.
General relative environmental condition control.

- **Special applications:**
  Aid for archaeological, medical etc. data acquisition.

In the following we give some examples for the various types of regional monitoring. It should be noted that all examples are based on remote sensing, as one of the most effective data acquisition tools for *EIS*.

2. Environmental monitoring with raster data for EIS

In the following, the possibilities of remote sensing are presented through the *EIS* data acquisition created by the Department of Photogrammetry (Technical University of Budapest).

A small scale system (100 m × 100 m pixel size on ground) was developed for the Fertô lake region sample area. Our purpose was a general description of the environment, aiming to aid eco-system-related decision making. Therefore, it is not the quantity removed from the environment which is investigated, but the changes, the effects on nature. In close relation with the monitoring methods *EIS* layers were created as follows:

- **Air quality layer:**
  Relative air pollution based on remote sensing (recording from space) meteorological and relief data;
- **Vegetation layer:**
  Change and status of qualified categories (forest, meadow) on the basis of remote sensing and archive data;
- **Human effects layer:**
  Categories based on data from remote sensing and from ranging over the site (landscape injuries, built in area, etc.);
- **Ecology status layer:**
  Categories based on remote sensing and archive data (natural ecosystems, protection areas, etc.) (see Colour-Page 11);
- **Soil condition layer:**
  Relative status of soil based on photos from space and soil maps (soil erosion, etc.);
- **Surface and ground water layer:**
  Pollution categories based mainly on measured and relief data.

It has been manifested during data acquisition that by means of remote sensing technology relatively large areas can be covered, when the purpose of the *EIS* is not an "overall" database (Fig. IV.3-1: Radar image, pixel size 15 m × 15 m). Furthermore, it should be taken into consideration that the information layers should be comparable, so the involved data heaps must be based on similar dimensions and concepts.

When the comparison of the created layers is not the only purpose of *EIS* the, but the deduction of new relations and environment modelling are expected on the basis of the analysis, the following elements should be taken into consideration when developing the environmental monitoring:

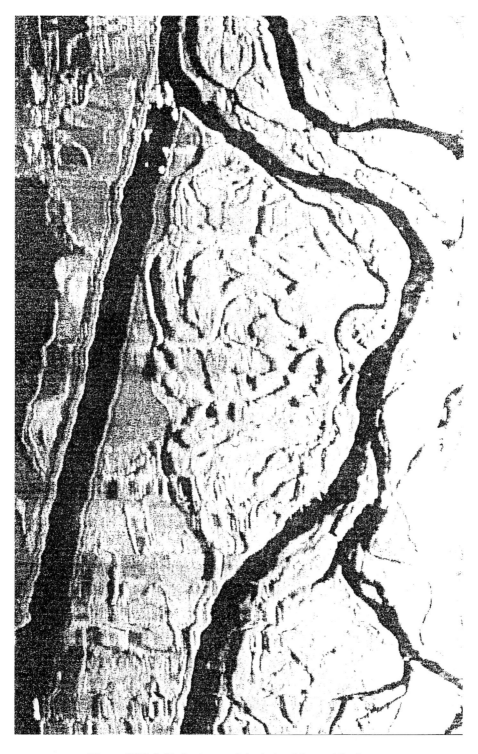

Figure IV.3-1: Radar image (pixel size 15 m × 15 m)

- Layer structures characterized by discrete geometrical positions should be transformed into 3D data supply. The corresponding remote sensing method should be found (see Fig. IV.3-2);

- The corresponding voxel size should be selected for the small scale (pixel size 100 m × 100 m) and for the large scale (pixel size 10 m × 10 m). Due to the vertical sensibility of the environment a voxel size of 100 m × 100 m × 25 m or 10 m × 10 m × 5 m are suggested, respectively.

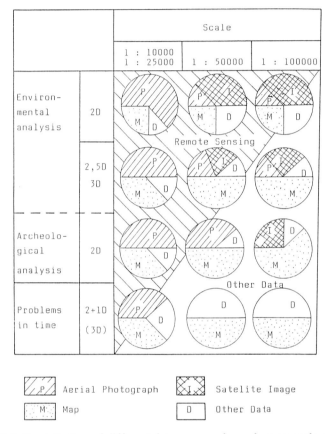

Figure IV.3-2: Contribution of different data sources dependent on scale and application

Some remarks should be made upon the integration of other non-remote sensing data into the raster system. In this case, the major role corresponds to the extension of point-related information sources. Due to the overlay of spatial and aerial photos not only data with "engineering character" are identified exactly, but various kinds of biology, health etc. data (localisation of habitats) may be obtained and displayed locally.

3. Data acquisition by vector systems

The medium scale environmental monitoring is presented by means of a vector based Data acquisition for **EIS**. The task was the collection of information in the Szombathely region. According to the requirements, the site measurement database (site monitoring) had to be integrated into a single system together with remote sensing data. According to the multilayer measurement data and the thematic map, the remote sensing data (aerial photos) were used for the following objectives (the used digital basic map was made on the basis of these data):

- **Ulterior identification of locations of site measurements.**
  The various sources have given the locations of measurement or description. Their integration into a homogeneous geometrical system was an important objective.

- **Area extension of measurements by aerial photos.**

- **Vector map interpretation of the results of remote sensing.**
  As the automatic layer comparison was not required as a function of *EIS*, the vector data logging has not lead to any further difficulties.

- **Correction of thematic maps and their geometrical adjustment.**

Based on these aspects, a layer system consisting of 12 subjects was generated. During it's creation remote sensing, map data as well as other data were used. In our case a 3D data collection was not required, so the system consists of plane layers having their own geometry.

4. Special purpose envionmental monitoring (temporal monitoring)

Above we have discussed the spatial relationships of environmental monitoring. However, there are other tasks where the temporal variation of the condition of objects should be considered, or the object analysis should be related to the past. In these cases a temporal monitoring can be carried out, when the third dimension of the *EIS* is 'time'. With respect to these elements, the environment monitoring can have the following types:

- **Investigation of variations of objects in time:**
  Generally a set of aerial photos covering a period of 40-50 years, photos from space over about 20 years and a set of maps over at least 100 years are available;

- **Investigation of archaeological objects:**
  Aiming to integrate various layers from long term (remote) past, on the basis of current status value and archive documentation;

- **Event or object construction.**

Presented also is the investigation of environmental elements as an example for the reconstruction of an event (Colour-Page 12: Battle of Gyôr, in 1809). The objective of remote sensing was the collection of auxilliary data related to this event. This means the common investigation of old documents and environment. The current state of environment relations and retrieved traces of the past have contributed to the investigation of this event.

Questions

1.   What is the difference between environmental monitoring and environmental description ?

2.   What is a voxel ?

3.   What method would you choose for updating a EIS at a scale 1:100.000 ?

4.   What is temporal monitoring ?

References

Bähr, H.-P., Vögtle, T. (Hrsg.): Digitale Bildverarbeitung. Wichmann Verlag, Heidelberg, 1991

Detrekői, A.: The importance of GIS/LIS for Hungary. Computers, Environment and Urban Systems, Vol. 17, Number 3, p. 213-216

Knyihar, A: User Interface - a procedure for a rasterbased GIS Software. Periodica Politechinca ser. Civil Eng., Vol. 36, No. 2, pp. 149-154

Knyihar, A., Winkler, G.: Environmental Information-System and Remote Sensing. Computers, Environment and Urban Systems, Vol. 17, Number 3, p. 217-221

Kraus, K.: Welche Umweltparameter kann man mit Photogrammetrie und Fernerkundung erfassen? *ZfVPh*, 79. Jahrgang, Heft 3, 1991

Winkler, P. (ed.): Remote Sensing for Monitoring the Changing Environment of Europe. 12th EARSeL Symposium, Hungary 1992, Publ.: A. A. Balkema, Rotterdam, 1993

Winkler, G. et al: Environmental Qualifizier System by Department of Photogrammetry of TUB. 16th. European Regional Conference ICID, Vol. III, 1992

Winkler, G.: Fernerkundung und 3D-Interpretation für ein Umwelt-Infprmationssystem (UIS). Zeitschrift für Photogrammetrie und Fernerkundung, Heft 6/92, Wichmann Verlag Heidelberg, 1992, S. 177-181

Winkler, G.: Remote Sensing in Space and in Time. Periodica Politechnica ser. Civil. Eng., Vol. 36, No. 2, p. 243-250

IV.4 Modelling of Hydrological Processes

M.J. van Dijk, R.H. Boekelman, T.H.M. Rientjes, Delft

Goal:

This chapter gives an outline of the different problems involved in modelling of hydrological processes, and the use of GIS as a data processor for these hydrological models.

Summary:

Hydrological models constitute an important tool for analysis of many environmental processes, like erosion and transport of contaminants. There are many models available which differ in their process schematization and conceptualization. Some existing methods for the modelling of qualitative and quantitative processes are presented.

Pre- and post-processing of input and output data of models can be done within a GIS framework. Problems resulting from this linkage of a model with a GIS are different data formats of the two systems and the inability of a GIS to handle time-variant data.

Keywords:

Hydrological models, empirical models, conceptual models, physically-based models, scale of representation, qualitative models, quantitative models, GIS-Models linking

1. Introduction

This chapter deals with modelling of hydrological processes, in particular physically-based rainfall runoff modelling at catchment scale. A model can be interpreted as a collection of mathematical equations to describe the behaviour of a system. The purpose of a model is "to replace reality, enabling measurement and experimentation in a cheap and quick way, when real experiments are impossible, too expensive, or too time consuming" ([Eppink, 1993]).

For modelling of rainfall-runoff processes, different approaches of conceptualization of the system are followed: empirically, conceptually, or physically-based ones. Except in the empirically-based approach, a loss function and a routing function is needed to model rainfall-runoff ([Beven, 1989]). The level of complexity of the models is independent on the applied schematization and conceptualization of the system, although empirically-based models tend to be much simpler than the physically-based ones. The difference between simple and complex models is the way they describe non-linearities in the loss and routing functions. A very complex model, however, is not a guarantee of a good representation of the real processes. Even very complex physically-based models have to be treated with great care since they are still extreme simplifications of the real world. This simplification is due to the impossibility to include all possible parameters and variables determining the system, the immense number of necessary (time-variant) data, and the misunderstanding of the complex system.

In correspondence to the model complexity are the model data requirements. Some simple models, for example, require only standard design rainfall distributions to calculate runoff responses. Distributed physically-based models, on the other hand, require large amounts of data when all parameters determining the system are to be included. Developments in remote sensing, GIS, data processing, and computing in recent years are of considerable importance in this aspect of data gathering and manipulation. However, the problem remains that very often data are not available, because data collection is expensive and difficult.

The data problem, also known as "parameter crisis", is one of the most important problems in hydrological modelling because the accuracy of the model outcomes is not only dependent on the degree to which the model structure "correctly" represents the hydrologic processes but mostly on the accuracy of the input data. Therefore, it makes more sense to use a simple model for which the input data requirements can be satisfied, than to use a complex model for which the input data requirements can not be satisfied.

2. Schematization of the process

A first step in modelling is the schematization of the hydrological system. When doing so, all processes influencing the modelled system can be drawn. A hydrological schematization of a catchment is presented in Fig. IV.4-1.

Precipitation falling on a catchment may evaporate or be discharged as ground and/or surface water. That part of precipitation leaving the catchment as ground and/or surface water is called excess rainfall. In a hydrological model two types of processes are described: the process by which the amount of excess rainfall is calculated from the precipitation (loss function), and the way in which this excess rainfall leaves the catchment (routing function). The total precipitation loss may be the result of different processes: interception, evaporation, neutralization of soil moisture deficiency, etc. Rainfall excess can usually be split up into groundwater runoff and direct runoff (overland flow, groundwater or subsurface flow, channel flow). When modelling single events (floods), the main cause is direct runoff. In this case percolation of rainfall into the groundwater reservoir is considered a loss function.

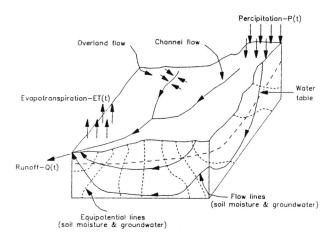

Figure IV.4-1: Schematization of a hydrological process (after [Freeze, Harlan, 1969])

3. Types of models

Models are simplifications of reality for the benefit of a special purpose; a universal model does not exist. The model designer decides which parameters/processes will be included in the system conceptualization, and how these parameters/processes will be modelled.

Models can be classified according to their different conceptualization of the modelled processes and their way of treating the spatial variability of these processes. An overall classification identifies empirically-, conceptually-, or physically-based models in a distributed or lumped environment.

3.1 Conceptualization

Black box, or empirically-based models:

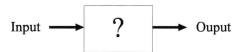

Figure IV.4-2: Scheme of a 'black box'

Empirically-based models usually result from time-series analyses. This type of models depends on finding a correlation between input values (mostly rainfall) and output values (discharge, sediment concentration, etc.) of time series, without taking the physical processes into account (Fig. IV.4-2). The parameters in empirical models thus have nothing to do with the physical properties of the processes.

Conceptually-based models:
Conceptually-based models consider some understanding of the underlying physical processes. However, the models are more or less empirical and their parameters do not depend on direct measurements. Therefore, calibration with concurrent input and output series is necessary. Most conceptual models are based on the storage concept or on the translation concept (Fig. IV.4-3) to model the physical processes of the system, e.g. infiltration, flow routing, etc. Sometimes a model consists of a combination of these two approaches.

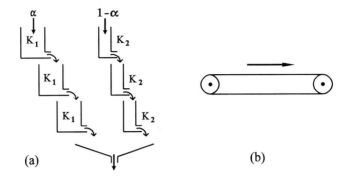

Figure IV.4-3: Storage concept (a), and translation concept (b)

Physically-based models:
The latest activity focuses mostly on physically-based modelling of catchment hydrology. The models are based on the understanding of the physics of the processes involved and describe the system by incorporating equations grounded on the laws of conservation of mass and energy.

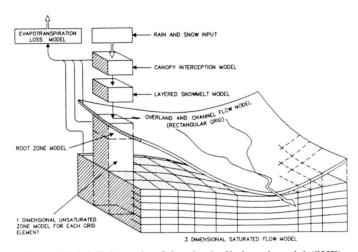

Figure IV.4-4: Schematic of the physically based model (SHE)

An assumed advantage of physically-based models is that they do not require long meteorological and hydrological records for calibration, since their parameters have physical meaning. Unfortunately, also physically-based models are mere representations of the true physical processes, and calibration is thus essential. This limitation of representation is caused by the equations underlying the models. These equations are good descriptors of spatially homogeneous processes, but in real world problems heterogeneity is introduced which leads to uncertainties. This implies that even physically-based models should be verified, by comparing predicted and measured values, to evaluate the modelling error. A standard method for calibration is the comparison of simulated and observed hydrographs. This is a necessary test but cannot be considered a sufficient test for models that purport to simulate the internal responses of a catchment ([Beven, 1989]). Therefore, calibration and verification of a physically-based model should not only be performed on a large scale with hydrographs, but also on the discretized process scale, which is very difficult to attain. The choice of variables and the choice of the temporal and spatial scales forms the structural basis of a physically-based model. These choices are fundamental and can affect the nature and reliability of the produced modelling results.

3.2 Scale of representation

Besides the model conceptualization, there is another distinction to be made between models: the schematization of spatial and temporal variabilities of the processes described. The scale at which process variations occur in the real world is infinite; for modelling purposes, however, we must average this infinite variability to some degree into finite elements. Such a finite element (often a square grid) is then assumed to be homogeneous and to have uniform parameters (altitude, slope, crop, soils, etc.).

Models can be split into two levels of scaling discretization: distributed models and lumped models, although there is no clear distinction between these two approaches; it only gives an indication of the level of discretization of the process in space and in time. This level of discretization is dependent on the spatial and temporal scales of the processes described.

It can be stated that a model is lumped if the spatial and temporal scales of the processes described in the model are smaller than the scales in which they are discretized. And a model is distributed when the spatial and temporal scales of the processes described in the model are of the same order as the spatial and temporal scales a catchment is discretized in. This introduces some problems to analyse physically-based models. The equations used in such models are based on the physics of small-scale homogeneous systems (e.g. Darcy). But these equations are applied at the scale of the chosen level of discretization, implicitly assuming that the process being modelled is homogeneous and isotrope at this scale. The assumption that small-scale equations can be applied to describe processes on this larger scale is made in every model, although there is no theoretical framework for this lumping ([Beven, 1989]). Therefore, each element can be seen as a small lumped model within which all conditions/parameters are spatially uniform/homogeneous.

We normally say that a model is distributed when the catchment is discretized in finite elements; and a model is lumped when all parameters (A, B, C...) are assumed to be homogeneous within the whole catchment (Fig. IV.4-5). The scale of discretization is not considered to be dependent on the process scales. Finally we can say that a distributed physically-based model is distributed at the catchment scale, but lumped at the element scale.

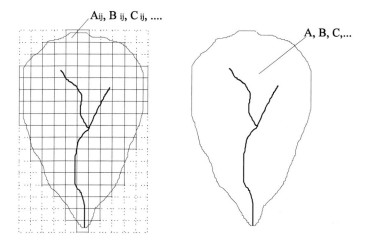

$A_{ij}, B_{ij}, C_{ij}, ...$

$A, B, C,...$

Figure IV.4-5: Distributed and lumped catchment parameters

One of the primary advantages of distributed modelling is that output results can be obtained for each element in the catchment. The accuracy of the element results of a distributed model are, however, dependent on the scale of discretization of the process, i.e. the size of the finite elements. If the element size is large, the inhomogeneity in such an element will be too big, thus leading to inaccurate estimates. The disadvantages of distributed models are:

- excessive information is required; this information is often not available.
- problems arise due to the size of the finite elements that should be chosen to describe the spatial variability of the phenomenon (which level of discretization gives an appropriate representation of the spatial behaviour of the different parameters and processes?)
- problems arise about the time interval that should be chosen.

4. Existing models

As explained in the preceding chapters, the two fundamental components of a rainfall-runoff model are a loss function and a routing function. The routing function of this hydrological model is not only needed to represent the processes of water flow to the catchment outlet, but also to represent the detachment and transport of sediment and agricultural chemicals. In this chapter, some models are presented and classified into quantitative and qualitative models. Quantitative models only deal with volumes of water; qualitative models estimate erosion, load of chemical substances, sediment load, and nutrient load that degrade water quality. Before choosing a model, the following few questions should be answered:

- What is the purpose of the model?
- What is the degree of accuracy required for the model results?
- Which are the temporal and spatial scales of the model?
- How many data are available?
- Which are the computing hardware requirements of the model?

4.1 Quantitative models

The primary interest of these models is to obtain a hydrograph at a predefined point of the draining river, mostly at the catchment outlet. A distinction can be made between a flood hydrograph for a single event, or a continuous hydrograph covering a whole range of events. There are special models for single events and continuous stream flow, but models for continuous stream flow can usually deal with flood hydrographs, too. Continuous stream flow simulation models are more complex and require more input data due to the greater amount of processes to be accounted for (evapotranspiration, soil moisture exchanges, etc.).

An example of a physically-based continuous stream flow model is the Système Hydrologique Européen (SHE) developed by the Danish Hydraulic Institute, the U.K. Institute of Hydrology, and SOGREAH in France. The model considers the primary hydrologic processes. Additional modules available are those for erosion and sedimentation, and irrigation ([Abbot et al., 1986]).

The UK Institute of Hydrology Distributed Model (IHDM) ([Beven et al., 1987]) is an example of another style of model, originally formulated for application to steep upland catchments. This model divides the catchment along the lines of greatest topographical slopes using one-dimen-sional (downslope) overland flow elements, one-dimensional (downstream) channel reaches and two-dimensional (vertical slice) groundwater (subsurface) flow components.

All these models take the surface and groundwater (subsurface) flow components into account. It is also possible to subdivide these two processes and to calculate the surface flow components and the groundwater flow components with separate models and link them with a GIS or a special interface. A groundwater model, able to simulate full three-dimensional systems, is MODFLOW developed by the U.S. Geological Survey. It is a very popular groundwater model, provided with a post-processor and particle tracking modules.

A well known single event model is the HEC-1 Flood Hydrograph Package developed by the Hydrologic Engineering Center of the U.S. Army Corps of Engineers. This model divides the catchment into many sub-basins, then routes the runoff of the separate sub-basins to the catchment outlet, eventually taking reservoirs and diversions into account.

4.2 Qualitative models

A wide range of qualitative models has been developed, which range from simple empirical models to complex simulation models. A widely used empirical erosion model is the Universal Soil Loss Equation (USLE), developed in 1977 at the Purdue University ([Wischmeier et al., 1978]). This model predicts the erosion potential for any defined area, without requiring runoff information. More detailed information about the USLE can be found in Chapter IV.1.

The USLE is often used in a modified form in other models to predict the erosion component. Such a model is the AGricultural NonPoint-Source model (AGNPS), a distributed model developed by the U.S. Agricultural Research Service in cooperation with the Minnesota Pollution Control Agency and the U.S. Soil Conservation Service. The purpose of this model is to analyse and to provide estimates of the runoff water quality of agricultural watersheds. The model predicts the runoff volume and the peak rate, the eroded and delivered sediment, the nitrogen, phosphorus and oxygen concentrations in the runoff and the sediment for single-storm events for all points in the watershed ([Young et al., 1987]).

A physically-based distributed model for predicting the effects of land use, management, and conservation practices on water quality and quantity is the Areal Nonpoint-Source Watershed Environment Response Simulation model (ANSWERS). This model is intended to predict erosion, sedimentation, and nutrient transport in catchments where agriculture is the primary land use, during and immediately following a rainfall event ([Beasley et al., 1981]).

5. Model-GIS linkage

Geoinformation Systems (GIS) provide a functional framework for coupling hydrological models with spatial entities. Distributed models make the greatest use of a GIS, because such models require heterogeneous model parameters over a grid-cell structure. A GIS can be used to obtain model parameters for each grid cell from maps or points. Until now the majority of research has been directed towards using GISs for input preparation (pre-processing) of model parameters and for output presentation of model results (post-processing). These model input and output data are all space and time-invariant data. When utilizing distributed hydrological models, very often tools are required which handle not only spatial data but also time-variant data. A major limitation of a GIS is its inability to deal with time series data.

In general, there are two ways in which hydrological models can be linked with a GIS:

a) The loosely coupled approach:
This is a loose coupling of a GIS and a model. The GIS is used to produce the model input; to effectively derive the parameters needed by a hydrological model and export them in the required model data format. The (time-dependent) hydrologic simulation calculations are performed by traditional hydrologic codes of the model. Afterwards, the results of the model are exported to the GIS and processed for analysis and visualization (Fig. IV.4-6). Interfaces, mainly provided by a GIS, allow data coupling/conversion, necessary for data exchange between the systems. This approach is time consuming but each system is used for the purpose it was designed for. GIS procedures for model input and output processing can consist of:
- georeferencing of variables and rescaling,
- establishment of relations between variables by overlaying maps,
- interpolation of discontinuous data,
- conversion of maps to a standardized format,
- output display, etc.

LOOSELY COUPLED APPROACH

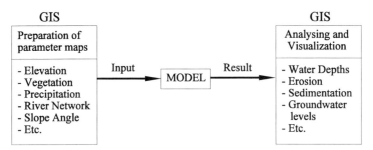

Figure IV.4-6: Loosely coupled approach

b) The tightly coupled approach:
A hydrological model embedded in a GIS is also called a "tight coupling between a GIS and a model". The model, being an integral part of the GIS, uses GIS procedures to perform hydrological analysis in space and time (Fig. IV.4-7). An advantage of this approach is that the user wastes less time on the mechanics of data transfer.

TIGHTLY COUPLED APPROACH

GIS

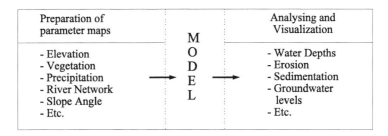

Figure IV.4-7: Tightly coupled approach

In the future, one of the research topics should focus on the tightly coupled approach which is the most limited one because a GIS, at present, cannot handle the time dimension which is required by a large number of hydrological models. The UK Institute of Hydrology at Wallingfort in collaboration with ICL made some advances in development of a tightly coupled approach. The collaboration resulted in the development of the Water Information System (WIS). WIS is a database which stores both spatial and temporal data on catchment characteristics as well as hydrological and water quality variables ([Romanovicz et al., 1993]). They implemented TOPMODEL within WIS, and included automatic optimization routines and sensitivity analysis options to support the analysis of model structures.

6. Critical notes

During the past decades there has been a change in direction of development of hydrological models, from empirically-based models towards physically-based models. There are a number of authors who are very critical about the use of physically-based models. They argue that the complex physically-based models are not able to model hydrological systems more reliably than simpler empirically-based models. The distributed modelling of parameters and variables in the distributed physically-based models incorporates problems like the validity of applying small scale equations at the grid scale.

The development of GIS's which has occurred during the same period as hydrological computer models has been of great help for complex distributed physically-based models, which demand a large amount of data. The procedures available in a GIS have helped to prepare the input parameter maps required by the models. This integrated approach of GIS and modelling has also brought some consequences: it can seriously limit the quality of modelling results when a GIS is treated as a "black box" without understanding the fundamental data manipulations. Future developments in hydrological modelling and GIS will focus on data collection, expansion of databases, improvement in computer power, and advances in the integration of models in a GIS. For hydrological models the improvements in data collection is crucial, since an understanding of the processes through data collection is the basis for every modelling effort.

Questions

1. What is a physically-based model?

2. What is a conceptually-based model?

3. Why is the scale in which the processes are discretized so important for physically based models?

4. Name the differences between loose coupling of a model and a GIS and tight coupling of a model and a GIS.

5. What is the "parameter crisis"? How can the introduction of remote sensing in hydrological practices influence this problem?

References

Abbott, M.B., Bathurst, J.C., Cunge, J.A., O'Connel, J.A., Rasmussen, J.:
An introduction to the European Hydrological System "SHE": 1. History and philosophy of a physically-based, distributed modelling system; 2. Structure of a physically-based, distributed modelling system. Journal of Hydrology 87, 1986, pp. 45-77

Bathurst, J.C.: Framework for erosion and sediment yield modelling. In: Bowles, O'Connel (eds.): Recent advances in the modelling of hydrologic systems. Chapter 13, 1991

Baxter, E., Vieux, E., Needham, S.: Nonpoint-pollution model sensitivity to grid-cell size. Journal of water resources planning and management, Vol. 119(2), 1993, pp 141-157

Beasley, D.B., Huggins, L.F.: ANSWERS - Users manual. U.S. Environmental Protection Agency Region V, Chicago, 1981

Beven, K.J., Calver, A., Morris, E.M.: The Institute of Hydrology Distributed Model. Report No. 98, Institute of Hydrology, Wallingfort, U.K., 1987

Beven, K.J.: Changing Ideas in Hydrology - The case of Physically Based Models. Journal of Hydrology, Vol. 105, 1989, pp 157-172

Beven, K.J.: Prophesy, reality and uncertainty in distributed hydrological modelling. Advances in Water Resources 16, 1993, pp 41-51

DeVantier, B.A., Feldman, A.D. (B): Review of GIS applications in hydrological modelling. Journal of Water Resources Planning and Management, Vol. 119(2), 1993, pp 246-261

DeVries, J.J., Hromadka, T.V.: Computer models for surface water. In: Maidment (ed.): Handbook of Hydrology, McGraw-Hill, 1993

Eppink, L.A.A.J.: Processes and models in erosion and soil and water conservation: Water erosion models: An overview. Lecture Notes, Wageningen, 1992

Heikens, D.L.J.: Modelleren voor het waterbeheer. Nieuwegein, November 1993

Mandl, P.: A conceptual framework for coupling GIS and computer models. EGIS, pp 431-437

O'Connell, P.E.: A historical perspective. In: Bowles, O'Connel (eds.): Recent advances in the modelling of hydrologic systems. Chapter 13, 1991

Romanowicz, R., Beven, K., Freer, J.: Topmodel as an application module within WIS. Hydrogis 93, Application of GIS in Hydrology and Water Resources, IAHS No. 211, 1993

Stuart, N., Stocks, C.: Hydrological modelling within GIS: An integrated approach. Hydrogis 93, Application of GIS in Hydrology and Water Resources, IAHS No. 211, 1993

TNO: Recent trends in hydrograph synthesis. Proceedings of the Technical meeting 21, Proceedings and Information No. 13, Committee for Hydrological Research T.N.O., The Hague, 1966

Wischmeier, W.H., Smith, D.D.: Predicting Rainfall Erosion Losses. Agricultural Handbook 537, USDA, Washington D.C., 1978

Young, R.A., Onstad, C.A., Bosch, D.D., Anderson, W.P.: AGNPS, agricultural nonpoint-source pollution model: A watershed analysis tool. Conservation Res. Report 35, US Dept. of Agric. Res. Service, Morris, 1987

IV.5 Hydrological Measuring Network Design

M.J. van Dijk, T.H.M. Rientjes, R.H. Boekelman, Delft

Goal:

Hydrological measuring network design is very important for analysing hydrological phenomena. This chapter explains the basics of network design and presents some methods for network design. With the use of these design methods an optimum network of gauging stations can be achieved.

Summary:

This chapter treats the importance of hydrological networks. The basic steps in the process of designing a network are discussed and some well known approaches to network design are given. In particular the methods of Rodriguez-Iturbe and Mejia, and the Kriging method are discussed and presented together with an example.

Keywords:

Rainfall networks, WMO minimum networks, Rodriguez-Iturbe and Mejia, Kriging

1. Introduction

At present, growing importance is given to the collection of environmental information. Environmental modelling and the definition of measures related to environmental protection policies are usually taken on basis of collected information. Hydrologists and meteorologists are in this respect privileged, especially in more developed countries, because hydrological records often exist for a long time. Hydrological information like precipitation data, runoff data, groundwater table data, piezometric data, etc., is collected and stored throughout the years by monitoring gauging stations of a (well designed) network. In developing countries, however, gauging stations are often poorly spatially distributed, not well managed, and frequently do not integrate a network configuration. Thus, some regions are "over-gauged" by independent (interrelated) agencies or, inversely, there are some regions which sometimes are not monitored at all. Many water resources management projects are (still) designed with inadequate and incorrect data or even with virtually no data. As a consequence, wrong water management decisions may be made, wrong criteria may be selected, and inappropriate and uneconomic designs may be developed and operated.

Therefore, the design of efficient and economic networks is an important issue for hydrologists and civil engineers. The system or network of hydrological gauging stations provides the information necessary to understand and to describe the hydrological phenomena and processes under study.

2. Design considerations

The common goal in monitoring any network of gauging stations is to understand and to describe the behaviour of the phenomenon or process under study. Hydrological phenomena and processes like precipitation, surface runoff, water levels, etc. have very different characteristics and depend on many factors such as climate, topography, geology, etc.. In general, one can say that any phenomenon can be characterized by a spatial and temporal distribution over the area under study. Network design deals with the understanding of these spatial and temporal distributions, in order to understand and to describe the dynamics of the phenomenon. The ideal situation from a monitoring point of view would be the operation of a very dense gauging network supplied with continuously registering equipment. The operation of such a network, however, would be very expensive and for many study objectives it would be over-dimensioned. In practice, many networks have been designed with special emphasis on minimum network operational costs and conveniences regarding network operation. It is clear that many networks are unbalanced not only in the number of stations but also in their spatial distribution within the network configuration. It is also clear that a true understanding and description of the phenomenon under study will never be achieved.

The most important questions in network design for any hydrological study deal with:
- the number of stations (i.e. network density), and
- the spatial distribution (locations) of stations within the area under study.

To answer these questions many aspects of network design must be taken into account. Aspects relating to physical, practical, and economical considerations have major influence on the designed network. For example, the spacing of a raingauge network must be based on: climatic characteristics, topography of the study area, raingauge site accessibility, availability of observers, costs of network operation, data processing capabilities, monitoring goals, etc.

Very important issues like network performance and data accuracy have been ignored for many years. During the past decade special attention has been paid to the necessary network density and to the spatial distribution of the individual gauging stations in relation to acquired gauging accuracies and defined performance criteria for specific hydrologic investigations. In this way, the questions concerning the optimal network density also relate to scientific considerations (required accuracy).

When (re)designing a hydrological network it is necessary to collect as much knowledge as possible about the characteristics of the phenomenon under study and about the physical properties of the study area. Complete network design should involve the specification of some operational aspects:

1.   Sampling variables (rainfall, evaporation, water levels, etc.).
2.   Number and location of sites (site accessibility, available observers, costs, etc.).
3.   Frequency of measurement (continuous, per minute, hourly, daily, weekly, etc.).
4.   Duration of measurement (project period, 1 year, 10 years, 25 years, etc.).
5.   Techniques and instrumentation (manual/automated, telemetric, digital/analogous, data logger, etc.).
6.   Data processing system (manual, computer, centralized, decentralized, real time processing, data distribution, data format, etc.).
7.   Measurement service (consultant, state department, local service, transport, etc.).

3. Social and economic aspects of networks

Social and economic aspects have often major influence on the design of hydrological networks. Two underlying questions relating to network design are: "*Is a society willing to invest in a gauging network?*" and "*How much can policy makers invest?*". Although trivial, in general this depends on the social benefits (i.e. "the value") that information of hydrological phenomena can have. An important question relating to network operation is: "*Can a network be operated by the local people?*"; people must be able to operate the stations, to carry out the observations and to maintain the equipment even in extreme conditions.

Designing a network is often confronted with contradictory intentions. Very often it is difficult to extend a network although the need of additional stations is obvious. On the other hand, situations can occur that make it difficult to stop the measurement at a station even if the obtained data are not longer required from the hydrological point of view. In this situation, the question is whether the information produced by a network is balanced by the cost of network construction and network operation (maintenance, management, and data processing). In order to be able to judge the value of the collected data it is important to review the information content in relation to objectives (goals) for which these (additional) data will be used.

Unless used for specific study purposes, the establishment of a too dense network should be avoided. The additional information gained by highly correlated (neighbouring) stations is not worth the financial efforts involved in maintaining such stations. On the other hand a too sparse network can give only indicative information about the phenomena of interest. In each case the'"optimum" network depends on the monitoring accuracy one is satisfied with. This accuracy must reflect and balance the social benefits (i.e. practical purposes) for which the data are collected.

A lack of information about the phenomenon can lead to wrong managing strategies resulting in economic losses. Compared to the ideal situation whereby all possible information is available one can speak of the economic loss or also called "information loss" ([v.d. Made, 1986]). The information loss can be diminished by an extension of the network which in its turn increases the costs of operation. If the network is designed and constructed in such a way that both the costs of network operation and the total information loss are minimal, then the economically optimum network configuration is found. The economic loss due to information loss is often difficult to define since the quantification of the social benefits can be based on subjective criteria.

The commonly applied strategy in designing an economically optimal network configuration, is:

- to build the most efficient network in terms of number of stations, station distribution, gauging interval, site accessibility, etc. within the budgetary limits and,
- to design a network configuration in such a way that the errors of spatially interpolated data are smaller than a (arbitrary chosen) fixed criterion.

These strategies have both their shortcomings. The first is simple to achieve, but does not take into account a performance criterion, i.e. the errors of estimate and thus the extent of the information loss. In the second approach the errors of estimate are fixed and so is the presumed information loss although the number of stations may not be a limiting factor for describing the (spatial) behaviour of the phenomenon under study. In practice it is difficult to assess a design criterion as a 'set point' for the standard error.

4. Objectives of networks and network types

Conscientious design of a network requires a clear definition of the objectives and goals of the network. The objectives and goals relate to the requirements of data quality and the necessary knowledge to understand the phenomenon under study. There are three major uses of hydrological data:
- Planning;
    Planning usually focuses on describing the behaviour and the natural variability of phenomena; extensive, high quality data are required on a long time base. An example of data used for planning purposes is the design and dimensioning of a storage reservoir to overcome dry periods.
- Operational management;
    Operational management requires less data than planning but data must preferably become available on a "real time" base. Data quality must be high for adequate management and it must also become available on a continuous time base. Operational management can be for hourly and daily water management and/or for very near future forecasting.
- Research;
    For research purposes data of high quality are needed. Data are often used for a better understanding of certain phenomena and/or processes. For specific studies, mostly in small areas like a single catchment, data are needed at all time levels. Trends over longer time periods as well as real-time behaviour of events may be studied.

Gauging criteria accompanying these data uses have different levels of accuracy. This results in networks with different station distributions and densities where each network is designed for a specific domain and its own objectives. The design of three independent networks is not appropriate, an integration of approaches should be aimed at. Rodda suggested that an optimal situation can be reached when the network is split into three levels of accuracy ([WMO, 1969]):
- First order (principal or base station) network:
    designed to cover a large region. The information obtained by the base stations is often applied for broad-scale national planning purposes. The first order network furnishes the basis for statistical studies and thus should be observed on a continuous time base for a certain period of time.
- Secondary order network:
    a densified network designed over a single sub-region, normally a basin. Stations should be operated for a limited number of years, although long enough to establish a good correlation between the secondary stations and the base stations
- Third order network:
    the most dense network and especially designed to collect data for a specific, clearly defined objective. The length of operation of additional stations is determined by the monitoring purposes.

5. Network optimization

Once an initial network is established it is important to optimize this network as soon as possible, and create a base network. However, the base network is vulnerable, since the loss of a station can have serious consequences regarding the accuracy of network operation and continuity of the time series. Therefore, the (re)design of the base network should be carried out with high precision. Optimization of a base network also includes the installation of a densified network of secondary stations in one or more representative areas for studying spatial and temporal process variability in more detail. When satisfactory correlation links exist between the base stations and the secondary stations, exclusion of the secondary order stations is acceptable since the prolonged time series are still guaranteed by the established correlation link between one or two base stations and the secondary order network stations.

The aim of optimizing networks is to achieve a network with minimum costs, with which it is possible to determine - with sufficient accuracy - the characteristics of the basic hydrological phenomenon elements, anywhere in the region, for practical purposes. By characteristics, all quantitative data, averages and extremes are meant that define the statistical distribution of the element studied ([WMO, 1974]). Especially in more developed countries optimum networks are required due to the high costs of network operation.

6. Network redesign

On the basis of data produced by an initial network, indicative information on the variability and the correlation (coherence) of processes can be determined. The network performance of an initial network can thus be analysed by different methods, like (geo)statistical analysis and rational reasoning. During the redesign process, the initial network configuration has to be optimized based on the performance analysis and the predefined design criteria of network accuracy and reliability, e.g. a given maximum standard error of estimate. The design criteria, on their turn depend very much on the objectives and goals for network establishment and operation and thus on the socio-economic aspects of network design. The techniques used for network redesign result in a theoretical optimum network. Such a theoretical network must be modified to satisfy practical boundary conditions and possible gauge locations. With the changes of socio-economic needs and local conditions in time the need to measure data changes too. Then, networks have to be revised which makes network redesign a continuous process. Fig. IV.5-1 presents a flow diagram of permanent network redesign (after [van der Made, 1987]).

7. Design of raingauge networks

The problem of optimal rainfall data collection has been a concern of hydrologists all over the world for many years. This concern is reflected in the vast amount of literature that has been published on different aspects of the matter. Many principles and techniques for network design have been proposed in the last decades, but few claim universal validity and applicability. The network design principles have evolved and will continue to evolve in time. To show the different approaches to network design, different authors will be reviewed:

7.1 Hershfield

[Hershfield, 1965] used a correlation of 0.9 as a criterion for obtaining the average spacing between raingauges for 24-hour and 1-hour duration rainfalls with a statistical two year recurrence interval.

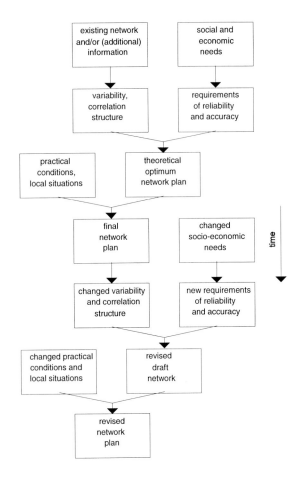

Figure IV.5-1: Flow diagram of permanent network design

7.2 WMO

[WMO, 1974] recommended for general hydro-meteorological purposes the following minimum densities for precipitation networks which are based on field experiences with (and performances of) comparable networks:

1. For plains in temperate, mediterranean, and tropical zones, 600 to 900 km² per station.
2. For mountainous regions of temperate, mediterranean, and tropical zones, 100 to 250 km² per station.
3. For small mountainous islands with irregular precipitation, 25 km² per station.
4. For arid and polar zones, 1500 to 10.000 km² per station.

7.3 Rodriguez-Iturbe and Mejia

[Rodriguez-Iturbe, Mejia, 1974] formulated a method for the design of precipitation networks that incorporates a correlation structure of the rainfall process in time and space. They developed a general framework to estimate the variance of:

-     the long-term mean areal rainfall,
-     the mean areal rainfall of a single storm event.

The variance is expressed as a function of correlation in time, correlation in space, length of operation of the network, and geometry of the gauging stations. The number of raingauges needed in a domain are estimated through this variance, the best locations for the raingauges, however, still need to be defined through additional (e.g. topographic and/or social) analyses.

Long-term mean areal rainfall:

The rainfall process is considered as a multidimensional random field function $Z(x,t)$ describing the total precipitation at location x and in time t. For determination of the mean value of this process it is assumed that the process is stationary and, furthermore, that its correlation function is separable in terms of its spatial and temporal structure.

$$C\left[Z(x_i,t),Z(x_j,t')\right] = \sigma_p^2 \, r\,(x_i\text{-}x_j) \; r^*(t\text{-}t')$$

$$(IV.5\text{-}1)$$

where σ_p^2 =   point variance of $Z\,(x_i,\,t)$,
                 $r\,(x_i\text{-}x_j)$ =   spatial correlation structure,
                 $r^*(t\text{-}t')$ =   temporal correlation structure.

The temporal correlation structure can be approximated by a simple Markovian scheme:

$$r^*(t-t') = \rho^{|t'-t|}$$

$$(IV.5\text{-}2)$$

where ρ =   the first auto-correlation coefficient (≤ 0.25).

The spatial correlation structure is a function of the distance between points, v. For isotropic, homogeneous random fields correlation is only a function of v, and decreases as the distance between points increases.

$$r\,(v) = e^{-h\,v}$$

$$(IV.5\text{-}3)$$

where $h\,[km^{-1}]$ =   the parameter of the correlation function, depending on the "characteristic correlation distance".

This correlation distance is the mean distance between two randomly chosen points in the area, characterized by the size and shape of the area. The mean distance between two points of a unit area is calculated by [Matern, 1960] (Tab. IV.5-1).

    In case of rectangled watersheds the mean distance between two points of a study area is now defined as the ratio of the diagonals of the rectangled watershed and the unit area of such a rectangle, multiplied by the mean distance v of the unit area.

Table IV.5-1: Mean distance of unit area

rectangle ($\lambda=a/b$)	v	other shapes	v
$\lambda = 1$	0.5214	circle	0.5108
$\lambda = 2$	0.5691	hexagon	0.5126
$\lambda = 4$	0.7137	equilateral triangle	0.5544
$\lambda = 16$	1.3426		

Geographisches Institut
der Universität Kiel

where λ =   the ratio of the sides a, b of a rectangle which approximates the shape of the watershed.

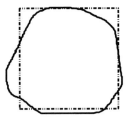 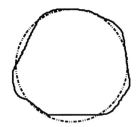

Figure IV.5-2: Examples characteristic correlation distance

Example: A watershed has an area of 64 km². If this area is approximated by a square (see Fig. IV.5-2), the diagonal of the watershed is 11.3 km. A unit area with the same shape has a diagonal of 1.41. In this case the characteristic correlation distance is 11.3 / 1.41 · 0.5214 = 4.18 km.
If the approximation is a circle (Fig. IV.5-2) we obtain for the characteristic correlation distance 9.03 / 1.13 · 0.5108 = 4.09 km.

The hydrologists want to estimate the long term mean areal rainfall by the arithmetic mean of N measurement points over T years. The precision of the estimate is measured by the variance. This variance has to be evaluated as a function of correlation structure of the process in both time and space, the number of stations in the network, the sampling geometry of the network, and the length of operation of the stations:

$$Var = \sigma_p^2 \left[F_1(T)\right]\left[F_2(N;Ah^2)\right]$$ (IV.5-4)

where Var =   the variance of the regional mean, expressed as a function of the point variance of the process multiplied by two reduction factors, F_1 due to sampling in time and F_2 due to sampling in space.

Fig. IV.5-3 shows F_1 as a function of the lag-one auto-correlation of the process ρ, and T.
The variance reduction due to spatial sampling F_2 depends on the correlation structure in space, the sampling geometry, and the number of stations. Two types of sampling schemes are considered in this study: simple random-sampling and stratified random-sampling:
-   In the simple random-sampling type of network each station is located with a uniform probability distribution over the whole space A independently of the other stations.
-   In the stratified random-sampling case the space A is divided into a number of non-overlapping comparable subareas. From each subarea, k points are randomly chosen where the rain gauges will be located.

Fig. IV.5-4 shows the variance reduction factor F_2 as a function of the number of stations and a non-dimensional area Ah^2 for random sampling.
Fig. IV.5-3 and IV.5-4 provide an analytical tool for trading time versus space in the estimation of long-term spatial averages of precipitation.

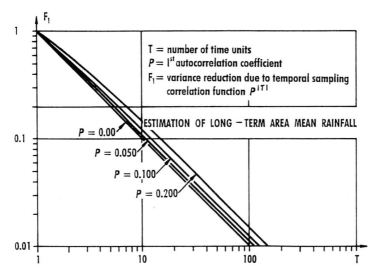

Figure IV.5-3: Variance reduction factor F_1

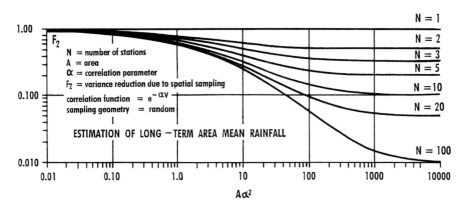

Figure IV.5-4: Variance reduction factor F_2 due to spatial sampling with random design used in the estimation of long-term mean area rainfall with $r(v) = e^{-\alpha v}$

Mean areal rainfall events:

In this case the mean rainfall over a certain area A must be estimated by the arithmetical mean of N observations. The depending factors are the spatial correlation coefficient, the number of stations, the total area, and the point variance (σ_p^2):

$$\sigma_N^2 = \sigma_P^2 \, F(N; Ah^2) \qquad \text{(IV.5-5)}$$

where σ_N^2 = the areal mean variance.

Fig. IV.5-5 shows the variance reduction factor F for random sampling as a function of the number of stations and an areal entity, Ah^2.

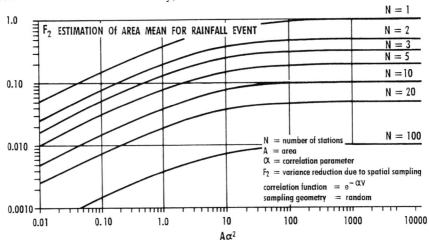

Figure IV.5-5: Variance reduction factor F due to spatial sampling with random design

7.4 Kriging

Application of Kriging in network design is favourable because the variance and interpolation error can be estimated for fictitious networks. The kriging variance can supply a tool for defining the confidence of estimation which in turn can be used as an indication in order to:
- decide on a more dense network,
- determine the optimal location of an additional measurement point,
- determine the optimal elimination of measurement points.

A property of the kriging variance is that, besides the spatial variability of the phenomenon under study, it depends only on the geometric configuration of the data points and of the domain to be estimated, but it does not include the actual data values $Z(x_i)$. So we can:
- consider an additional point x_{n+1},
- build the kriging system corresponding to the set $n+1$ data points,
- compute the corresponding Kriging variance without having to make any hypothesis about $Z(x_{n+1})$.

Global estimation:
When the purpose of a network is to estimate an average value over a given domain A, we shall try to determine where to locate the $n+1^{st}$ measurement point in order to get the smallest resulting estimation variance for the average value over A. The estimation variance will be denoted as σ^2 with the n already existing data points and as $\sigma_0^2(x)$ when the fictitious point x has been added to the set of data points. The relative variance reduction $R(x)$ can be defined as:

$$R(x) = \frac{\sigma^2 - \sigma_0^2(x)}{\sigma^2} \qquad (IV.5-6)$$

If the choice is limited to a certain number of points selected according to a given criterion, the fictitious measurement point is successively located at each of them and the corresponding variance reductions are determined.

If no a-priori conditions exist the fictitious point x is moved across the domain A and its adjacent surroundings, so that one can contour the relative variance reduction and locate the best (greatest variance reduction) area for an extra measurement point.

Local estimation:
When the purpose is to obtain local estimates (point-related values or values over meshes), the problem is slightly different: there are as many kriging variances as there are nodes or meshes on the grid. An additional measurement point will influence the kriging variance of any grid node (or mesh) in the kriging neighbourhood of which this point is located. In this case, we will only try to remove local maxima from the variance map by locating the additional points there.

Optimal deletion of a measurement point from the network:
For operational, economic, or political reasons a site has to be deleted from the current network of sites. Define $Z(x_{n-1})$ to be the network without the nth site. A sensible statistical criterion for the deletion of a site is to choose the site that achieves the minimum kriging variance when predicted, i.e. delete the site that can be predicted best from the remaining n-1 sites.

It is clear that the estimated variance is an important factor for determining the location of rain gauges since it depends only on the geometrical location of the measured points. Obviously, the choice of the variogram model and of the parameters is conditioned by the particular set of available data. But once the variogram model is chosen, the variance can be regarded as depending exclusively on the location of the rain gauges. Hence it becomes possible to compute the error variance associated with any set of hypothetical data points without getting actual data at these points.

Networks can be (re)designed with the different maximum estimate variances as criterion. A minimum total amount of stations should be used to achieve this, taking into consideration the existing stations which should be deleted if necessary. After this the best network configuration should be chosen with respect to other criteria like costs, available observers, accessibility, etc.

8. Example of long-term areal mean rainfall network design

As an example, the rain gauge network of the island Cebu (The Philippines) will be taken to illustrate network design with the figures of the variance reduction factors. The region of central Cebu can be approximated by a rectangle of 30 km × 35 km, with a diagonal of 46 km. A rectangled unit area with a side ratio of 1 has a diagonal of 1.41, and thus the characteristic correlation distance is 0.5214 · 46 / 1.41 = 17 km (Tab. III.5-1). An exponential correlation structure in space must be fitted for distances of the order of 17 km, which were found to be equal to 0.42:

$$r(17) = e^{-17h} = 0.42$$

The correlation parameter h is found to be 0.051. Thus the equation to be used for describing the spatial correlation structure for the cental Cebu region is:

$$r(v) = e^{-0.051v}$$

With this correlation function and the area A, the variance reduction factors due to spatial sampling with random design F_2 can be estimated from Fig. IV.5-3. Next we need to estimate the variance reduction factor due to temporal sampling F_1. For this we will have to estimate the auto-correlation coefficient representative for the whole area. For the annual data in this example the obtained auto-correlation coefficient was 0.21. The variance reduction factors F_1 and F_2 are presented in Tab. IV.5-2 and IV.5-3.

Table IV.5-2: Variance reduction factor F_1

T	F_1	T	F_1
1	1.00	15	0.100
2	0.60	20	0.075
5	0.28	50	0.030
10	0.15	100	0.050

Table IV.5-3: Variance reduction factor F_2

N	Random	Stratified
1	1.00	1.00
2	0.73	0.68
3	0.65	0.59
5	0.58	0.52
10	0.50	0.49
20	0.46	0.47
100	0.45	0.47

Combining these tables, an estimate of the efficiency of different network schemes will be obtained. Stations in operation during 10 years, the expected total variance reduction factor is $0.15 \cdot 0.46 = 0.069$. In other words, this network will produce an estimate of the long-term areal mean with a variance of the order of 7% of the variance of the point rainfall process. For central Cebu the obtained point variance was 161 982 mm². For the case in which $F_1 \cdot F_2 = 0.069$, the variance of the estimate of the long-term areal mean is 11 177 mm², or a standard deviation of ±106 mm.

9. Network design in combination with GIS

This example deals with the network of the basin of the Mananga river, one of the main rivers on Cebu-island (Philippines). The characteristics of the rainfall occurrences within the basin can be analysed with the help of the rain gauges located in the basin. However, in the Mananga basin there are only four operational rain gauge stations, a too small number to describe the characteristics of them. Therefore, eight other stations lying in the surrounding mountain ranges are included in the analyses of the rainfall process.

The Mananga basin will be used to demonstrate the use the GIS software package ILWIS for rain gauge network optimization through Kriging. The raster maps of the different parameters will be made on a raster with a grid spacing of 100 m × 100 m, resulting in a raster of 106 columns and 136 rows (pixel size = 100 m × 100 m).

With the Kriging interpolation method a map is made of the estimation variances over the basin (Fig. IV.5-6). It is clear that the estimated variance is an important factor to determine the location of rain gauges, the variance depending only on the geometrical location of the rain gauges. Obviously, the choice of the variogram model and its parameters is conditioned by the particular set of available data. But once the variogram model is chosen, the variance can be viewed as depending exclusively on the location of the rain gauges. Hence it becomes possible to compute the error variance associated with any set of hypothetical data points without getting actual data at such points. The estimation variance is thus a very suitable tool for network optimization. The existing network of rain gauges can be replaced or supplemented by the best representative new set of rain gauges at specified locations. In our case, the standard deviations are highest in the northern part of the area. Therefore, the best places to add a new rain gauge would be on the particular place where the estimation variance is highest.

If two stations are added, the estimation variance in the northern part is considerably reduced (Fig. IV.5-7). The two added stations are only known by coordinates from the rainfall map. A topographic map should be analysed to check if it is possible to place the new stations at or close to the chosen locations in order to satisfy other criteria like accessibility. For this exercise we have at our disposal a road map of the Mananga basin from which the distances to the new locations can be calculated. If the distance map and the estimation variance map (without extra stations) are combined, the best locations for additional stations become visible.

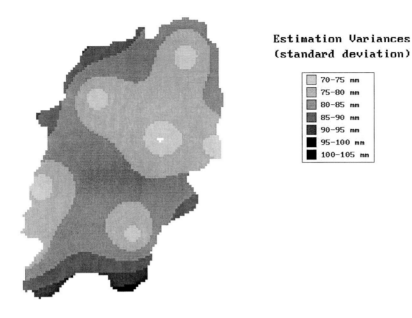

Figure IV.5-6: Kriging estimation variances

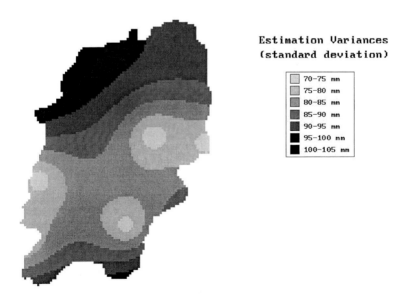

Now the problem is what weight has to be assigned to the distance and to the estimation variance for optimization. This is very subjective and will be different for every hydrologist. A possible combination of the two maps is shown in Fig. IV.5-8.

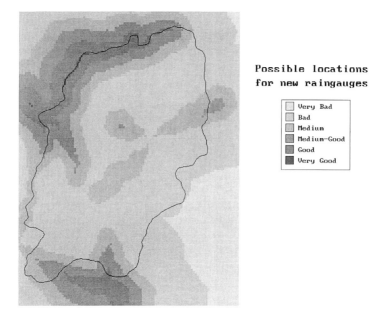

Figure IV.5-8: Optimal locations for additional rain gauges

Questions

1. Explain why a first order network is so important.

2. What is understood by "information loss"?

3. Why do researchers often use the correlation function for network design?

4. Are the design methods of rain gauge networks also useful for evaporation pan networks?

5. Take the example of Section 8: How many stations, with an operation time of how many years, are needed to estimate the long term areal mean with a variance of 5%?

References

Bastin, G., Lorent, B., Duqué, C., Gevers, M.: Optimal Estimation of the average areal rainfall. In: Bogardi, Bardossy (eds.): Multicriterion Network Design using Geostatistics. Water Resources Research, Vol. 21, No. 2, 1985

Bras, R.L.: Hydrology, An Introduction to Hydrologic Science. Addison-Wesley Publishing Company, 1990

Bras, R.L., Rodriguez-Iturbe, I.: Random Functions and Hydrology. Addison-Wesley Publishing Company, 1985

Delhomme, J.P.: Kriging in the hydrosciences. Advances in Water Resources, Vol.1, No.5, 1978

Dijk, M.J. van, Kappel, R.R. van: Optimization of the Cebu rainfall network. M.sc. Thesis, Hydrology Section, Delft University of Technology, 1992

Dijk, M.J. van, Rientjes, T.H.M.: Geostatistics and Hydrology. Part 3: Hydro-Meteorological Measurement Network Design. Hydrology Section, Delft University of Technology, 1994

Eagleson, P.S.: Optimum density of rainfall networks.Water Resources Research, Vol. 3, No. 4, 1967, pp. 1021-1033

Made, J.W. van der: Analysis of some criteria for design and operation of surface water gauging networks. Van Gorcum, Assen, 1987

Made, J.W. van der: Design aspects of hydrological networks. TNO, The Hague, 1988

Matern, B.: Spatial variation. Medd. statens Skogsforskningsinst. Sweden, 49(4), 1960

Moss, M.E.: Integrated Design of Hydrological Networks. US Geological Survey, Water Resources Division, Reston, Virginia 22092, USA, IAHS Publication No.158

Rodriguez-Iturbe, I and Mejía, J.M.: The design of rainfall networks in time and space. Water Resources Research, Vol. 10, No. 4, August 1974, pp. 713-728

WMO (World Meteorological Organization): Hydrological network design and information transfer. WMO No. 433, Operational Hydrology Report, No. 8

WMO (World Meteorological Organization): Hydrological network design -needs, problems and approaches. Operational Hydrology Report, No. 12, 1969

WMO (World Meteorological Organization): Hydrological network design - needs, problems and approaches. Operational Hydrology Report, No. 12, 1982

IV.6 Runoff Irrigation in the Sahel Zone [1]

Thomas Vögtle, Karlsruhe

Goal:

This chapter will give an example of GIS application in large areas. The whole planning process of such a project will be explained step by step: data pre-processing, modelling and visualization. After compilation of required basic information, the selection of suitable parameters and alternative methods for data acquisition/data extraction will be focused. Another main aspect deals with the creation of a decision structure - i.e. the algorithmic combination of the extracted parameters - to determine the required GIS results.

Summary:

The main parameters to determine suitable areas for runoff irrigation in the Sahel-Zone may be *storage capacity of soils*, *accessibility* and *type of irrigation system*. The first one can be derived from soils, vegetation and land use of the surface of the earth. The evaluation of satellite images by computer-assisted multispectral classification delivers this information. The accessibility of an irrigation system will be expressed by the distance to the neighbouring settlements. The type of such a system (micro-/macro-catchment) depends on its location: in flat areas (slope < 10%) micro-catchments are built, slopes > 10% require macro-catchments. A slope model is calculated by means of a digital terrain model (DTM). Both data - distance model and DTM - can be derived by digitizing a topographical map (centres of settlement and contour lines resp.). A specific decision structure based on this information will classify favourable, possible and unsuitable areas for runoff irrigation.

Keywords:

Irrigation, remote sensing, multispectral classification, digital terrain model, digitizing maps, decision structure, GIS application

---

[1] This chapter is a translation of [Vögtle, Tauer, 1991] by kindly permission of WICHMANN Verlag, Heidelberg

1. Introduction

In collaboration with the Institute of Hydraulic Structures and Agricultural Engineering, University of Karlsruhe, an EC-project was defined to determine potential sites for runoff irrigation in huge areas ([Tauer, Vögtle, 1990], [Tauer, Humborg, 1992]).

For a better understanding of the selection process of relevant parameters and suitable methods, basic information about the natural and social environment of a semi-arid zone (Sahel-Zone, North Africa) will be presented in a short introduction.

2. Principles of runoff irrigation (RI)

Runoff irrigation employs rainwater without temporary water storage. Consequently, runoff irrigation can only be accomplished in semi-arid areas during the rainy season by directing surface runoff to selected (agricultural) fields. The plants receive their supply of water during a dry period following a rainfall event exclusively from the moisture thus stored in the soil. Aside from sufficient water availability, adequately deep soils with high water storage capacity must exist on sites where a RI system is to be installed.

This type of irrigation, which uses surface runoff induced by rainfall, has been practised in semi-arid areas for more than 2000 years. Surface runoff is diverted by means of simple earth or rock dams and directed to fields that have been surrounded by ridges or may be terraced. The catchment areas, also called water harvest areas, can vary in size from a few square meters to several square kilometres.

Two different types of RI systems are employed depending on the slope of the terrain. Locations where catchment area and field lie adjacent to each other on the same plain are called *microcatchment systems*. On the other side, *macrocatchment systems* are located at a slope, where the (normally terraced) fields are placed at the bottom of the slope (Fig. IV.6-1).

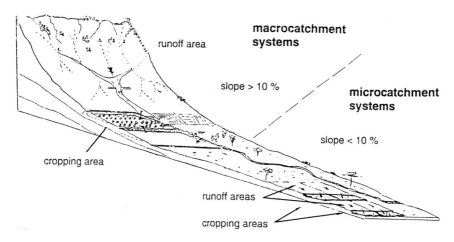

Figure IV.6-1: Types of RI systems utilized on terrain of various slopes

3. Environmental conditions of the Sahel-zone

3.1 Climate

The climate in the Sahel-Zone is characterized by significant seasonal changes between dry and rainy periods. Additional to this aspect, the duration and intensity of the rainy season increase from north to south with its culmination in July/August. The amount of precipitation depends on the global passat winds and is influenced by the periodic dislocation of the inner-tropical conveyance zone (ITC). Throughout the year there are high temperatures with a maximum of about 50 °C before the rainy season (April/May). The extreme fluctuations of rainfall, both its annual total and its distribution within the rainy season, are a primary constraint for agriculture.

3.2 Hydrology

The runoff process in the Sahel-Zone is highly influenced by the extreme climatic conditions. With exception of a few small rivers in the mountainous region in the southern part there are no permanent watercourses. Also permanent natural lakes are only an exception in the Sahel-Zone, while during the rainy season large swamp or pool areas may appear, but most of them fall dry a few weeks after the end of the rainy season. Due to a low groundwater level, which decreases about 5 m to 10 m during the dry season, almost no springs exist.

3.3 Agriculture

Small-scale farming is the most prevalent system in the Sahel for maintaining both a subsistence economy and an extensive grazing system. The types of plants to be grown depend on the soils and the available amount of water. The mechanization of agriculture is minimal, the field work is mostly done manually. The important crops are sorghum, groundnut, maize and beans.

No fixed cropping sequence is employed because each crop is grown on suitable sites (e.g. groundnuts on sandy locations, maize on deep soils). Only in regions where alternative fields are available, fields are allowed to lie fallow.

The most important factor regarding the location of suitable fields is the availability of water. Favourable locations get more and more rare with the expanding population, and the farmers must resort to plots with poorer soil. Techniques for small irrigation systems, such as RI to retain rapidly flowing surface water, are only practised where they have been introduced by development agencies. Evaluations of the quantities harvested in such regions have proved that RI systems offer some advantages.

3.4 Soils

In the Sahel-Zone, soils are affected by climate, stony subsoils, relief and supply of moisture. The soils are primarily shallow and at an immature stage of development. Soils must have a sufficient water storage capacity to be suitable for RI. The plants must obtain moisture exclusively from the water stored in the ground during the dry period. Therefore, loamy or clayey soils can be used for RI. These can be found at locations with a great density and height of vegetation. Thus, it is possible to derive soil characteristics from the type of vegetation.

3.5 Sociology and infrastructure

With exception of some urban centres and a few geographically favourable locations, the population density in all Sahelian countries is low, less than 40 inhabitants per square kilometre. Even though the rural population can still survive on the yields of their fields in climatically favourable periods, an over-exploitation of the environment has begun by expansion of agriculture to erosion-prone sites and the use of shorter fallow periods. Furthermore, the population growth in Sahel countries is about 2% to 3% per year. The establishment of RI systems will not only ensure the production of food, but also reduce the necessary amount of agricultural fields, which takes the protection of environment and natural resources into consideration.

Besides the aspect of population, the distances between potential irrigable sites and settlements become an important limiting factor. Traditional agricultural fields are normally located close to the villages. Distances up to 4-5 km - about one-hour walk - are acceptable. At particularly favourable locations, where an optimal water supply is available for the fields, also greater distances are sometimes negotiated or small temporary (seasonal) dwellings are set up.

4. Discussion of suitable methods

Despite the simplicity of runoff irrigation, a number of parameters must be taken into consideration in determining potential sites for RI systems. The aim of this study is to design a method for pinpointing the favourable areas on the basis of the relevant factors, e.g. soils, topography, sociology and others (see Section 3.). For this application a suitable method has to be developed for creating and combining (modelling) relevant data layers. An important aspect will be the applicability of the method to large territories, and not only to local ones.

In principle, the following methods can be applied:

- Field visit: data collection by ground check and in-situ measurements

- Interpretation of available maps and statistical data

- Aerial survey and visual interpretation or analytical/digital evaluation of aerial photographs

- Evaluation of satellite images (remote sensing)

4.1 Field visit

Field visits, involving the measurement, recording and mapping of relevant data directly during a ground check - usually in cooperation with local experts - are advantageous, particularly if RI sites are to be established for small, local areas only, e.g. at the district level. For larger areas, at the regional level or greater, this method is obviously too time-consuming and too expensive.

4.2 Interpretation of available maps

New and actual map series - especially in digital form - are a very good basis for planning and modelling. But even in developing countries the lack of actual and detailed maps - e.g. topographic and thematic maps for topography, pedology, vegetation and others - introduces great problems in the data acquisition and data treatment process for detecting suitable RI sites. In many cases we will find no country-wide information, with exception of small-scale topographic maps (scale ≤ 1 : 200.000). In addition, it must be taken into consideration that these maps often are out-of-date. Also other statistical data which may be important for planning purposes - like climatic or agricultural data - are mostly not documented adequately or are not available.

4.3 Aerial survey

Aerial colour photographs have a satisfactory spectral and a very good geometrical resolution (colour/infrared film with approx. scale S ≥ 1 : 80.000). These images can be considered, in principle, for more extensive regional areas. The procedure for the evaluation of aerial photographs, whether as a visual interpretation or a complete photogrammetrical evaluation of stereo images, is generally known; it is explained in detail in standard reference textbooks, e.g. [Moffitt, Mikhail, 1980], [Slama, 1980], [Konecny, 1984]. Although the necessary field work is minimal, three other aspects can limit the applicability of aerial photogrammetry: high-resolution photogrammetrical equipment is very sophisticated, and costly precision instruments are sensitive to the extreme climatic conditions. Therefore, they can only be conducted by a few centralized institutions. Furthermore, the quality of the available regional infrastructure (e.g. accessibility and conditions of (local) airports or landing strips, availability of surveying aeroplanes) can be limiting factors. The third aspect is that, even though aerial surveying is possible for areas larger than regional level, the analogous evaluation, which is the only suitable method at the moment in developing countries, is expensive and time-consuming due to the enormous amount of data. Therefore, aerial photographs were judged to be not very suitable for this study.

4.4 Evaluation of remote sensing data

By increasing the area under investigation, the evaluation of remote sensing data in form of satellite imagery has proved to be a better approach than those mentioned above. This method is much more rapid and provides a cheaper determination of relevant parameters for the area examined. Up to now satellite images provide information about the characteristics and use of the surface of the earth. Key topographic information in terms of contours and elevation data - e.g. roads or populated areas, topography and others - can be ascertained with current, more sophisticated earth observation systems. Only a small effort (acquisition of ground truth, i.e. training samples for classification) is required to process remote sensing data, and it is nearly independent on the size of the territory under investigation. Taking different illuminations into consideration, the data gained from one satellite scene can be transferred to adjacent scenes without necessity of field visits.

There are two main methods for the evaluation of satellite images, which use digital image processing features:
  * Visual interpretation via image optimization
  * Computer-aided multispectral classification

By selecting a suitable method for the specific task of this application the visual interpretation has to be excluded. On the one hand (colour-)optimized satellite images offer a more sensitive discrimination in land use and vegetation classes, but on the other hand a valid interpretation needs a good knowledge of the local conditions and a great experience of the interpreter.

Apart from this the method is very time-consuming and - of course - influenced by subjective aspects and the individual perception of the operator (and therefore, not free from errors), especially for large satellite scenes.

On the other hand computer-aided or automatic image classification will require less interactive work by the operator (see Chapter III.3), and the results are stored directly in digital form for further processing and combining with other data layers, e.g. in a GIS.

In this study it has to be taken into consideration that only specific land-use classes are of fundamental interest. Therefore, an unsupervised classification which can only distinguish between different *reflection classes*, but not between the necessary *semantic classes*, has to be excluded. For a supervised classification we need a-priori knowledge about the land-use of selected, representative training areas (=ground truth), which takes about 2 to 3 weeks for an area of approx. 2000 km^2 to 3000 km^2 (presuming experience and a well prepared mission).

4.5 Conception of the evaluation procedure

If we take into account all information mentioned above, it is obvious that most parameters (as far as possible) should be extracted from remotely sensed data. For completion, available data and topographic maps will be used while field visits are only necessary for the acquisition of ground truth. The input data (digitized or directly in digital form) will be preprocessed and stored as data layers in a GIS. By superimposing and logical combination of these data layers in a GIS the final results for determining potential sites for RI can be extracted. For this application, a raster-based GIS was preferred because most information should be extracted from digital images which are already in raster format. Therefore, all other data - especially vector data - have to be converted into raster data which allows the combination of different data layers to be done in an easy way ([Tauer, Vögtle, 1990]).

5. Selection of relevant parameters

It has already been explained in Section 3 that the available amount of water is the most important factor for the determination of RI systems. The water storage capacity of soils - mostly validly described by the field capacity - has a great variability in the Sahel-Zone and can not be determined for large areas in a simple way. Therefore, the storage capacity of the soil should be derived indirectly from secondary effects (e.g. vegetation type and their spatial distribution). This method obviously cannot achieve a high accuracy, so the description of the storage quality has to be limited to three main classes: *unsuitable - possible - favourable.*

The choice of a RI system (micro- or macro-catchment) depends on the slope of the terrain. Only in areas with higher slopes (hills, mountains) macro-catchment RI systems with terraced fields on the bottom of a slope can be built up, while micro-catchments are used in flat areas. In the present study, a slope of 10% has been defined as threshold between the two types of RI systems. The inclination data can be derived from topographical maps or - if available - from a Digital Terrain Model (DTM).

The distance between suitable fields and settlements has an enormous influence in the acceptance of potential RI sites. Therefore, all centres of settlements were digitized either directly from the satellite image or from a topographical map. Around these centres concentrical circles were drawn digitally, that means, every point of the study area belongs to a distance class (distance to the nearest settlement).

Table IV.6-1: Methods for determining parameters relevant to Runoff Irrigation

parameter	derived from	method
storage capacity of soil	soil cover, natural vegetation, land use	evaluation of satellite images by computer-assisted classification
accessibility	distance between cultivated area and settlements	manually from topographic maps or by digitizing the centres of settlements and creating comprehensive distance model
type of system	slope of terrain	manually from topographic maps or derived from an inclination model based on a DTM (from digitized iso-lines or directly from aerial or satellite data)

These three parameters which are used for determining potential sites for RI systems are shown in Tab. IV.6-1 together with their determination methods.

In the next sections it will be shown how to build up the data layers for this specific GIS as well as the further data processing by means of a practical application in a test area of Mali (North Africa), northern Kayes region.

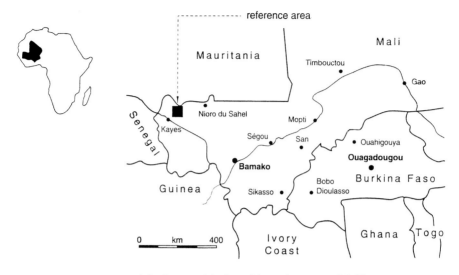

Figure IV.6-2: Geographical position of test area 'Mali'

6. Creation of data layers for the test area 'Mali'

6.1 Test area 'Mali'

The geographical position of the test area 'Mali' is shown in Fig. IV.6-2. Its area of approximately 50 km × 64 km is limited in the north by the *Térékolévalley,* in the east by the *Kaarta* plateau which is about 200 m to 300 m higher than the surrounding areas, in the south by the *Garivalley* and in the west by a large plain till the *Lac Magui.*

Table IV.6-2: List of semantic classes for TM and SPOT

class	semantic meaning (TM)	semantic meaning (SPOT)
2	water surfaces	shades
3	shades	swamps
4	swamps	green vegetation, orchards
5	green vegetation, orchards	dense grass, bushes
6	grass, single bushes	sparse vegetation, scree slopes
7	grass, shrub savanna	rocks, sparse vegetation
8	burned areas	shrubs, trees, scree areas (plateau)
9	rocks, scree slopes	vegetation, bare soils (plain)
10	sparse vegetation	dense tree, shrub vegetation
11	vegetation, bare soils (plain)	as 5, but different plant species
12	green vegetation	medium tree, shrub vegetation
13	dense tree, shrub vegetation#	as 8, but in shaded areas
14	sand, dry grass	grass, bare soils
15	bare soils, sparse vegetation	as 4, but different soil moisture
16	as 14, but different soil moisture	reject
17	as 14, but different soil moisture	
18	as 14, but different soil moisture	
19	reject	

6.2 Results of the multispectral classification

On the base of a field mission for the acquisition of the required *ground-truth,* a supervised classification of two different satellite images was calculated to investigate the suitability of various sensor systems. Therefore, the American Landsat-TM and the French SPOT system were used. Landsat-TM has a better spectral resolution (7 spectral bands vs. only 3 bands of SPOT) but has a poorer geometrical resolution (pixel size on the ground about 30 m × 30 m vs. 20 m × 20 m of SPOT). Due to a different reflectance behaviour of wet and dry vegetation, the discrimination of those objects in the Sahel-Zone, i.e. the assignment to specific classes, depends highly on the date of the satellite image. Therefore, scenes of nearly the same date shortly after the rainy season were used.The classes which could be discriminated are shown in Tab. IV.6-2, separately for Landsat-TM and SPOT. The poorer spectral resolution of SPOT causes a smaller number of classes.

The separability of the chosen (semantic) classes can be shown by means of feature ellipses in a graphical way. As an example, the feature ellipses of 12 classes (built up with the mean reflectance values (brightness) and their standard deviations in band 4 and 7 of Landsat-TM) can be seen in Fig. IV.6-3.

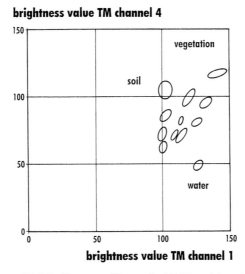

Figure IV.6-3: Feature ellipses for TM band 1 and 4

The quality of such a classification can be described with the classification rate based upon training areas, i.e. the ratio of correctly classified pixel to the total number of pixel. This can be visualized in form of a confusion matrix, where the diagonal elements are the ratio of correctly classified pixel, the other elements contain the values of incorrectly classified pixel and their related (wrong) classes (Tab. IV.6-3). The overall classification rate - weighted by the area of each class inside the satellite image - was calculated 97.4% for the TM-scene and 89.1% for the SPOT-scene.

As it can be derived from the amount of wrong classifications we can define some typical problems in multispectral classification process. The reasons for and the treatment of these problems in a further iteration process will be shortly explained:

Settlement: It can be derived from positions of the feature ellipses in Fig.IV.6-3 that the classes *settlement* and *bare soils* can not be separated optimally, i.e. many locations with bare soils have been misclassified as *settlement,* on the other hand, *bare soils* often appear inside *settlements.* The reason for this phenomenon may be similar materials (soils) of the simple buildings and their surrounding areas, additionally there are mostly no paved roads. If we take into consideration the integrating effect of a large pixel size (in relation to the dimension of that houses and roads), the spectral reflectance values will be nearly the same. Therefore, these two classes can not be distinguished. A solution of this problem may be the use of additional information from topographic maps (e.g. to mask the settlement areas) or the use of additional spectral features like *texture* or *structure*.

Table IV.6-3: Classification result for 18 classes (in [%]) for TM-scene 'Mali' (class description see Tab. IV.6-2)

class	2	3	4	5	6	7	8	9	10	11	12	13	14	15	16	17	18
2	**100**	0	0	0	0	0	0	0	0	0	0	0	0	0	0	0	0
3	0	**100**	0	0	0	0	0	0	0	0	0	0	0	0	0	0	0
4	0	0	**100**	0	0	0	0	0	0	0	0	0	0	0	0	0	0
5	0	0	0	**98**	0	0	0	0	0	0	0	2	0	0	0	0	0
6	0	0	0	0	**99**	0	0	0	0	0	0	0	0	1	0	0	0
7	0	0	0	0	0	**89**	0	1	0	0	0	0	0	1	0	0	9
8	0	0	0	0	0	0	**98**	2	0	0	0	0	0	0	0	0	0
9	0	0	0	0	0	1	2	**94**	0	0	0	0	0	0	0	0	3
10	0	0	0	0	0	0	0	0	**100**	0	0	0	0	0	0	0	0
11	0	0	0	0	0	1	0	0	0	**99**	0	0	0	0	0	0	0
12	0	0	0	0	0	0	0	0	0	0	**100**	0	0	0	0	0	0
13	0	0	0	3	0	0	0	0	1	0	0	**96**	0	0	0	0	0
14	0	0	0	0	0	0	0	0	0	0	0	0	**99**	1	0	0	0
15	0	0	0	0	0	0	0	0	0	0	0	0	0	**100**	0	0	0
16	0	0	0	0	0	0	0	0	0	0	0	0	0	0	**100**	0	0
17	0	0	0	0	0	3	0	0	0	0	0	0	0	0	0	**97**	0
18	0	0	0	0	0	2	0	1	0	0	0	0	0	0	0	0	**97**

Shadows: Especially in hilly or mountainous areas, great differences in reflectance can appear (dependent on the sun angle) even if the land coverage is the same. Those parts of the relief which are oriented towards the sun will get brighter, while those which are oriented away from it will get reduced in reflectance or even lie in shadow areas. But these poorer reflectance values are similar to other classes like *swamp* or *water*, so misclassifications may appear. An improvement of the classification results can be achieved by the knowledge of topography, e.g. in form of a Digital Terrain Model (see Chapter III.10), where radiometric corrections can be calculated and applied. At the Institute for Photogrammetry and Remote Sensing (IPF), University of Karlsruhe, such a software package has been developed, but not applied in this study, due to economical aspects (relation between realistic improvement and necessary expense). As an alternative, the reflectance-reduced areas were treated as own (sub-)classes.

Water: To detect water areas within a satellite image, training samples (see Section 4.4) were measured in large, continuous water areas (i.e. mostly lakes) which have a uniform reflectance. Therefore, we get only a small (statistical) variance in the grey values. But small, shallow pools where surrounding areas fall additionally into the same pixel, have not been classified as *water*. This information would be absolutely important for irrigation activities in the Sahel-Zone. Additional training areas were measured for those pools. Afterwards, all water areas were classified correctly.

6.3 Storage capacity of soil

As mentioned in the sections above, storage capacity of soil cannot be determined directly and quantitatively. But the available amount of water can be derived roughly (*very good, sufficient, bad*) from the typical vegetation coverage of a classified image (Colour-Page 13a).

Table IV.6-4: Suitability of semantic classes regarding water storage capacity

Suitability	TM-Classes	SPOT-Classes
unsuitable	2, 3, 4, 5, 8, 9, 10, 12, 13	2, 3, 4, 6, 7, 8, 10, 12, 13
possible	6, 14, 15, 16, 17, 18, 19	5, 11, 16
favourable	7, 11	9, 14, 15

6.4 Accessibility

To determine the accessibility of potential sites for runoff irrigation, a distance model with the settlements inside our test area was created. In a first step, the centres of towns and villages were digitized from a topographic map (1:200.000). Afterwards, 4 concentrical circles were drawn around each centre with a radial distance of 2 km, 4 km, 6 km and 8 km respectively, resulting in 5 different distance classes (Colour-Page 13b). An acceptable distance between settlements and irrigation fields can now be defined dependent on the quality of soil, i.e. if a location has a favourable soil quality, a longer distance to the next settlement will be accepted (e.g. up to 8 km), but the suitability of this field will decrease from *favourable* to *possible*.

6.5 Type of RI-system

The choice of the type of RI-system depends on the slope of the surface. An inclination model can be derived easily and automatically from a *Digital Terrain Model* (DTM, see Section 5 and Chapter III.10). A manual determination of slope information for such a large test area would be too time-consuming and therefore, too expensive. On the other hand, a considerably great effort is also required to produce a DTM, if it is not yet available in digital form (e.g. as in most parts of Central Europe. Dependent on the accuracy required for a specific application, there are different methods for determination of a DTM. Subsequently, the most important ones will be explained shortly (in order of decreasing costs):
- Terrestrial tachymetry (surveying method)
- Stereoscopic evaluation of aerial / satellite images
- Digitizing of topographical maps

Geodetic surveying by means of terrestrial tachymetry (see Chapter II.2) is the method with the highest accuracy, but also the most time-consuming one, if the data acquisition has to be done for large areas. Nowadays, modern geodetic instruments like electronical tachymeter with digital data storage and automatic data transfer to computer systems are used in practical work. In this method, only the representative relief points (e.g. points with a significant change in slope, top of a hill etc.) are measured, therefore, a good approximation to the terrain can be attained by a minimal amount of data. However, the whole project area has to be traversed and inspected by a surveying team, so the data acquisition for a large area - like in this study - would be too time-consuming and too expensive.

An alternative can be the evaluation of stereoscopic aerial or satellite images. Here we attain elevation data by means of photogrammetric analogous, analytical or digital workstations (e.g. analytical plotter). The main advantage of this method is a fast data acquisition of large areas with only a minimal investment of field work (measuring of control points). With this method, representative terrain points (like described above) can be measured as well as elevation profiles or rasters, because the necessary measuring time per point is usually quite shorter than for terrestrical surveying. However, precondition is the availability of (actual) aerial or satellite images or the possibility to produce new ones. The evaluation of stereoscopic satellite images (e.g. SPOT, MOMS, see Chapter II.4) delivers only a reduced accuracy (about ± 5 m to ± 10 m standard deviation); therefore, we have to judge the suitability of this method for every specific application.

Digitizing topographic maps is performed manually in most cases, scanning maps with an automatic raster/vector-conversion is still under investigation (see Chapter III.7). The contour lines - i.e. lines of the same elevation - as well as important terrain points (e.g. top of a hill, depressions etc.) are digitized with a cursor, connected to a digitizing table (e.g. A1- or A0-format). Each contour line and each single terrain point will be marked by a specific label which represents its elevation value. Special algorithms are able to create a DTM (for instance a regular elevation grid) out of this amount of single X-, Y-, Z-points. By digitizing maps, data acquisition of large areas can be done in a short time within an acceptable budget.

Due to these reasons and lower requirements of accuracy for DTM determination (only to choose the type of RI system), the digitizing of a topographic map (scale 1:200.000) was applied in this study, larger scales were not available. From this DTM a slope model was derived by calculating the maximal negative slope (=minimal slope) between every grid point and its 8 neighbour points:

$$S_{gp} = MIN \left(\frac{(H_{gp} - H_i)}{d_i} \right) \qquad i = 1, 2, ..., 8$$

where S_{gp} - Slope to a grid point
       H_{gp} - Elevation of the grid point
       H_i - Elevation of the neighbour point i
       d_i - Distance to the neighbour point i

With this equation, top of hills, upper terrain edges and similar terrain phenomenons which indeed produce higher slopes but do not sample water, can be excluded. As a result, we again get a raster matrix which contains the slope values. Two different slope classes were defined, class 1 for slopes lower than 10% and class 2 for slopes higher than 10% (Colour-Page 14a).

7. Combination of different digital data layers in a GIS

For a superposition and data modelling of the three layers
- classified satellite image for water storage capacity
- distance model for accessibility
- inclination model for type of RI system

a specific decision structure had to be created (Fig. IV.6-4). This structure represents the data modelling process, i.e. it is dependent on the application. The algorithm was implemented on a raster-oriented GIS, because most data layers (land use, slope model) are acquired in raster format. In the upper data layer (Fig. IV.6-4) the result of the classification process is stored, in the middle layer the distance model and in the lower one the slope model.

We have to take into consideration that all data layers must be related to the same geo-reference system (e.g. coordinate system, see Chapter II.1). Data layers derived from topographical maps are still georeferenced, while satellite images contain geometrical distortions (satellite orbit not parallel to the coordinate system, local distortions caused by the topography etc.). Thus the images have to be corrected (geo-referenced) first by means of a geometrical transformation (based on control points or by using the methods of orthophoto generation, see Chapter III.2).

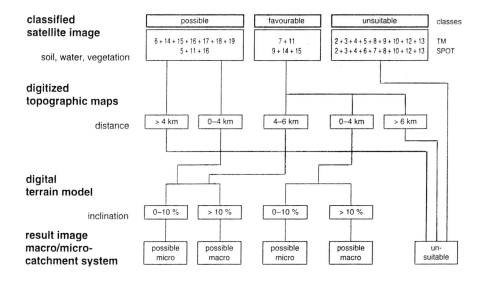

Figure IV.6-4: Decision structure for determination of runoff irrigation sites in Mali

The control points - i.e. identical points in image and map - have been digitized from a topographic map 1:200.000. However, it was difficult to find a sufficient number of those control points, because the Sahelian landscape has not enough significant structures/objects which can be identified unambiguously in both, map and image (e.g. intersections of roads, great buildings etc.) and which can be measured with an adequate accuracy.

Besides *Bool Operations,* the method of *Look-up Table* (LUT) computations can be used to reduce processing time. A special LUT was generated for each data layer in the GIS to assign a specific key number or code to each class. The (simple) sum of the three codes (one per layer) will lead to a new result code which can be related to one of the result classes. Therefore, the calculations per raster point can be reduced to 3 additions, and large quantities of data can be processed in a very short time, even on smaller computers ([Vögtle, Tauer, 1991]).

As result, a GIS image is generated which contains the 5 pre-defined classes of suitability for runoff irrigation (Colour-Page 14b). For a final output (e.g. on film, raster plotter etc.) this image has to be completed by a map frame, legend, title and other descriptions (see Chapter III.1).

It could be seen during the realisation of this application, that digital processing of numerous data layers in a GIS has some advantages in comparison with classical planning methods:

- Use of additional information/data layers (e.g. satellite images) in a GIS

- Possibility of an automated processing, modelling and analysing of the data

- Realisation of numerous variations and/or alternatives without significantly increasing necessary effort

- Subsequent extensions, additions and changes in the data base without great effort

- Objective and repeatable results

- No fatigue of operator

Questions

1. What is the most suitable method to acquire land use in very large areas ?

2. What are the advantages/disadvantages of Landsat-TM and SPOT sensors for this task?

3. How can an accessibility model be created ?

4. Which method for determination of a digital terrain model is suitable for large areas ?

5. How can an algorithm for modelling of data layers be implemented ?

References

Bill, R., Fritsch, D.: Grundlage der Geo-Informationssysteme 1. Wichmann Verlag, Karlsruhe, 1991

Burrough, A.: Priciples of Geographical Information Systems for Land Resources Assessment. In: Monographs on Soil and Resources Survey. Oxford, Calendron, 1986, 194 p.

Chrisman, N.R.: The role of quality information in the long-term functioning of a geographic information system. Cartographica (21), 1984, p. 79-87

Encarnacao, J.L.: Computer Graphics. R. Oldenbourg Verlag, München-Wien, 1975

Konecny, G., Lehmann, G.: Photogrammetrie. Walter de Gruyter Verlag, Berlin, 1984

Moffitt, F.H., Mikhail, E.M.: Photogrammetry. Harper & Row Publishers, New York, 3. Edition, 1980, 648 p.

Slama, Ch. C. (ed.): Manual of Photogrammetry. American Society of Photogrammetry, Falls Church, 4. Edition, 1980, 1056 p.

Tauer,W., Humborg, G.: Runoff Irrigation in the Sahel Zone, Remote Sensing and Geographical Information Systems for Determining Potential Sites. Verlag Josef Margraf, Weikersheim, 1992, 204 p.

Tauer, W., Vögtle, T.: Das Potential an Sturzwasserbewässerungsflächen in der Sahelzone. Abschlußbericht zum EG Forschungsprojekt CEE R&D Programme TS2*-0018-D(BA), Karlsruhe, 1990, 130 p.

Tauer, W., Prinz, D., Vögtle, T.: The potential of runoff-farming in the Sahel region: Developing methodology to identify suitable areas. Water Resources Management, Kluwer Academic Publ., Dortrecht, Boston, London (in preparation)

Tomlinson, R., Boyle, A.R.: The state of development of systems for handling natural resources inventory data. Cartographica (18), 1981, S. 65-95

Vögtle, T., Tauer, W.: GIS-Anwendung: Sturzwasserbewässerung in der Sahel-Zone. In: Bähr, Vögtle (eds.): Digitale Bildverarbeitung - Anwendung in Photogrammetrie, Kartographie und Fernerkundung. Wichmann Verlag, Karlsruhe, 1991, p.271-284

IV.7 Hydrological Modelling of Floods

M.E. Boomgaard, R. Petter, H.R. Vermeulen
R.H. Boekelman, M.J. van Dijk, T.H.M. Rientjes, Delft

Goal:

The goal of this model is to show how to use a GIS for data pre-processing for a physically-based rainfall runoff model.

Summary:

The process of rainfall runoff will be explained and demonstrated through water-depth maps made with the GIS software package ILWIS, and with hydrographs. A physically based distributed model, DAGUDO, was applied to the catchment of the upper Progo river in Central Java. Depending on land use the model applies two different approaches to describe surface runoff. The model simulates floods resulting mainly from this flow component. A connection is made between two-dimensional overland flow and a network of one-dimensional river segments.

Keywords:

Physically based hydrological modelling, floods, kinematic wave equation, DAGUDO, digital elevation model

1. Introduction

Floods are one of the major causes of loss of life and economic damage from natural disasters. Therefore, it is not surprising that the prediction and analysis of floods is of great interest in the hydrological science. However, even today our knowledge about the occurence and behaviour of floods is not too clear. The introduction of digital computers for hydrological prediction of flood events, both for gauged and ungauged sites, made it possible to analyse the inundation problem further.

GIS can be used as a pre- and post-processing tool for hydrological modelling of floods since it can link topographic data to other properties like description of soils, land use, groundcover, ground water conditions or rainfall amounts. A problem with all these properties is that an accurate description of even a small watershed makes GIS a memory intensive and compu-tationally complex system.

The prediction of surface runoff is one of the most useful hydrological capabilities of a GIS system. The prediction may be used to assess or predict aspects of floods. GIS is further an important tool for education. Therefore, GIS can not only be used as a database for digital maps but it can also produce nice presentations of the factors affecting the runoff process.

If a physically based model (see Chapter IV.4) is considered, a large amount of spatially distributed information is needed. Whether this information is partially lumped or fully distributed, depends on the user. If time or available data for modelling are limited, a rough distribution may be sufficient to acquire reasonable results. This chapter gives an example of efficient use of a physically based model, DAGUDO, by its application to the upper Progo catchment in Central Java.

The catchment covers an area of 420 km^2 and is characterized by two volcanos which reach an altitude of 3000 m. The Progo river springs from the volcano slopes and provides sawahs[1] in the lower reaches of the basin with irrigation water. The catchment outlet is located at an altitude of 500 m.

2. Theoretical background of DAGUDO

Primary goal of research is flood estimation in tropical areas with steep slopes. In such areas high rainfall intensities allow the assumption that overland flow is the major runoff source. Infiltration and storage components are regarded as losses which do not affect the flood hydrograph resulting from high intensity storms.

2.1 Surface runoff

The model DAGUDO has been developed for rather steep areas where the flow velocity is higher than the wave velocity (critical flow). As a consequence of this, downstream distortions do not influence the upstream water levels and the term of inertion can be neglected. Therefore, the *Saint-Venant equations* used in the model can be simplified to the kinematic wave equation. This kinematic wave equation can properly be used when the parameters of inertion and pressure are negligible compared to those of gravity and friction ([Eagleson, 1970]). This simplified kinematic wave equation is an equation of continuity:

$$\frac{\partial h}{\partial t} + \frac{\partial q}{\partial x} + \frac{\partial q}{\partial y} = 0 \qquad\qquad (IV.7\text{-}1)$$

---

[1] rice fields surrounded by bunds

where q = discharge [m²/s]
         h = thickness of the water layer
         t = time.

The overland flow for one-dimensional surface runoff is described by the *Strickler-Manning formula*:

$$Q = k_m \ b \ i^{\frac{1}{2}} \ h^{\frac{5}{3}}$$

(IV.7-2)

where Q = discharge [m³/s]
         k_m = Manning roughness coefficient [m$^{1/3}$/s]
         b = width of the stream flow [m]
         i = slope
         h = water depth [m].

Because of the equilibrium between gravity and friction, the flow acceleration can be set to zero. This results in an equal energy gradient and slope gradient. As the model simulates a two dimensional flow, the calculation of the flow rate has to be carried out in two directions, x and y. By substituting the Strickler-Manning formula in the equation of continuity, the two-dimensional flow is obtained.

To implement this in a computer program an implicit calculation scheme has been used. The advantage of an implicit scheme is the possibility of using longer time intervals than in an explicit calculation scheme, before instability takes place. However, the algorithm for the solution of the unknown variable is much more complicated.

The factor $h^{5/3}$ in the Strickler-Manning formula can be divided into a linear and a non-linear part:

$$Q_j = k_{m(j)} \ b \ i_{(j)}^{\frac{1}{2}} \ h_{(j)}^{\frac{2}{3}} \ h_{(j)}$$

(IV.7-3)

where j = index of a pixel.

The linear part h_j can be used to create a time dependent equation. Implicit calculation then requires the water depth at the end of each time step (h^{t+1}). The non-linear part $h^{2/3}$ is determined explicitly at the beginning of each time step.

When the Strickler-Manning formula is discretized with a central method, the following expression for the discharge is obtained:

$$Q_j^t = 0.5^{\frac{5}{3}} \ k_j \ b \ i_j^{\frac{1}{2}} \ \left(h_{j+1}^t + h_j^t \right)^{\frac{2}{3}} \ \left(h_{j+1}^{t+1} + h_j^{t+1} \right)$$

(IV.7-4)

In the model, the flow rate from or to a pixel is determined by the values in this pixel and the values in the four surrounding pixels. This is a so-called "5 points method" (Fig. IV.7-1). For this scheme the equation of continuity can be written as follows:

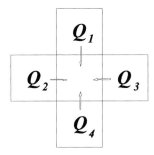

Figure IV.7-1: Five point approach

$$O \cdot \frac{\partial h}{\partial t} = Q_1 + Q_2 + Q_3 + Q_4 \tag{IV.7-5}$$

where O =   the pixel area (the same for all pixels).

By substitution of the discretized Strickler-Manning equation in the continuity equation, an expression appears containing the variable h^{t+1}.

The solution of the implicit scheme is an iterative process which is very time-consuming. The ADI (Alternating Direction Implicit) method, developed by [Peaceman, Roachford, 1955], is used to solve the extensive equation matrices. The main characteristic of the ADI method is the division of a matrix in a vertical and a horizontal part. Every iterative step requires the solution of equation matrices in the vertical and horizontal direction applying the double sweep method or Gauss elimination.

By this iterative process a matrix of new water-depths is formed which is the input for the next computation step. During a computation run the water levels or discharges in all pixels can be stored at different times, and afterwards they can be visualized by means of a GIS.

2.2 River runoff

Major river arms have been represented schematically by a network of segments. Similar to the overland flow, the river flow within one segment has been modelled using the kinematic wave approach with the Manning equation. Flow to an adjacent downstream segment, however, is determined explicitly, which limits the extent of the time step. Overland flow operates without such a limitation.

A special module has been developed to describe the transition from overland flow to river flow. The separation of these processes has the advantage that both can be computed using different time steps; as explained before, river flow requires shorter time intervals than overland flow but computing time for the river network is relatively short: therefore, the time step used for overland flow is decisive when these processes are split.

The model is able to transform the water level into flow rate (m³/s) at any given point in the river for every time step. A hydrograph can thus be obtained.

3. Data processing

The model uses digital input maps concerning elevation, land use, soil texture, rainfall distribution and river network. For the generation of these maps the GIS ILWIS (Integrated Land and Watershed management Information System, developed by ITC Enschede, The Netherlands) was used. Digital maps were produced at a 250 m × 250 m grid size comprising approx. 6700 elements.

At first a digital elevation model (DEM) is produced (see Chapter III-10). Since the topography of the watershed has to be modelled and because surface water tends to flow downhill, the hydrological importance of terrain modelling is clear. The DEM (Fig. IV.7-2) was interpolated from contour lines of a 1:50.000 elevation map. The problem in modelling DEM data is the production of non-existing depressions due to noise in the elevation data affecting the interpolation schemes to describe variations in elevation between raster points. The application of a smoothing filter can be a solution to facilitate "drainage" of small enclosed depressions.

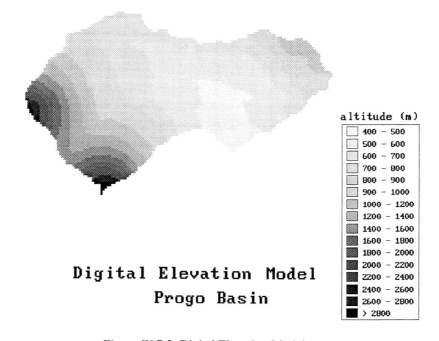

altitude (m)

	400 – 500
	500 – 600
	600 – 700
	700 – 800
	800 – 900
	900 – 1000
	1000 – 1200
	1200 – 1400
	1400 – 1600
	1600 – 1800
	1800 – 2000
	2000 – 2200
	2200 – 2400
	2400 – 2600
	2600 – 2800
	> 2800

Digital Elevation Model

Progo Basin

Figure IV.7-2: Digital Elevation Model

Roughness of land surface was directly related to land use; a map of *effective* roughness values was derived from a 1:50.000 land use map, using values proposed by [Engman, 1986]. The digital land use map is displayed in Fig. IV.7-3. Many gullies have evolved, particularly on the volcano slopes, due to heavy rainfall. As a result, runoff is faster than would be expected from the land use roughness coefficients.

Soil texture was considered to determine infiltration rates and for each soil type an infiltration function was defined.

The digital river map (DRM) was obtained by digitizing a 1:50.000 river map. The same grid size (250 m × 250 m) was used, allowing only modelling of the main river network. The influence of the dense network of small streams was accounted for by the effective roughness values of land surface. The DRM was composed of over five hundred river segments and four types of river branches were distinguished. Each branch type was assigned a value for bed width and river roughness. A minimum and a maximum bed slope were also assigned.

Rainfall is one of the most important processes, particularly the spatial distribution of rainfall. Sometimes an assumption of uniformly distributed rainfall in time and space is a very crude approximation. Therefore, DAGUDO has been designed to deal with spatially distributed rainfall, calculated by means of an interpolation scheme, and theoretically, every desired *temporal* rainfall distribution can thus be used in the model. Tab. IV.7-1 shows the set-up of different values applied to the standard case.

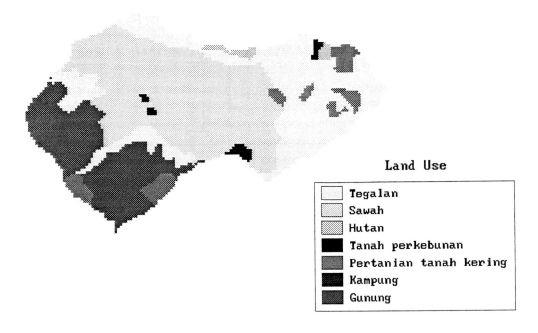

Figure IV.7-3: Digital land use map

Table IV.7-1: Standard values for sensitivity analysis

number of columns / rows	149 / 84
number of river segments	549
time step	120 s
grid size	250 m
k_m-value	distributed
precipitation	uniform, 30 mm during one hour

4. Results

The resulting computations and observations are shown in Fig. IV.7- 4. The analysed storm rainfall occurred on January 29th, 1977 and was considered to last three hours with constant intensity. Fig. IV.7-5 shows the spatial rainfall distribution. As base flow is not accounted for in the model, the magnitude of this phenomenon was estimated from the records and subtracted from the measured hydrograph. It can be seen that discharge becomes significant only after more than two hours. This is caused by the area's storage capacity which is considerable. Deviations from measured discharges mainly occur during the rising time; Fig. IV.7-4 clearly shows that the bulging shape of the hydrograph is caused by the spatial distribution of rainfall.This example shows that peak discharge is reasonably approximated by the model, whereas the shape of the hydrographs may deviate. Our attention was primarily focused on the simulation of peaks.

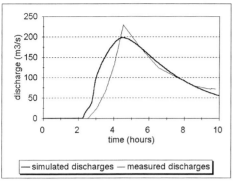 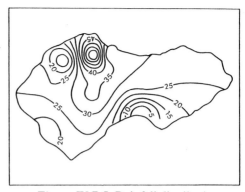

Figure IV.7-4: Hydrograph   *Figure IV.7-5:* Rainfall distribution

In Fig. IV.7-6 a uniformly distributed storm was applied. Every ten minutes the water depth at the pixels was stored in files. Afterwards the process of rainfall runoff could be shown by presenting the water depths at the different time intervals.

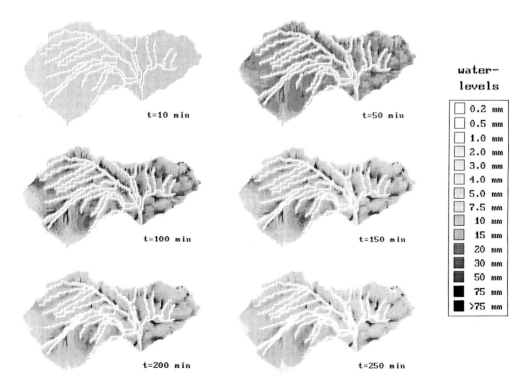

Figure IV.7-6: Water depths at different time intervals

References

Abbott, M.B. et al.: An introduction to the European Hydrological system "SHE": Structure of a physically based, distributed modelling system. J. Hydrol. 87, 45-77, 1986

Boomgaard, M.E., Petter, R.: A physically based rainfall-runoff simulation model: case study of the upper Progo basin, Central Java. Delft University of Technology, Faculty of Civil Engineering, Delft, The Netherlands, 1993

Chow, Ven Te: Handbook of Applied Hydrology, 1964

Dado, E., Franssen, G.S.: Een model voor de oppervlakteafvoercomponent in hellende gebieden in combinatie met het geografisch informatiesysteem ILWIS (Translated from Dutch: "A model for surface runoff in sloping areas in combination with the geographic information system ILWIS"). Delft University of Technology, Faculty of Civil Engineering, Delft, The Netherlands, 1992

DeVantier, B.A., Feldman, A.D.: Review of GIS applications in Hydrological modelling. Journal of water resources planning and management, Vol. 199 (2), 1993, pp. 246-261

Engman, E.T.: Roughness coefficients for routing surface runoff. J. Irrig. and Drain. Eng. 112 (1), 39-53, 1986

ITC: ILWIS 1.3 user's manual. ITC Enschede, 1992

Mc Donald & partners: Hunting technical services LTD (1971): Kali Progo basin study. London, Yogyakarta, 1971

Peaceman, D.W., Roachford, H.H.: Numerical calculation of multidimensional miscible displacement, 1962

Seyhan, E.: Mathematical simulation of watershed hydrologic processes. Geogr. Inst. of R.U. Utrecht, 1977

Sri Harto: A study of the unit hydrographs basic characteristics of rivers on the isle of Java for flood estimation. UGM Yogyakarta, 1985

Stelling, G.S., Booy, N.: Computational modelling in open channel hydraulics. 1991

IV.8 Regio-Climate-Project (REKLIP)

Klaus-Jürgen Schilling, Karlsruhe

Goal:

This chapter will give a short introduction in a practical application of environmental monitoring in the field of climatology. The production of a climatological atlas for the Regio-Climate-Project REKLIP by means of digital cartography will be shown.

Summary:

The "Regio-Climate-Project" (REKLIP) is a tri-national project (France, Swizerland, Germany) dealing with the analysis of the climate and its change during the last 40 years in the area of the upper Rhine valley. The result will be a climatological atlas comprising 80 maps produced with methods of digital cartography within a Geoinformation System (GIS). For this purpose a digital data base containing topographic and climatologic data was established using a software combination of ARC/INFO as GIS and ORACLE as data base.

This chapter shows the concept of the GIS-solution, especially the connection of ARC/INFO with PC-based systems for cartographic layout (ALDUS Freehand and ADOBE Photoshop).

The map examples demonstrate the combination of digital topography (hill shading) with thematic information (overlayed raster matrices). Based on these results further climate modelling will be done concerning the above mentioned Rhine valley region.

Keywords:

GIS, ORACLE-Database, cartography, map creation, map production

1. Introduction

The "Regio-Climate-Project" (REKLIP) is a trinational project of France, Swizerland and Germany. The area of interest has been located in the upper Rhine valley which is a unique landscape unit, but devided by the national frontiers of the three participants. Therefore, the principal aim of this project was to organize a joint acquisition and analysis of climatological data which should illuminate the spatial and temporal distributions of the most important climatic parameters ([Schweinfurth, Laing, 1992], [Fiedler et al., 1987]). Its scientific research program began in 1989 with an expected duration of 8 years.

Moreover, the effect of climate upon civilization has to be considered. Beside the climate proper, other parameters such as "air quality" and "energy transport in the lower atmosphere", the so called *microclimate*, will be included during the process of analysis.

The most important aims of REKLIP are:

1. Creation of a climate atlas in which climatic parameters (wind, temperature, clouds, fog, humidity, etc.) for the project area are represented.

2. Provision of criteria (climatic models) for the assessment of hygienic air conditions after chemical or nuclear accidents.

3. Analysis of the impact of human activities (urbanization, traffic, industry, agriculture, generation of energy, etc.) on both, air quality and microclimate.

On one side, the results are supported by modern observation methods and on the other side they are based on highly developed numerical analyses ([Schweinfurth, Laing, 1992], [Schweinfurth, Laing, 1993], [Schilling, Wiesel, 1994]). To meet these requirements, a GIS based on the software system ARC/INFO and the database system ORACLE has been established.

The geographical unit of the upper Rhine valley, which is shared by France, Swizerland and Germany, is limited by the Vosges mountains in the west, the Swiss Jura in the south and the German Black Forest in the east. Earlier research stopped at the political frontiers but the climate does not care about these artificial limitations. REKLIP is the first attempt to pursue research activities in the upper Rhine valley taking this region as a climatic unit.

The institutions involved in REKLIP are several institutes of universities located in the project area (Strasbourg, Basel, Freiburg and Karlsruhe) as well as some public organizations like national meteorological services and other national or municipal agencies (Fig. IV.8-1). The central location for data collection and site of the GIS is the *Institut für Photogrammetrie und Fernerkundung (IPF)* in Karlsruhe ([Schilling, Wiesel, 1994], [Schilling, 1993]).

Some of the institutes have smaller subsystems at their disposal to control and prepare data which are then transmitted to the main system. Communication for data exchange occurs at local and regional (data) networks. Simultaneously the institutes may work directly at their subsystems with data stored in the central main system or exchange data with other participants.

2. Data base

Task of the central data base is to supply climatological data to all participants. This data mainly consist of long term observations (from 1951 up to 1980) of the national meteorological services ([Schilling, Wiesel, 1993]). With this information, the different working groups will do statistical analyses to derive maps for the climatic atlas for all the meteorologic parameters. The supply of additional information, like a digital terrain model, digital topographic maps and the results of three land use classifications based on scenes of Landsat TM, are used for the computation of models, for the analyses with special thematic backgrounds and for the production of maps for the climatological atlas ([Schilling, 1993], [Weindorf, 1993]).

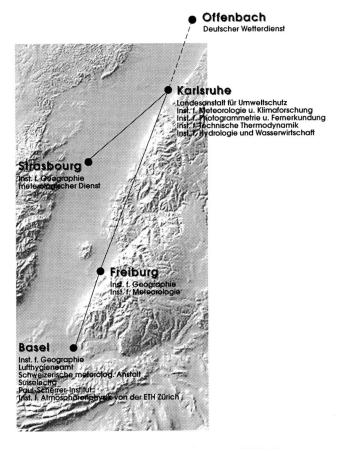

Figure IV.8-1: Institutions involved in REKLIP

2.1 Meteorological data

Besides the already mentioned continuous recordings of national weather services for the period 1951-1980, there are 36 permanently working meteorological stations for data adquisition which record information about different meteorological parameters in intervals of thirty minutes. To enter data about these measured parameters into the data base (working with ORACLE), a standardized data format has been established and each station has to send its data to the central data base in that format. Each station produces nearly 1 MB of information per month (a yearly total of 400 MB). These continuously recorded data were supplemented by two intensive observation campaigns in spring and autumn of 1992. During this two weekly campaigns there has been a data capture at the meteorological stations in a one minute cyclus at 5 selected days. Furthermore there were other mobile measurement systems (cars, balloons, airplanes). Starting from the data base of the permanent meteorological stations and the intensive observation campaigns in the following operational steps of the project, metereological models for forecast and prediction should be created. These models should not only be used for meteorological forecasts, but also in case of chemical or nuclear accidents with effects on the atmosphere (see also aims of REKLIP) ([Schweinfurth, Laing, 1992], [Schweinfurth, Laing, 1993], [Schilling, Wiesel, 1994], [Schilling, 1993]).

2.2 Topographic data

Topographic information of the surface are helpful for spatial analysis. On one side, they are useful for the maps in the climatological atlas, on the other side, they are important for some numerical calculations (emission of streets and urbanized regions, cascading, etc.). The capture of geometry was made by digitizing topographic maps in scales of 1:100.000 and 1:200.000 ([Leiber, 1993], [Moll, 1992], [Müller, 1992]). A listing of the maps used can be seen in Tab. IV.8-1:

Table IV.8-1: Used maps in REKLIP

Sheet Number	Sheet Name	Scale
cc7110	Mannheim	1 : 200.000
cc7118	Stuttgart-North	1 : 200.000
cc7910	Freiburg-North	1 : 200.000
cc8710	Freiburg-South	1 : 200.000
12	Vogesen-North	1 : 100.000
31	Vogesen-South	1 : 100.000
1	Land Map Switzerland	1 : 200.000

With help of the ARC/INFO data structure the data were devided in the layers *highways, federal streets, roads, water, urbanized regions and railways.*

After digitizing it was necessary to carry out geometric corrections because of the resolution of the digitizer and the distortions of the maps (see Chapter III.7). This is inevitable if different data layers have to be overlayed. In further work thematic completitions took place, for example a classification of hydrology and settlement. At last all digitized layers have been transformed into the UTM coordinate system. This step made it possible to combine identical thematic layers of neighbouring maps into one data layer for the whole project area.

The produced maps consist of topographic (geometric) information, digitized at the *Institute for Photogrammetry and Remote Sensing* (IPF) and thematic information delivered by the other REKLIP-participants to the IPF.

3. Maps in the Climatological Atlas

Based on the digitized topographic data we produced a great amount of maps (Fig. IV.8-2). The topography - here represented by cultural features, hydrological features and relief - constitutes the cartographic base map of the (thematic) meteorological data. Just those features of culture and hydrology, which have close links with the thematic contents, were represented in the map. Thereby it should be avoided to overload the map with too much background information because otherwise the thematic contents loose their significance. Compulsory to all maps is the representation of the relief, either by contour lines, by hill-shading or in a combined form ([King, 1993]).

Most important in REKLIP is the "*base map*" with a scale of 1:500.000 (Colour-Page 15) and the "*general map*" with a scale of 1:1 Mio. They obtain their thematic contents by overlaying raster matrices whose resolution came from a sensible deduction of the measurements. Depending on the number and availability of measurement points, grid cell sizes between 500 m (precipitation) and 5000 m (clouds, fog) resulted.

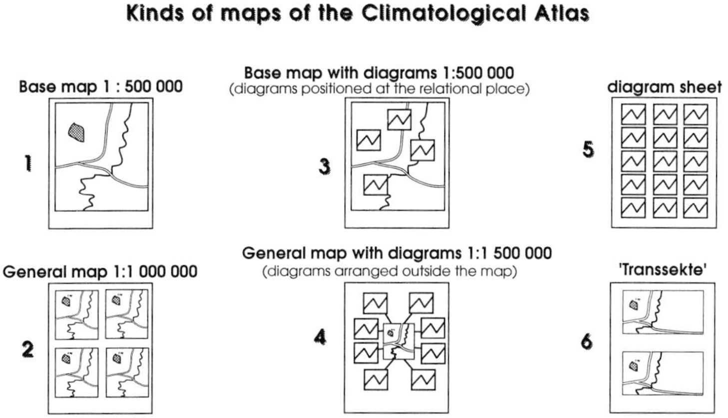

Kinds of maps of the Climatological Atlas

Figure IV.8-2: Different kinds of maps in REKLIP

Moreover selected parts can be illustrated in a detailed map at a scale of 1:200.000. Because of the necessary high density of the measurements this scale is applied to smaller geographic parts of this project like the precipitation distribution in the basin of Freiburg.

These previously mentioned forms of expression matter the areal subjects (continua). For representation of point features there are diagram maps within the REKLIP- map- series in form of "*base maps with diagrams*" at a scale of 1:500.000 (Colour-Page 16), where diagrams are directly positioned in the topographic background map and as "*general maps B*" at a scale of 1:1.5 Mio. In the center of this sheet there is a topographic map at a scale of 1:1.5 Mio. for purpose of orientation. Around this map the illustrations and diagrams are arranged. The topographic relation between diagrams and measurement points is available to greater local representation.

The extension of the REKLIP area and the given paper size for printing fix the cutting of the map sheets. At a scale of 1:500.000 the extension of the project area of 180 km × 250 km leads to a size of the mapsheet of 36.0 cm × 50.0 cm. In this case it is possible to print the map in the DINA2 (42 cm × 59.2 cm) size. Maps at a scale 1:1 Mio. are printed in thematical order at one paper sheet DIN A2. Also the "*Diagrammblatt*" map and the "*Detail*" maps must be in DIN A2 format ([Schilling, 1993]).

The digitized topographical objects must be checked for geometric accuracy and completeness. After this check they can be used as map elements. The aim of the REKLIP climatological atlas was to plot the topographic elements much prettier than the official topographic maps. In case of this challenge the digitized streets have had a signature consisting of two lines. It was possible to create these two lines for every street in ARC/INFO. In case of the different scales in digitizing (1:200.000 and 1:100.000) and plotting (1:200.000 and 1:500.000) there arised different problems, e.g. overlapping. These problems could only be solved by manual generalization. A lot of time has to be spent on this. At the plotting scale of 1:200.000 we mostly have to separate the rivers from the streets by displacement. At a scale of 1:500.000 it was necessary to reconstruct the automatically generated two line signature of the streets. Because of the geometrical density of some digitized objects the generation of double lines failed in some cases.

In the first plannings of the climatological atlas we wanted to create all maps out of one data base. Due to generalization effects we have one identical data base for the maps at 1:200.000 and 1:500.000. For the other maps at 1:1 Mio. and 1:1.5 Mio. the manual generalization work was too comprehensive, so it was more convenient to digitize new ones. Therefore, a new data base for 1:1 Mio. and 1:1.5 Mio. results on the base of the IWK (Internat.World Map). The advantage of this work is a reduced data base at these scales. We can handle the new data base easier and quicker than the old one (1:200.000) but on the other side, two different data sets are more difficult to handle: changes and completions have to be carried out in every scale, which turned out to be a time consuming work. Furthermore you must pay attention on pursuing all changes in each map. Five different kinds of maps are developed for the climatological atlas. Two examples can be seen at Colour-Pages 15 and 16.

3.1 Map Production

First it was planned to produce all maps by using only one system, namely ARC/INFO. In the course of time some different problems arised forcing us to think over. In ARC/INFO (at that time Version 5.0.1) the following demands are difficult or not to realize (in Version 6.1.1 these problems are solved), e.g.:

- hill-shading
- import of diagrams
- lettering
- cartographic symbols

It was planned to use the hill-shading as background information to the maps of kind 1, 2, 3 and 4. The production of hill-shading was not solvable satisfactory in ARC/INFO (Version 5.0.1).

With other software systems like ADOBE Photoshop (for hill-shading) and ALDUS Freehand (for other points) better results can be obtained. Another important fact is the data import which can only be done indirectly. There are graphic systems which has solved this problem in an easier way. The third and most important point is the lettering task in ARC/INFO which do not provide the high quality that cartographic maps require. Special cartographic pograms are better qualified for typography. Therefore, we have thought about how to produce maps for this climatological atlas. ARC/INFO (Version 5.0.1) ([Laing, Schilling, 1993]) by itself cannot realize all the requirements. There are two possibilities to gain control of these problems: Either you have to evolve your own program for best map production or commercial programs have to be used.

The first possibility was out of question due to time problems. Thus we decided to use the commercial systems 'Aldus Freehand' and 'Adobe Photoshop'. Adobe Photoshop is a pixel based image processing program. It is used to produce the hill-shading (as background information) ([King, 1993]). The vector based program Aldus Freehand is wide spread in cartographical applications. Freehand is used to design the maps concerning cartographical aspects. Lettering, data import and positioning the diagrams is realized by Freehand. Also printing films are made by Freehand. The following steps are realized:

1. Digitization of the maps; production of thematic data by the meteorologists
2. Export of the digitized maps out of ARC/INFO
3. Production of hill-shading in Photoshop
4. Import of digitized topographic data (ARC/INFO export-files), meteorological data and hill-shading in Freehand
5. Lettering of the maps and arranging the diagrams (Freehand)
6. Map output (Freehand)

In the working process shown above the maps are produced in digital way. The advantage is the digital existence of all features which are available in this way for further projects and calculations. All data can be used out of one data base and can be processed and updated. The disadvantage may be the use of three different systems (ARC/INFO, Aldus Freehand and Adobe Photoshop). But if there are interfaces between different systems, it is acceptable to use them to get the best results. And this is successfully realized in this way.

This stage of development is at the same time a new direction to a digital type of climatological atlases. Therefore, the thematic information layers have just to be updated, so they can be overlayed with the topographical data. Afterwards all layers can be represented and processed together. In addition to this pure visualization of data, in a next step data base queries and access to the data base can directly be done supported by special analysis functions. Beside of statistical results, analysis of time series can be carried out. The further result of the present climatological atlas in its traditional analogous form will be a complete *"meteorological information system"*.

4. Conclusion

The use of geoinformation systems with a variety of applications is steadily increasing. The common base of all applications can be defined (cartographic) data updating and output. Unfortunately, the modern computer-aided means are in opposition to the ideas of the classic cartography. This leads to unsatisfactory results which don't reach the quality of traditionally produced maps, though a lot of techniques and material are available. This is probably caused by the fact that the costs of simple systems are decreasing, so an increasing number of people or companies have such systems to their disposal. At first sight, the performance of these systems are impressive, but on frequent examination or intensive use, it turns out that one or another component is missing. As a consequence, the users start looking for an additional system complying the missing tasks. The result is an amount of different systems in use, with whom the tasks can rarely be fulfilled in a correct and efficient way (not mentioned the problems of passing data from one system to the other). The different systems can only be used correctly if the interface between the systems works without any problems. In this case an optimal and acceptable result can be achieved. As it can be seen, a system to produce maps completely in digital way still is missing.

Questions

1. How can you combine topographic information with thematic information?

2. Which tools/progrrams are neccessary for producing maps in a digital way?

3. What are the main contents of the REKLIP-maps?

4. How many different kinds of maps are in the REKLIP-atlas and which are the main differences between these maps?

5. Point out the problems of producing "digital maps" and show the solutions in REKLIP!

6. What are the problems from "maps in a digital form"?

References

Fiedler, F., et al.: Regio-Klima-Projekt (REKLIP) - Wissenschaftlicher Plan. Institut für Meteorologie und Klimaforschung, Universität Karlsruhe, Karlsruhe, 1987

King, R.: Analytische Schummerung mit Hilfe von digitalen Bildverarbeitungsalgorithmen. Unveröffentl. Diplomarbeit, Institut für Photogrammetrie und Fernerkundung, Fachhochschule Karlsruhe, 1993

Laing, R., Schilling, K.-J.: ARC/INFO - Anwenderhandbuch. Institut für Photogrammetrie und Fernerkundung, Universität Karlsruhe, Karlsruhe, 1993

Leiber, B.: Rechnergestützte Herstellung topographischer Übersichtskarten in den Maßstäben 1 : 1 MIO und 1 : 1.5 MIO für REKLIP unter Verwendung von ARC/INFO. Unveröffent. Diplomarbeit, Institut für Photogrammetrie und Fernerkundung, Karlsruhe 1993

Moll, C.: Rechnergestützte Herstellung einer topographischen Karte im Maßstab 1:500.000 für das Regio-Klima-Projekt unter Verwendung des Geo-Informationssytems ARC/INFO. Unveröffentl. Diplomarbeit, Institut für Photogrammetrie und Fernerkundung, Karlsruhe 1992

Müller, A.: Herstellung topographischer Karten für das Regio-Klima-Projekt im Maßstab 1:200.000 unter Verwendung des Geo-Informationssytems ARC/INFO. Unveröffentl. Diplom arbeit, Institut für Photogrammetrie und Fernerkundung, Karlsruhe, 1992

Schilling, K.-J.: Regio-Climate-Project, GIS and Cartography exemplary shown by REKLIP. Euroconference GIS, Roussillon, September 1993

Schilling, K.-J.: Regio-Klima-Projekt - Die ORACLE-Datenbank. Interner Bericht über die REKLIP-Datenbank unter ORACLE, Institut für Photogrammetrie und Fernerkundung, Karlsruhe, 1994

Schilling, K.-J., Wiesel, J.: REKLIP-Jahresbericht 1993, Projekt: Einrichtung, Verwaltung und Weiterentwicklung des zentralen Geo-Informationssystems für REKLIP. Arbeitsbericht 1993, Karlsruhe, Januar 1994

Schilling, K.-J., Wiesel, J.: Zugriff auf die REKLIP-Datenbank. Informationsschrift zum Arbeiten mit der REKLIP-Datenbank, Institut für Photogrammetrie und Fernerkundung, Karlsruhe, 1993

Schleich, A.: Bereitstellung eines Systems zur automatisierten Ausgabe von raumbezogenen Daten eines GIS unter Berücksichtigung kartographischer Aspekte. Unveröffentl. Diplomarbeit, Institut für Photogrammetrie und Fernerkundung, Karlsruhe, 1994

Schweinfurth, G., Laing, R.: REKLIP-Jahresbericht 1991, Projekt: Einrichtung, Verwaltung und Weiterentwicklung des zentralen Geo-Informationssystems für REKLIP. Arbeitsbericht 1991, Karlsruhe, Januar 1992

Schweinfurth, G., Laing, R.: REKLIP-Jahresbericht 1992, Projekt: Einrichtung, Verwaltung und Weiterentwicklung des zentralen Geo-Informationssystems für REKLIP. Arbeitsbericht 1992, Karlsruhe, Januar 1993

Weindorf, M.: Erstellung eines Programmes zur Umformatierung der REKLIP-Daten in ein von ORACLE lesbares Format und Import der Daten in ORACLE bzw. Integration/Ausgabe in MS-EXCEL. Unveröffentl. Studienarbeit, Institut für Photogrammetrie und Fernerkundung, Karlsruhe, 1993

a)

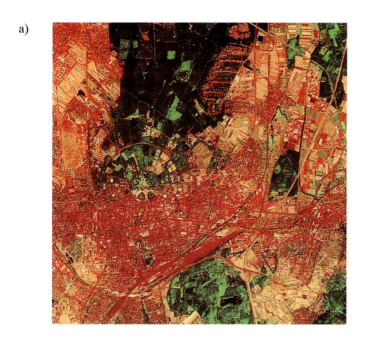

b)

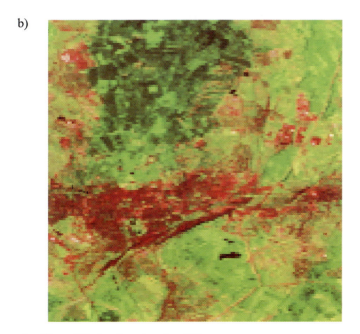

Colour-Page 1: City of Karlsruhe taken by different sensors. a) photographical picture taken by *KFA-1000* camera; b) multispectral scanner *Landsat MSS*.

a)
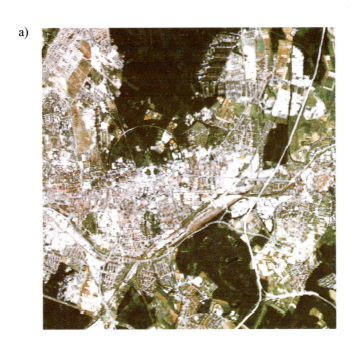

b)
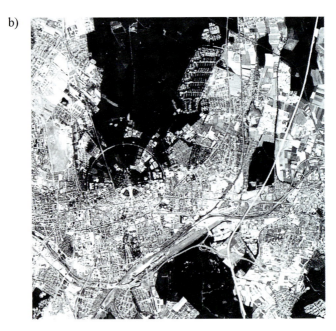

Colour-Page 2: City of Karlsruhe taken by different sensors. a) multispectral scanner *Landsat TM*; b) multispectral scanner *SPOT* (panchromatic band).

Data : SPOT HRV1 Multispectral

KJ : 50-252 ;   date : 28/6/86 ;
Extrait (560*450) ;   localisation : Strasbourg
(Fig. 4 & 6)

FACTORIAL COLOR COMPOSIT 3, 2, 1 (CMY)

Channel 1
Minimum : 98.Maximum : 220.
Mean : 158.30St. dev. : 26.06

Channel 2
Minimum : 98. Maximum : 220.
Mean : 158.30 St. dev. : 26.06

Channel 3
Minimum : 98. Maximum : 220.
Mean : 158.30   St. dev. : 26.06

TRAINING SETS - SAMPLING LAYERS

o Water surfaces

▣ Inhabited areas
    ● Housing areas
    ☐ Industrial zones

▲ Vegetation areas

Author : Christiane Weber - URA 902 CNRS
- Université Louis Pasteur –   France. 1994

Colour-Page 3:   Multispectral SPOT HRV1 image of the city of Strasbourg, including
training sets.

SUPERVISED CLASSIFICATION - Stepwise discriminant analysis

1 Water : Canal
2 Water : Pound
3 Water : Pound
4 Water : Pound
5 Down town buildings
6 Down town buildings
7 Down town buildings
8 Densified surbub unit
9 High-rise buildings with some vegetation surfaces
10 Individual house unit
11 Densified built unit on large parcels with vegetation
12 Individual house unit with garden
13 Individual house unit with garden
14 Industrial zones
15 Station sorting out surfaces
16 Harbour surfaces
17 Exhibition surfaces
18 Shaving site
19 Mixed Rhin valley forest
20 Mixed Rhin valley forest
21 Wooded surfaces
22 Wooded surfaces
23 Cultivation
24 Cultivation
25 Cultivation

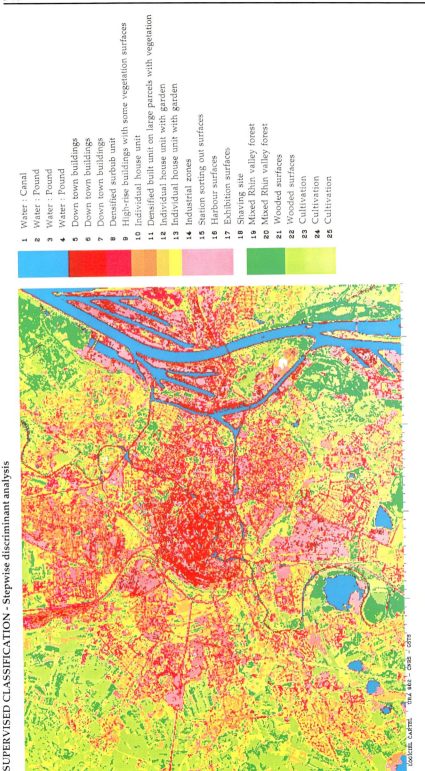

LOGICIEL CARTEL   URA 902 - CNRS - GSTS

Data : SPOT HRV1 Multispectral

KJ : 50-252 ;   date : 28/6/86 ;
Extrait (560*450) ;   localisation : Strasbourg
(Fig. 8)

Author : Christiane Weber - URA 902 CNRS - Université Louis Pasteur - France. 1994

Colour-Page 4: Supervised classification — stepwise discriminant analysis (SPOT HRV1).

SUPERVISED CLASSIFICATION - Gathered categories
Stepwise discriminant analysis

1 Water
2 Down town buildings and surbubs
3 High-rise buildings with some vegetation surfaces
4 Mixed unit on large parcels with vegetation
5 Individual house unit with garden
6 Mixed fringes unit
7 Industrial zones
8 Mixed Rhin valley forest
9 Cultivation

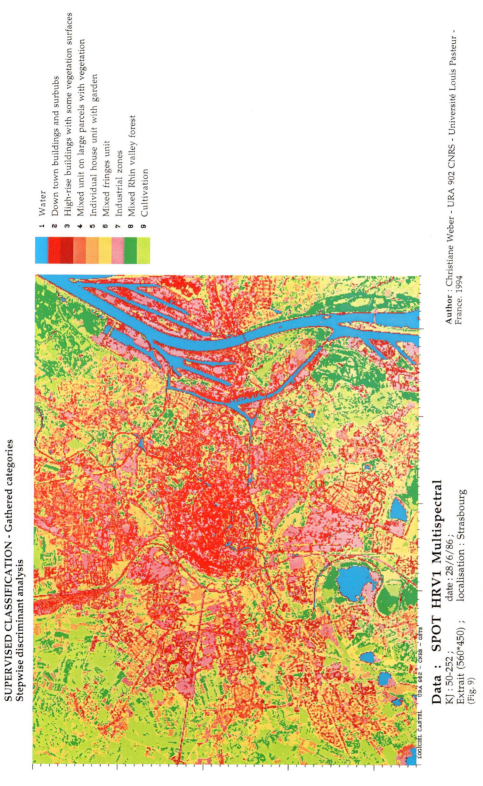

Logiciel CARTEL   URA 902 - CNRS - GSTS

Data :   SPOT HRV1 Multispectral
KJ : 50-252 ;   date : 28/6/86 ;
Extrait (560*450) ;   localisation : Strasbourg
(Fig. 9)

Author : Christiane Weber - URA 902 CNRS - Université Louis Pasteur - France. 1994

Colour-Page 5: Supervised classification – gathered categories (SPOT HRV1).

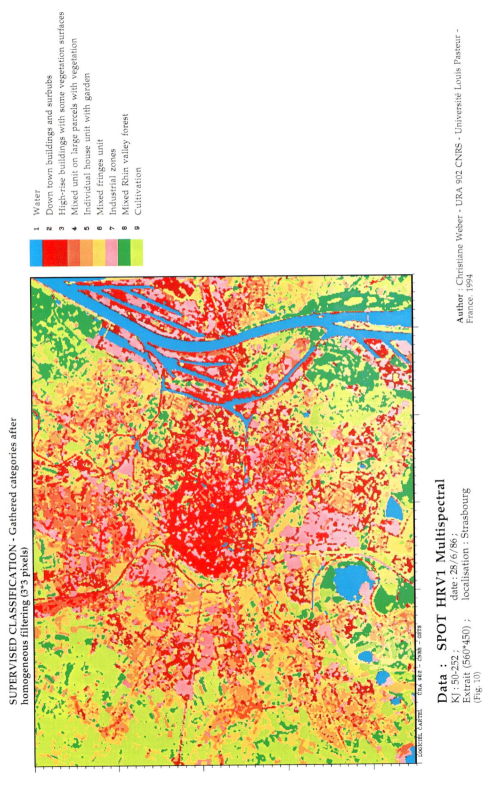

SUPERVISED CLASSIFICATION - Gathered categories after
homogeneous filtering (3*3 pixels)

1 Water
2 Down town buildings and surbubs
3 High-rise buildings with some vegetation surfaces
4 Mixed unit on large parcels with vegetation
5 Individual house unit with garden
6 Mixed fringes unit
7 Industrial zones
8 Mixed Rhin valley forest
9 Cultivation

LOGICIEL CARTEL   URA 902 - CNRS - GSTS

Data : SPOT HRV1 Multispectral

KJ : 50-252 ;   date : 28/6/86 ;
Extrait (560*450) ;   localisation : Strasbourg
(Fig. 10)

Author : Christiane Weber - URA 902 CNRS - Université Louis Pasteur -
France. 1994

Colour-Page 6: Supervised classification – gathered categories after homogeneous filtering (3x3 pixel).

Colour-Page 7: Geological map of the Mananga basin (for construction of the K-factor map).

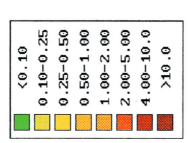

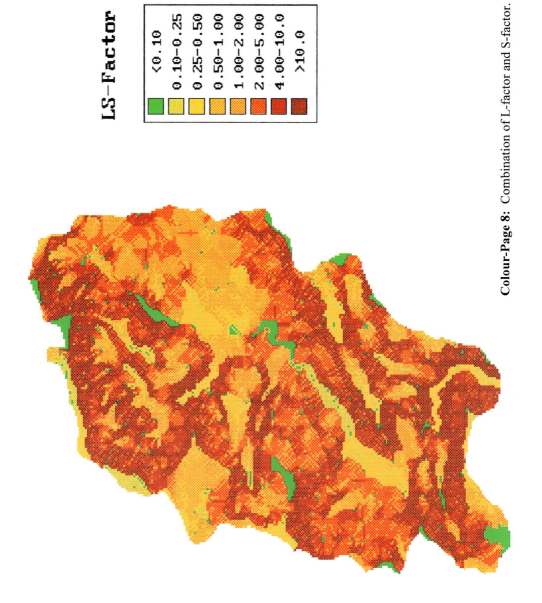

Colour-Page 8: Combination of L-factor and S-factor.

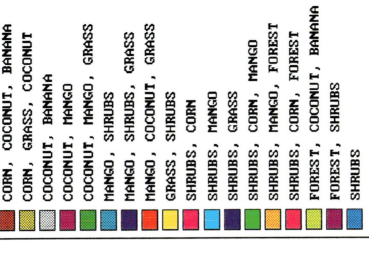

LANDUSE

- CORN, COCONUT, BANANA
- CORN, GRASS, COCONUT
- COCONUT, BANANA
- COCONUT, MANGO
- COCONUT, MANGO, GRASS
- MANGO, SHRUBS
- MANGO, SHRUBS, GRASS
- MANGO, COCONUT, GRASS
- GRASS, SHRUBS
- SHRUBS, CORN
- SHRUBS, MANGO
- SHRUBS, GRASS
- SHRUBS, CORN, MANGO
- SHRUBS, MANGO, FOREST
- SHRUBS, CORN, FOREST
- FOREST, COCONUT, BANANA
- FOREST, SHRUBS
- SHRUBS

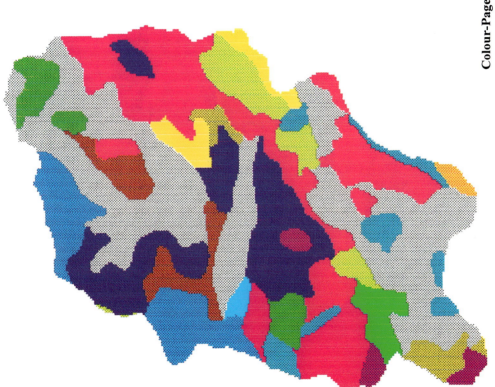

Colour-Page 9: Landuse map (for calculation of the C-factor).

SOIL LOSS

LOW

HIGH

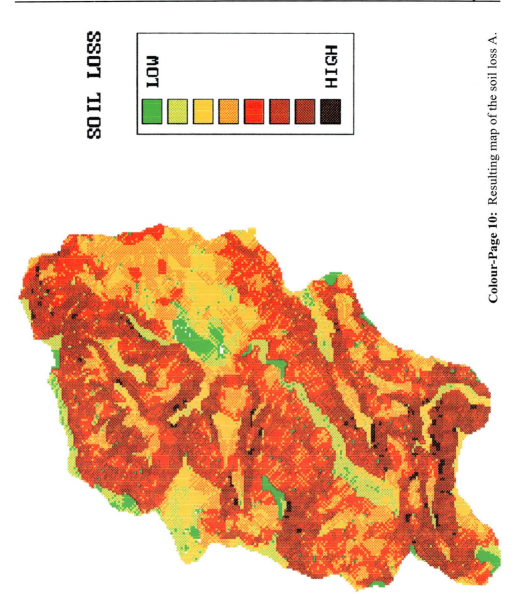

Colour-Page 10: Resulting map of the soil loss A.

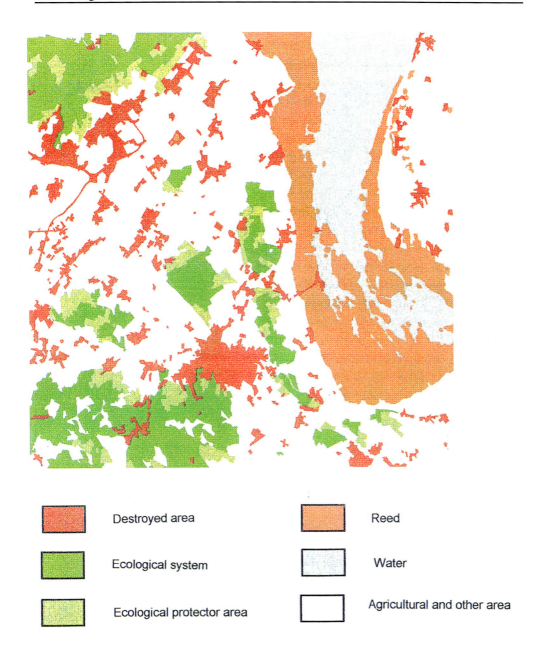

■	Destroyed area	■	Reed
■	Ecological system	■	Water
■	Ecological protector area	□	Agricultural and other area

Colour-Page 11: Ecology status layer.

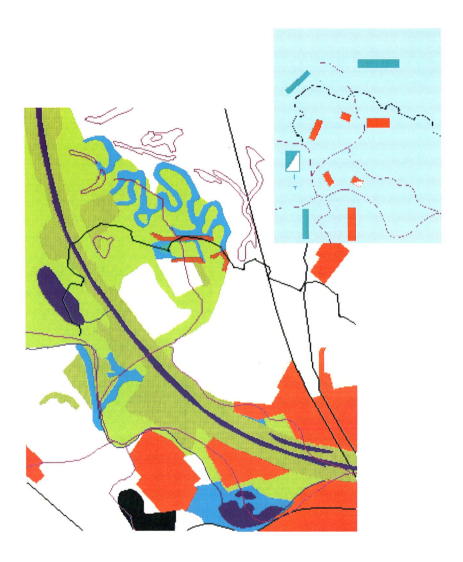

Colour-Page 12: Reconstruction of the Battlefield of Gyôr, in 1809, by combining satellite images, aerial photographs and old military maps. The dark line shows the old hungarian fort. Small picture (upper right): Positions of the hungarian (red) and french (blue) militaries before the Battle of Gyôr.

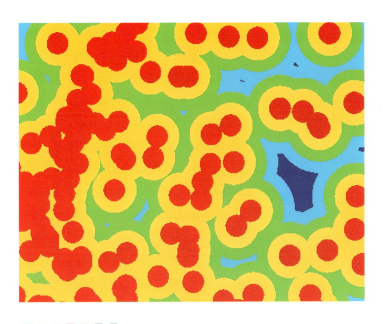

Colour-Page 13:

a) Multispectral classification (Landsat TM image) of test area *Mali*.

green: dense vegetation
yellow: shrub savanna
violet: sparse shrub savanna
red: sparse vegetation, bare soils
ochre: bare soils, rocks, sparse vegetation
blue: water surface, swamp
grey: burned areas

b) Distance model of test area *Mali*.

red: 0 to 2 km
green: 4 to 6 km
navy: > 8 km
yellow: 2 to 4 km
blue: 6 to 8 km

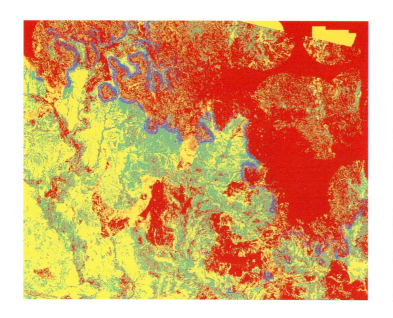

b) Resulting GIS image of suitable areas for ROI in test area *Mali*.

green: favourable (micro-catchm.)
blue: favourable (macro-catchm.)
yellow: possible (micro-catchm.)
orange: possible (macro-catchm.)
red: unsuitable

Colour-Page 14:
a) Slope model of test area *Mali*.

yellow: slope ≤ 10%
blue: slope > 10%

Colour-Page 15 (see page 351): Base map of the project area of REKLIP (1: 500 000).

Colour-Page 16 (see page 352): Base map with thematic diagrams (velocity of wind).

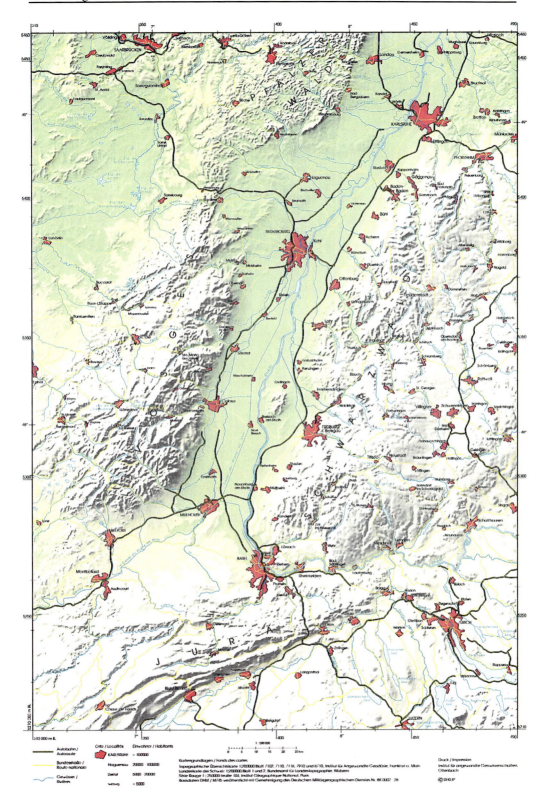

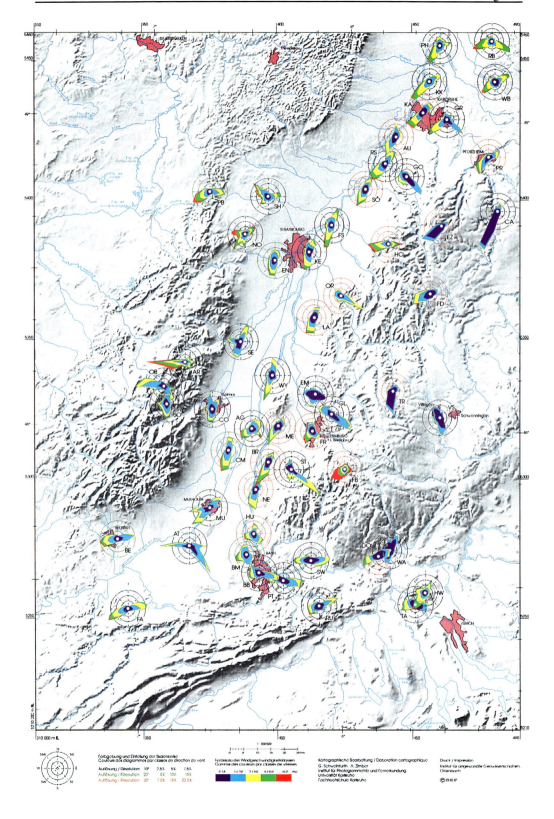

Authors

Henri Annoni
was born in Strasbourg in 1935. He studied physics with special emphasis on nuclear research in Strasbourg and as an invited research fellow at Berkley/USA and Geneva/Switzerland. He passed his D.E.A. in 1961 as well as the thése d'état in 1968.
From 1970 to 1975 Henri Annoni held a position at the University of Clermont-Ferrand and was invited to give conferences in Spain and in the USSR. In 1976 he was appointed scientific advisor to the French Embassy in Seoul/South Corea.
In 1979 he took on a Full Professorship at the "Ecole Nationale Supérieure des Arts et Industries" (ENSAIS) in Strasbourg. He lectured on Informatics for Surveyors both here and in an annexed university course in Syracuse/USA. At the ENSAIS, Henri Annoni was responsible for D.E.A. procedures together with the University of Strasbourg as well as for international collaboration, especially on the European level (ERASMUS, TEMPUS, Collége Franco-Allemand).
He died in 1994 under tragic circumstances.

Hans-Peter Bähr
Professor Dr.-Ing.habil., Head of Institute for Photogrammetry and Remote Sensing, University Fridericiana Karlsruhe, Germany (since 1983).
Main Research Fields: Digital Image Processing in Photogrammetry, Remote Sensing and Cartography.

Reinder H. Boekelman
M.Sc., Civil Engineer, Lecturer Department of Water Management, Faculty of Civil Eng., Delft University of Technology, The Netherlands.
Main Research Fields: Groundwater Modelling, Salt Water Intrusion.

Colette Cauvin-Reymond
Professor, Docteur d'état, Department of Geography, Université Louis Pasteur de Strasbourg, France (since 1968), Head of "Unité Associée au CNRS, 'Image et Ville'".
Main Research Fields: Thematic and Analytic Cartography, Spatial Cartographical Transformations (anamorphosis), Spatial Analysis, GIS, Spatial Cognition, Transportation/Accessibility, Urban Research.

Ákos Detrekői
Professor Dr. SC., Head of the Department of Photogrammetry,
Technical University of Budapest, Hungary (since 1980).
Main Research Fields: Mathematical Analysis of Measurements,
Photogrammetry, Image Processing, Remote Sensing, GIS.

Marc Jan van Dijk
M.Sc., Researcher at the Sanitary Engineering & Water
Management Division, Delft University of Technology, The
Netherlands.
Main Research Fields: (Geo-)Hydrological Modelling.

Béla Márkus
Associate Professor, Ph.D., Head of the Department of
Geoinformatics, College of Surveying and Land Management,
University of Forest and Wood Sciences, Hungary (since 1994).
Main Research Fields: GIS, Digital Elevation Modelling.

Gábor Mélykúti
Associate Professor, Dr.-Ing., Ph.D., Department of Photo-
grammetry, Technical University of Budapest, Hungary.
Main Research Fields: Analytical Photogrammetry, Digital
Terrain Model, Geoinformatics.

Tom Rientjes
Land and Watermanagement Engineer, Researcher/Lecturer at the
Department of Water Management, Delft University of
Technology, The Netherlands.
Main Research Fields: Modelling of Rainfall-Runoff Processes,
Groundwater Modelling.

Ferenc Sárközy
Prof., Ph.D., Professor at the Department of Surveying, Technical
University of Budapest, Hungary.
Main Research Fields: Network Optimization, Digtal Spatial
Data Capture, GIS Data Models, 3D and 4D Models.

Klaus-Jürgen Schilling
Dipl.-Ing., Surveying Engineer, Researcher at the Institute of
Photogrammetry and Remote Sensing, University of Karlsruhe,
Germany.
Main Research Fields: Semantic Modelling of Remote Sensing
Data, Semantic Networks, Development of Image Processing
Methods, Digital Topographical Database

Hans R. Vermeulen
Associate Professor, Civil Engineer, Department of Water Management, Faculty of Civil Eng., Delft University of Technology, The Netherlands.
Main Research Fields: Hydrological and Hydraulic Measurements, Geo-Statistics, Hydrology and Integrated Water-management in tropical countries.

Thomas Vögtle
Dr.-Ing., Surveying Engineer, Full-Time Researcher/Lecturer at the Institute of Photogrammetry and Remote Sensing, University of Karlsruhe, Germany.
Main Research Fields: Integrated Processing of GIS- / Remote Sensing Data, Digital Cartography, Land Use Mapping.

Christiane Weber
Dr., Full-Time Researcher in Centre National de la Recherche Scientifique, Université Louis Pasteur de Strasbourg, France (since 1985).
Main Research Fields: Study in Urban Modelling and Systemic Structure, Remote Sensing applied in natural and urban fields, GIS.

Joachim Wiesel
Dr.-Ing., Surveying Engineer, Full-Time Researcher/Lecturer at the Institute of Photogrammetry and Remote Sensing, University of Karlsruhe, Germany.
Main Research Fields: Visualization and User/Database-Interface of geo-related data in WWW, Building Information Systems, Integration of different Hard- and Software Systems, Digital Photogrammetry.

Gusztáv Winkler
Associate Professor, Dr.-Ing., Department of Photogrammetry, Technical University of Budapest, Hungary.
Main Research Fields: Photointerpretation, Remote Sensing and GIS in Environmental Monitoring.